Terrestrial Neutron-induced Soft Errors in Advanced Memory Devices

TERRESTRIAL NEUTRON-INDUCED SOFT ERRORS IN ADVANCED MEMORY DEVICES

Takashi Nakamura, Mamoru Baba
Tohoku University, Japan

Eishi Ibe, Yasuo Yahagi
Hitachi Ltd, Japan

Hideaki Kameyama
Renesas Technology Corporation, Japan

World Scientific

NEW JERSEY • LONDON • SINGAPORE • BEIJING • SHANGHAI • HONG KONG • TAIPEI • CHENNAI

Published by

World Scientific Publishing Co. Pte. Ltd.
5 Toh Tuck Link, Singapore 596224
USA office: 27 Warren Street, Suite 401-402, Hackensack, NJ 07601
UK office: 57 Shelton Street, Covent Garden, London WC2H 9HE

British Library Cataloguing-in-Publication Data
A catalogue record for this book is available from the British Library.

TERRESTRIAL NEUTRON-INDUCED SOFT-ERRORS OF ADVANCED MEMORY DEVICES

ISBN-13 978-981-277-881-9
ISBN-10 981-277-881-0

Printed by FuIsland Offset Printing (S) Pte Ltd, Singapore

PREFACE

Terrestrial neutron-induced soft-error (NISE) in semiconductor memory devices is currently one of the major concerns in issues of reliability. Understanding its mechanism and quantification of soft-error rates under environmental radiation on the ground are primarily crucial in the semiconductor technologies. Basic knowledge of NISE is required for scientists and engineers in the many relevant industrial and academic fields that are pertinent to the design and quality assurance of semiconductor devices, components, and systems. It should be noted that those devices, components and systems comprise the key technologies that are prevalent in modern IT societies.

This book covers the relevant up-to-date topics in terrestrial neutron-induced soft-errors, and aims to provide basic knowledge regarding NISE to the readers, with several valuable and unique features, as listed below.

(1) The terrestrial neutron spectrum and dose on the ground are continuously measured by several types of neutron detectors, and the variation of the spectrum and dose is discussed here. The differences between the terrestrial neutron spectrum and the spallation neutron spectrum used in accelerator-based testings are also discussed. Field test results of SRAMs and DRAMs in the terrestrial environment are shown, and are consistent with the variation in the measured terrestrial neutron spectra between the various sites used for field testing.

(2) Comprehensive and illustrative descriptions of Single Event Upsets (SEUs) in various SRAM and DRAM devices over a wide range of neutron energies (1-800MeV) are summarized. The first systematic descriptions of (quasi-) monoenergetic neutron test methods used to obtain SEU cross sections as a function of neutron energy are given.

(3) The threshold energies of SEU cross sections are given by fitting the data to a Weibull-fit curve. For the first time it is found that the threshold energies are as low as about 1 MeV, which is much lower than

the threshold energy for non-elastic nuclear reaction in silicon. The reaction mechanism for this threshold value is explained here.

(4) By introducing a novel non-linear unfolding method, it is demonstrated that the SEU cross sections can also be obtained as a function of neutron energy using results from SEU testing done with spallation neutron (continuous spectrum) beams. These energy dependent cross sections are very consistent with (quasi-) monoenergetic neutron test results. This unfolding technique enables one to extract a consistent set of SEU cross sections from test results taken at various spallation neutron facilities around the world, despite the fact the neutron spectrum varies from facility to facility.

(5) A Monte-Carlo simulation model for the entire physical process related to neutron-induced soft-error is described. The model can be applied to any nucleon (neutron/proton) radiation fields including accelerator and terrestrial radiation fields. The model is validated using the terrestrial-field and accelerator test results. Using realistic 3D device models in the simulations, unique results are presented for asymmetrical multi-cell upset, material effects in the components, and threshold energy.

(6) The most recent developments with respect to international standardization of testing methods are introduced, covering the USA standards JESD89A, 89-1, 89-3, and the Japanese guideline EDR4705.

This book consists of nine chapters as follows:

Chapter 1: Introduction

This chapter reviews the historical and current developments regarding terrestrial neutron-induced single event effects (SEEs) and the significance of SEU (Single Event Upsets) of memory devices, particularly SRAMs, together with an explanation of the physical mechanism of SEEs. The overall structure of the book is also described.

Chapter 2: Terrestrial Neutron Spectrometry and Dosimetry

The neutron detectors used for terrestrial neutron measurements are described, such as the Bonner ball, Phoswich detector, organic liquid scintillator, and rem counter. The sequential measurement of the terrestrial neutron spectrum and dose is given, along with the relevant environmental data. Altitude and geomagnetic variations in neutron dose are introduced. Other important parameters which affect the neutron dose are discussed.

Chapter 3: Irradiation Testing in the Terrestrial Field

The methods and equipment used to measure neutron soft-errors in the terrestrial field are described. The chi-square method used to estimate the soft-error rate (SER) is described. Correction methods to normalize SER rates are also introduced.

Chapter 4: Neutron Irradiation Test Facilities

The specifications of various accelerator-based neutron facilities now available in the world for irradiation testing of memory devices are given for two types of neutron facilities: (quasi)-monoenergetic neutron facilities using p-Li and other reactions, and spallation neutron facilities that produce a continuous energy spectrum.

Chapter 5: Review of Experimental Data and Discussions

General test methods using (quasi-) monoenergetic and spallation neutron beam facilities are described. Relevant analyses techniques such as tail correction, Weibull-fitting, unfolding, and multi-cell upset (MCU) classification are also described with a variety of the measured data. The authors' framework on SECIS (SElf Consistent Integrated System for neutron-induced single event) is also introduced.

Chapter 6: Monte-Carlo Simulation Methods

A Monte-Carlo simulation model, which is implemented in the program packages CORIMS/SEALER, is described over the entire physical process related to neutron soft-errors. The model can be applied to any nucleon (neutron/proton) radiation field, including accelerator and terrestrial radiation fields. A preliminary approach for composite material effects is introduced.

Chapter 7: Simulation Results and Their Implications

The model is validated with terrestrial field and accelerator test results. Unique results with realistic 3D device models are given for asymmetrical multi-cell upset, material effects in the components, and threshold energy.

Chapter 8: International Standardization of the Neutron Test Method

The most recent developments with respect to international standardization of testing methods are introduced, covering the USA standards JESD89, 89-1, 89-3, JESD89A and the Japanese guideline EDR-4705.

Chapter 9: Summary and Challenges

This chapter briefly summarizes the contents of the book. Topics which are not included in this book are discussed, along with very important items left for action in the future.

Acknowledgments

We would like to gratefully acknowledge Dr. Per.-Ulf Renberg, Dr. Alexander Prokofiev of the TSL of Uppsala Univ., Dr. Steve A. Wender and Dr. Bruce Takala of LANSCE and Prof. Yasuki Nagai and Dr. Nobuyuki Matsuoka of Osaka Univ. for their useful discussions and support of the experiments. Dr. Tatsuhiko Sato and Dr. Munehiko Kowatari of JAEA, Dr. Kazunori Nagaoka of JCAC, and Dr. Haruhisa Matsumoto of JAXA and Dr. Tomoya Nunomiya of Fuji Electric Systems, Co., gave us invaluable cosmic-ray neutron data. Dr. Robert Baumann of Texas Instruments, Dr. ShiJie Wen, Dr. Soon S. Chung of Cisco Systems, Inc., Mr. Charlie Slayman of Sun Microsystems, Inc., Dr. Norbert Seifert of Intel, Corp., Mr. Olivier Lauzeral of iRoc Technologies, Prof. Yukinobu Watanabe of Kyushu Univ., Dr.Yoshiharu Tosaka of Fujitsu, Ltd., Mr. Hiroshi Furuta of NEC, and Dr. Hajime Kobayashi of SONY made invaluable comments to improve our approaches. We also wish to thank Mr. Masatoshi Sato, Mr. Tasuke Tanaka, and Mr. Masato Ikeda of Renesas Technology Corp.

Mr. Mitsumori Hidaka and Mr. Atsushi Saito of Elpida Memory, Inc. Special thanks is devoted to Dr. Lawrence Heilbronn, Lawrence Berkeley National Laboratory for his intensive support in correcting the language.

Takashi Nakamura
Tohoku Univ.

Eishi Ibe
Hitachi Ltd.
December 5, 2007

ABOUT THE AUTHORS

Takashi Nakamura is now a Professor Emeritus of Tohoku University and works as an advisor to four Japanese companies. His e-mail address is nakamura@cyric.tohoku.ac.jp.

He was born in Nara, Japan, in 1939. In 1962, he graduated from the School of Nuclear Engineering at Kyoto University. In 1964, he received his master's degree from the Graduate School of Engineering at Kyoto University, and later that year became a research associate in the faculty of the School of Engineering at Kyoto University. In May of 1970, he received his Ph.D from the Graduate School of Engineering at Kyoto University. From April of 1973 to March of 1974, he was a guest researcher at the Atomic Energy Research Institute of Sweden. In 1975, he became an associate professor at the Institute for Nuclear Study of the University of Tokyo. In 1986, he accepted the position of professor at the Cyclotron and Radioisotope Center at Tohoku University, and in 1999 he became a professor in the Graduate School of Engineering at Tohoku University. During his time at Tohoku University, he was also a guest professor at the Institute of Space and Astronautical Science (1987–1990), and was a guest professor at the National Institute of Radiological Sciences. Since 2003, he has been Professor Emeritus at Tohoku University and a Visiting Professor at the Cyclotron and Radioisotope Center at Tohoku University. During his professional career, he has received numerous awards, including the Ichimura Award (Innovative Technological Invention) in 1986, the Best Paper Award of the Atomic Energy Society of Japan (1995), the Distinguished Service Award for Nuclear Safety by the Minister of the Science and Technology Agency of Japan (2000), and the American Health Physics Society William C. Morgan Lecturer Award (presented at the annual meeting in Cleveland, OH, USA, in 2001). From March of 2001 to March of 2003 he was the president of the Japan Health Physics Society and from October 2002 to October 2006 he was the President of the Asian and Oceanic Radiation Protection Association. He currently is the President of the Radiation Council of Japan.

Eishi Ibe received his B.S. degree in Physics from Kyoto University, Japan in 1975 and Ph.D degree in Nuclear Engineering from Osaka University, Japan in 1985.

He is currently a chief researcher in the Production Engineering Research Laboratory, Hitachi, Ltd. (e-mail: hidefumi.ibe.hf@hitachi.com). He was a research fellow at the California Institute of Technology in 1986–1987. He was a member of the advisory committee for the MIT nuclear reactor laboratory in 1990–1993. He is the Asia/Pacific chair of JESD89 Task Group in which international consensus is being made on testing methods for neutron-induced soft-error of semiconductor devices. He has experienced and led a wide variety of projects related to radiation effects on materials and components. He conducted a pioneering work on simulation techniques of water radiolysis in the coolant of nuclear power plants. He established a theoretical basis for hydrogen injection techniques which are now widely applied to Japanese boiling water reactors in order to mitigate stress corrosion cracking of component materials. He has led joint research on relevant projects at the MIT nuclear reactor laboratory in USA in 1990–1993, Halden reactor project in Norway in 1993–1996 and neutron radiation experiments at the Tandem accelerator of JAERI, Japan in 1989. He has received awards in 1986, and1990 from Japanese Atomic Energy Society and in 1996 from American Nuclear Society. He has authored chapters in three books and more than 100 technical papers and presentations in this field.

From 1998 up to now, he has focused his activity on radiation effects in semiconductors, in particular, those related to neutron-induced errors. He organized the first symposium in Japan on this subject in 2000 and led the JEITA (Japan Electronics and Information Technology Industries Association) project team from 2002–2005 on the Japanese standard EDR-4705 for testing methods of environmental radiation-induced soft-error in 2002–2005. He has also led the JESD89 revision task group in 2004 as the Asia/Pacific chair among 70 members from universities, governmental, and industrial organizations worldwide.

He has developed Monte-Carlo simulation techniques for terrestrial neutron-induced soft-errors where he has made outstanding contributions on scaling effects and multi-cell upsets. He also discovered a new soft-error mode MCBI (Multi-Coupled Bipolar Interaction) and has explained its mechanism by using a 3D simulation with a unique 4 partial memory cell model.

He has led a series of experiments in order to establish testing standards and validate simulation results by using neutron beam facilities around the world such as TSL (Theodore Sverdberg Laboratory, Sweden), LANSCE (Los Alamos Neutron Science Center), CYRIC (CYclotron and RadioIsotope Center), FNL (Fast Neutron Laboaratory) and RCNP (Research Center for Nuclear Physics).

He has been honoured as a IEEE Fellow as of Jan. 1, 2008 for contribution to neutron-induced soft-error analysis for semiconductor memory devices.

Mamoru Baba is a Professor of the Cyclotron & Radioisotope Center, Tohoku University, Aoba, Aramaki, Aoba-ku, Sendai 980-8658, Japan, babam@cyric.tohoku.ac.jp.

He received B.S. and M.S. degrees from Tohoku University in Sendai, Japan in 1967 and 1969, respectively, in the field of physics and nuclear science. He joined the teaching staff of the Nuclear Engineering Department (later Graduate School of Quantum Science and Energy Engineering), Tohoku University, Japan in 1970, and worked as research associate and associate professor until 1999. He received his Ph.D from Tohoku University in1991. He moved to the Cyclotron and Radioisotope Center in 1999 as a professor in the radiation safety division. His main focus has been radiation physics and measurements, experimental neutron nuclear data, and accelerator application to medicine and engineering. He has published more than 120 scientific papers on these fields. His current interest is in the application of accelerator and radiation measurements to space and medical fields, as well as radiation safety. He has been a member of the Atomic Energy Society of Japan (AESJ), and served as a head of the division of accelerator and beam science, as well as the nuclear data division. He is now serving chairman of the Japanese Society of Radiation Safety Management.

Yasuo Yahagi is currently a senior researcher in the Production Engineering Research Laboratory, Hitachi, Ltd. His e-mail address is yasuo.yahagi.rg@hitachi.com.

He received B.S. and M.S. degrees from the University of Tokyo in Japan in 1991 and 1993, in the fields of earth science, and high-pressure physics, respectively, and a Ph.D degree in Quantum Science and Engineering from Tohoku University, Japan in 2005. He was a visiting lecturer at the Graduate School of Science and Engineering, Tokyo Institute of Technology in Japan from 2004 to 2005.

After his considerable contribution to enhance the production yield of thin-film-transistor liquid crystal displays (TFT-LCDs) at Hitachi Ltd. by using various techniques for controlling electro-static discharge, suppressing chemical contamination, etc., he has moved his main area of research to radiation effects on semiconductor devices, in particular, those related to neutron-induced errors, since 2001.

He has carried out systematic experimental works to establish testing standards and validate simulation results by using neutron beam facilities around the world such as TSL (The Sverdberg Laboratory, Uppsala University, Sweden), LANSCE (Los Alamos Neutron Science Center, Los Alamos National Laboratory, US), CYRIC (Cyclotron and Radioisotope Center, Tohoku University, Japan), FNL (Fast Neutron Laboratory, Tohoku University, Japan) and RCNP (Research Center for Nuclear Physics, Osaka University, Japan). Conducting experiments at FNL, he showed for the first time in the world that high energy neutrons with 1 MeV cause SEU in advanced SRAMs.

He has contributed to establishing the Japanese standard of testing methods of environmental radiation-induced soft-error, EDR-4705, published in 2005 by the Japan Electronics and Information Technology Industries Association. He is a member of the Japan Society of Applied Physics.

Hideaki Kameyama is with Renesas Technology Corp., 20-1 Josuihoncho, Kodaira, Tokyo, 187-8588 Japan. He is manager of the System Solution Quality Assurance Dept. Quality Assurance Div. kameyama.hideaki@rks.renesas.com**.** Hideaki Kameyama received a B.S. (1981) in applied chemistry from the University of Tokushima, Tokushima, Japan. In 1981, he joined Hitachi microcomputer engineering Ltd. and worked for five years as a quality assurance staff for MOS devices. The next three years he worked as a reliability engineer at an oversea subsidiary company of Hitachi in US. In 1990, he returned to Hitachi Micon-systems Ltd. as a reliability engineer on SRAM, AND/NOR Flash memories and MNOS type EEPROM. At the 2000 IRPS symposium, he presented new data regarding the retention mechanism on Flash memory reliability. From 2000 to 2004, he explored system level soft error experiments as a project leader by using 130nm/180nm SRAM at several test sites both in USA and Japan. He served on the committee working on the Japanese soft error test standard EDR4705 of JEITA, until 2003.His current interests are SER reliability for advanced SRAM device, SIP PKG device qualification and SOC quality & reliability management oriented wafer process reliability. He is a member of EIA/JEDEC JC14.3 and also of IEEE/Electron Devices Society.

CONTENTS

Chapter 1

Introduction

Eishi Ibe

This chapter reviews the historical and current developments in the field of terrestrial neutron-induced single event effects (SEEs) and the significance of SEU (Single Event Upsets) of memory devices, particularly SRAMs. An explanation of the physical mechanism of SEEs is also included. Lastly, the overall structure of the book is described.

1.1 Background

As electronic devices are exposed to the harsh radiation environment from galactic cosmic rays or solar flares, a wide variety of single event effects (SEEs) in semiconductor devices caused by protons or heavy ions have been investigated for space applications over many years [1.1–1.5]. SEEs due to terrestrial neutrons have been also investigated in avionics applications [1.6, 1.7]. Although SEEs due to both alpha rays and terrestrial neutrons on the ground were pointed out in the late 1970s [1.8, 1.9], major research and countermeasures have focused almost solely on alpha ray-induced single event upsets (SEUs; soft-error) in memory devices until 1990s [1.10]. Terrestrial neutrons have been recognized as a looming source of SEUs since early 1990s, and concerns are now growing to a wider and deeper extent than ever before [1.11–1.16].

A number of new modes of SEEs (soft-errors and hard errors) are being proposed, and are summarized in **Table 1.1**. As can be seen in Table 1.1, logic [1.17, 1.18] and power MOSFET devices [1.19, 1.20] are also vulnerable to these apparent threats.

The critical level of SEU rate was initially fixed at 1000 FIT/chip (1FIT = 1 error in 10^9 hours) regardless of the type and density of the devices [1.21]. This requirement is almost equivalent to a rate as low as

Table 1.1 Error modes from terrestrial neutrons.

<table>
<tr><th>Category</th><th colspan="2">Mode</th><th>Memo</th></tr>
<tr><td rowspan="7">Soft Error (SEU)</td><td colspan="2">SBU</td><td>Single bit upset</td></tr>
<tr><td colspan="2">MCU</td><td>Multiple-bits upset for one event</td></tr>
<tr><td colspan="2">MBU</td><td>MCU in the same word (not correctable by ECC)</td></tr>
<tr><td colspan="2">Block error</td><td>Multiple bits errors along the BL or WL originally due to errors in peripheral circuits.</td></tr>
<tr><td colspan="2">MCBI</td><td>Multiple-bits upset due to parasitic bipolar action triggered by snapback in channel. Correctable by re-writing. Sometimes associates with low current.</td></tr>
<tr><td colspan="2">FBE</td><td>Main error mode of SOI. Mitigated by Body Tie.</td></tr>
<tr><td colspan="2">SET</td><td>Error mode of logic devices such as latch, inverter and clock.</td></tr>
<tr><td rowspan="4">Pseudo Hard Error</td><td colspan="2">SESB</td><td>Bipolar action in S/D channel. Impact ionization may affect.</td></tr>
<tr><td rowspan="3">PCSE</td><td>SEL</td><td>High current continues to flow due to parasitic cylister. Only power cycle can resume, but sometimes destructive (hard error)</td></tr>
<tr><td>SEFI</td><td>PCSE of logic devices.</td></tr>
<tr><td>Firm Error</td><td>Error mode of SRAM-based FPGA</td></tr>
<tr><td rowspan="2">Hard Error</td><td colspan="2">SEGR</td><td>Destruction of thin oxide layer mainly due to high-energy heavy ions. Power MOSFET, Flash memory.</td></tr>
<tr><td colspan="2">SEB</td><td>Destructive/explosive error of power MOSFET.</td></tr>
</table>

1 error in 114 years, but it also equivalent to a rate of 1 error in a month for larger server systems. While the error rate of 1000FIT in multimedia like movies or music obviously does not matter, it is definitely not acceptable for critical safety systems such as those found in automotive and emergency rescue systems. In addition, it is becoming very difficult to keep the 1000FIT/chip level as devices continue to be scaled down. The critical level of error rate, therefore, may be determined according to the error mode and applications, as summarized in **Table 1.2**. The soft error rate requirements may be given in terms of FIT/Mbit instead of FIT/chip, for example. For some critical safety applications or error modes, the requirements may be as low as less than 1FIT/system. At any rate, it is very difficult to quantify the susceptibility of a particular device with simple methods because the requirements for error rate are very low. Accelerator-based test methods and numerical simulation techniques, therefore, are indispensable for clarifying mechanisms and for the quantification of current and future trends in SEEs. Reliable solutions with (quasi-) monoenergetic and spallation neutron tests [1.11] and the numerical simulation package CORIMS (Cosmic Ray Impact Simulator)

Table 1.2 Requirements for terrestrial neutron-induced error rate.

Category	Error mode	Possible requirement	Remarks
soft-error	SEU (single event upset)	1000FIT/chip	MTBF*～100years/chip, <1day/super computer
		200-400FIT/Mbit or higher	· can be corrected by ECC · depend on application
	MCU (Multi cell upset/error)	A few to 10 % of total SEUs or higher	· fatal only if multiple bits are in the same word · depend on application
	SET (single event transient)	To be determined	· Soft error of logic devices
	MBU (Multi bit upset)	0-10 FIT/chip	· MCU in the same word · May cause system crash
Pseudo-hard error	SEFI (Single event functional interrupt)	0-1FIT/chip or system	· Error of logic system · May be recovered by reset
	SEL (Single event latchup)	0-1FIT/chip or system	· Continuous current flow · recovered by power cycle
Hard error	SEL, SEB, SEGR	0-1FIT/chip or system	· Permanent · May be fixed by re-mapping

*Mean Time Between Failure

have been developed within the framework of SECIS (SElf-Consistent Integrated System for neutron-induced soft-error [1.16, 1.22–1.25]. These methods are developed by the authors and introduced in this book.

Major attention is paid to SEUs of SRAM devices and multi-cell upset (MCU: multiple errors at one nuclear spallation reaction) because SRAMs are recognized as the most vulnerable device as device scaling proceeds. The typical trends for SRAM and DRAM data collected from the literature are summarized in **Fig. 1.1** [1.15]. Trends in SEU cross section (= number of errors/particle fluence (n/cm^2)) of SRAM and DRAM are totally different: SRAM is getting worse as the memory density increases, while DRAM is getting better. The origin of these conflicting trends will be elucidated in Chapter 7 with the help of results from simulations. Multi-bit upset (MBU: MCUs in the same word) is much more crucial than SEU for practical applications because MBUs cannot be corrected by the standard error correction code (ECC), and as a result can cause unexpected and costly shutdown of large server systems [1.13]. Unlike alpha particles, heavy secondary ions created by nuclear spallation reactions are believed to spark MBUs [1.23]. Because those heavy ions have relatively short ranges (typically less than 30 μm in silicon), interleave techniques, where bits in the same word are set apart with interval distances sometimes as large as 40–50μm, are applied to suppress MBUs. Such techniques, however, would make architecture design sophisticated or could have an adverse effect on performance (speed and power dissipation).

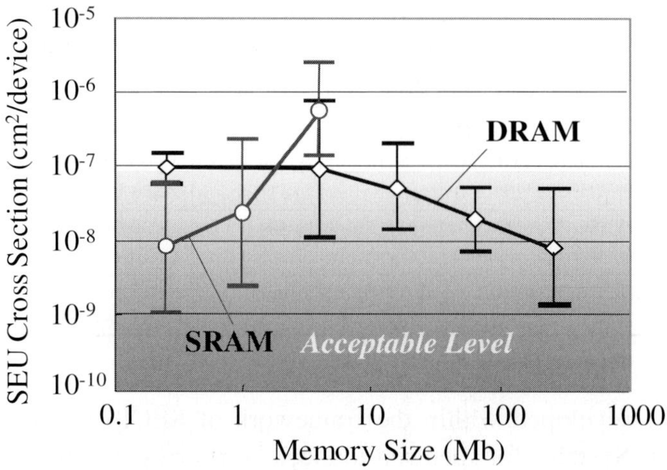

Fig. 1.1 Trends of soft-error susceptibility in SRAM and DRAM as a function of memory size. SEU cross section is defined as the number of errors divided by fluence (nucleons/cm^2) [1.15].

1.2 General Description of the SEE Mechanism

High energy neutrons, protons, pions, muons and neutrinos are produced in nuclear cascade showers created by nuclear spallation reactions between cosmic rays (mainly protons) and atmospheric nuclei (nitrogen and oxygen), as illustrated in **Fig. 1.2** [1.26]. Charged particles are halted in a relatively short range, but neutrons produce a cascade of spallation reactions (air shower) that eventually make terrestrial neutrons at the ground level, as illustrated in **Fig. 1.3**. The energy spectra at the sea level are shown in **Fig. 1.4 [1.26]**. Since charged particles twine around magnetic force lines, the geomagnetic and heliomagnetic fields act as shields against low energy cosmic rays. Air also acts as a shield against neutrons, so that neutron flux varies with the location on the Earth. The terrestrial neutron flux at the sea level is about 20 n/cm^2/h (E_n>1MeV). The nature and spectrometry/dosimetry of terrestrial neutrons are described in detail in Chapter 2.

A simplified bird's eye view of CMOS-SRAM (static random access memory) is illustrated in **Fig. 1.5**. When a nucleus undergoes a collision with a ballistic neutron, a nuclear spallation reaction, in which the nucleus breaks into secondary fragments, can take place with a certain probability. When the storage node (diffusion layer) is hit by a secondary ion, a certain amount of electrons/holes produced along the ion track are collected to the nodes, typically by the funneling mechanism [1.27].

Electrons and holes are also collected by the diffusion process. An SEU takes place when charge collected to the node exceeds the critical charge Q_c, over which the data "1(high)" in the node changes to "0(low)".

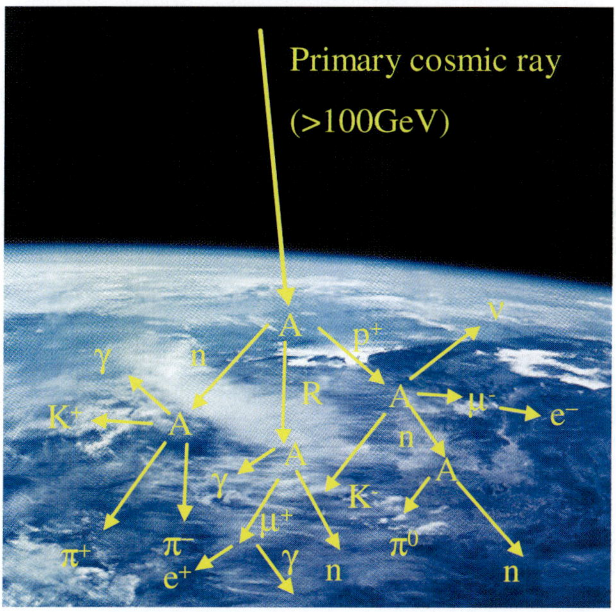

Fig. 1.2 Nuclear cascade shower in the atmosphere due to a primary cosmic ray. A stands for the nucleus of an air element and R stands for the residual nucleus from a nuclear spallation reaction.

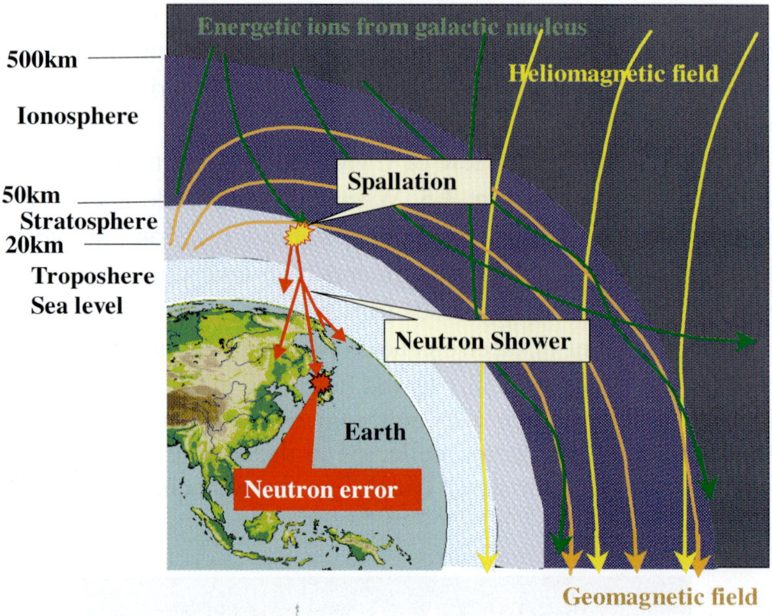

Fig. 1.3 Macroscopic mechanism of a neutron-induced soft-error.

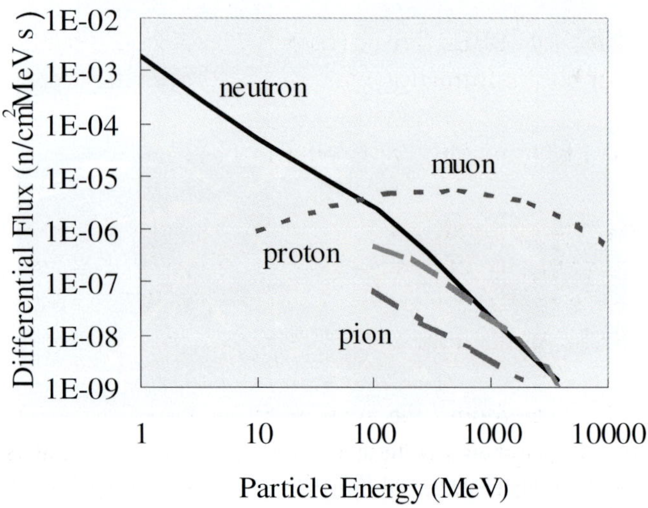

Fig. 1.4 Cosmic-ray energy spectra at the sea level [1.26].

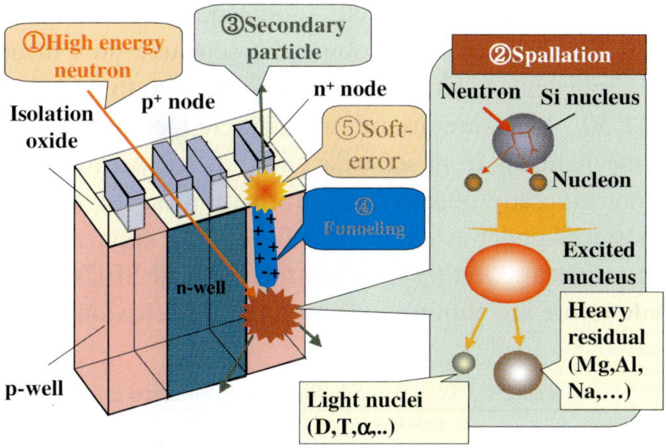

Fig. 1.5 CMOS SRAM structure and the microscopic mechanism of terrestrial neutron soft-error. Events proceed sequentially from ① to ⑤.

As device scaling proceeds, the charge collected due to alpha particle decreases because the charge density along alpha particle path is only about 5–10 fC/μm, whereas the charge density created by heavier secondary ions (C,N,O,...,Ta,W) is as high as 100–200 fC/μm. This is why neutron-induced SEUs are getting dominant as the device size shrinks.

Other types of SEEs are caused by either a charge collection mechanism or by the formation of parasitic thyristor/transistor effects.

1.3 Overview of Quantitative Evaluation Methods

Several basic test methods can be itemized to elucidate mechanisms and quantify susceptibility of semiconductor devices to neutron SEUs. The methods are categorized into four types according to radiation sources: proton, neutron, heavy ion, and focused laser beam, as summarized in **Table 1.3**.

Among these methods, the focused pulse-laser method is relatively simple and is applied to simulate the energy deposition that is needed to produce electron-hole pairs in the semiconductor devices [1.28, 1.29]. The quantitative relationship between actual ion energy and laser beam intensity, however, has not been validated. As such, only relative comparisons between tests or probing the sensitive area in devices for single event transients (SETs) may be possible with this method. Heavy-ion tests [1.30, 1.31] are useful to characterize the physical charge collection mechanisms from secondary ions at nuclear reaction sites, but they may not give any quantitative information on the susceptibility of electronic devices under actual neutron fields. The applicability of pulsed laser beam and heavy ion tests to neutron-induced SEUs are limited, as such, and therefore are eliminated from further discussions here.

Table 1.3 Test methods for neutron-induced error.

Probe	Name	Method	Merit/Demerit
Proton	Proton accelerator test	Irradiate DUT with mono-energetic proton	·Many facilities available ·Equality with neutron?
Neutron	Field test	Keep a number of devices at a certain location on the earth	·Costly, time consuming ·Reliable ·Corrections necessary
	(Quasi-) mono-energetic neutron test	Irradiate DUTs with (quasi-) mono energetic neutrons	·Facilities limited (TSL, CYRIC, FNL, RCNP) ·Versatile ·Correction s necessary (quasi-mono energetic neutron)
	Spallation neutron test	Irradiate DUTs with neutrons of broad energy range similar to terrestrial neutron spectrum	·High flux ·Facility limited (LANSCE, TRIUMF,RCNP) ·similarity with terrestrial neutron spectrum ·uncertainty in selection of energy range
	Thermal neutron test	Irradiate DUITs with thermal neutron by using an experimental nuclear reactor.	·Facility limited (KUR, MIT) ·Estimation of SER in the field is difficult due to difference in spectrum shape.
Heavy ion	Heavy ion test/ single ion test	Bombard mono-energetic heavy ion(s) on DUTs.	·Suitable to understand basic mechanism of charge collection ·No immediate correlation with neutron SEU.
Pulsed /Focused laser	Focused laser beam test	Pulsed laser beam is focused at a certain location and depth of interest.	·Easy access ·Pre-treatment of DUT ·Equality with neutron SEU

Proton beam tests have been a widely accepted method for SEU tests. In particular, these tests are useful for space applications where high-energy protons are major component of the radiation environment. Proton beams are also used as substitutes for neutron SEU tests at ground level because the proton's physical properties are quite similar to the neutron: the total nuclear reaction cross sections beyond 50 MeV are very close to those for neutrons, and the physical properties are almost identical between neutron and proton, except for isospin. Proton beam tests are useful as accelerator-based tests for rare error modes because a high flux can be obtained relatively easily. A certain number of experimental results show similarities between proton and neutron SEUs. However, the equality of proton to neutron in measuring SEU cross sections does not appear to be fully validated, since a substantial number of reports in literature show significant differences in SEU cross sections between proton and neutron beam measurements [1.32, 1.33]. Furthermore, protons with energies as high as 20 MeV and above, which are commonly used in atmospheric tests, have penetration depths deep enough to reach the sensitive volume of un-lid devices. This means that both ionization effects and nuclear spallation reactions take place simultaneously in proton beam testing. Both interactions are generally difficult to distinguish and, perhaps more important, ionisation effects may accelerate SEU rates through the total ionising dose (TID) effect [1.34]. Further study may be necessary to have more clear-cut conclusions with this regard. A neutron beam, therefore, is strongly recommended as the most straight-forward and reliable probe to quantify SEU cross section on the ground.

The methods where neutrons are utilized are further divided into four sub-categories. They are the field test, the (quasi-) monoenergetic neutron test, the spallation neutron test, and the thermal neutron test. Among these four sub-categories, thermal neutron effects [1.35, 1.36] are neglected in this book because they are due to (n, α) reaction of ^{10}B in BPSG. BPSG was used for a polishing technique but has been replaced by the CMP (Chemical Mechanical Polishing) technique. It may be noteworthy that thermal neutron effects may be visible when the boron concentration gets extremely high in PMOSFET. Details of the field test are described in Chapter 3. Details of accelerator-based neutron tests are also described in Chapters 4 and 5.

The physical processes described in **Fig. 1.5** can be simulated based on nuclear spallation reactions and semiconductor physics in an infinite memory cell array. Mote-Carlo simulation techniques and relevant results are described in Chapters 6 and 7.

International standardization of testing methods is one of the most important issues in the silicon industry because any reliable products/systems cannot be designed without SEE tolerance data of devices, using globally common gauges. The current status of standards in the USA/Japan, along with international standardization, are described in Chapter 8.

Future challenges are described in Chapter 9 as the summary of this book.

Chapter 2

Terrestrial Neutron Spectrometry and Dosimetry

Takashi Nakamura

This chapter reviews neutron detection methods used in terrestrial and space environments where neutrons and protons coexist. The experimental procedures using various neutron detectors in Japan, U.S.A. and Europe are described along with the experimental results. The altitude and latitude variations of the experimental results with solar modulation are discussed and compared with calculations.

2.1 Introduction

When primary cosmic rays composed of galactic cosmic rays and solar cosmic rays enter the earth's atmosphere, they produce secondary cosmic rays through many cascade reactions of primary cosmic-ray particles with atmospheric atoms. The primary cosmic-ray particles are about 90% protons, less than 10% alpha particles and about 1% heavier atomic nuclei. The earth's atmosphere consists of about 1033 g/cm^2 of oxygen and nitrogen, with a density that changes constantly with altitude. The atmosphere is so thick that the primary particles have many collisions with the earth's atmosphere and produce a lot of cascade particles before they reach sea level. These cascade particles consist mainly of pions, muons, nucleons (neutrons and protons), electrons and photons. At terrestrial altitudes, strongly penetrating particles such as muons and neutrons can reach sea level. At sea level, cosmic rays are about 90% ionizing radiation consisting mostly of muons with some photons and electrons, and about 10% neutrons.

The Sun has an 11-year solar cycle, with the most recent maximum in solar activity occuring in the period of 2000–2002. During a solar maximum, big solar flares frequently occur and can have a large effect on Earth's environment. Dynamic outbreaks of geomagnetic storms and radiation showers striking Earth occur sporadically, but with increasing intensity and frequency during the years around solar maximum. Protons are the main constituent among the solar event particles that reach the earth. When a proton comes into and interacts with the atmosphere, high-energy secondary neutrons and another electromagnetic waves are generated. There are several places in the world where neutron fluxes have been sequentially measured using moderated-type large BF_3 detectors. These detectors, however, can measure the neutron fluences, but not the energy spectra.

Over the past 10 years, there has been increasing concern about the exposure of air crews to atmospheric cosmic radiation. At aviation altitudes, the neutron component of the secondary cosmic radiation contributes about half of the dose equivalent; however, it has been difficult to accurately calculate or measure the cosmic-ray neutron spectrum in the atmosphere to determine an accurate dosimetry [2.1–2.5]. Dose rates from atmospheric cosmic radiation at commercial aviation altitudes are such that crews working on present-day jet aircraft are an occupationally exposed group with a relatively high average effective dose, following recommendations of the International Commission on Radiological Protection in Publication 60 [2.6]. Under these circumstances, many neutron measurements aboard the airplane and on the ground at different elevation have been done in Europe and North America [2.2, 2.7–2.16], mainly at high geomagnetic latitudes. However, only one measurement has been done aboard an airplane over Japan at low geomagnetic latitude on February 27, 1985 [2.17].

The characteristic feature is that the depth dependence of total neutron flux has a maximum at an altitude of 15–20 km (55–125 g/cm^2 depth) above sea level. This maximum occurs at all latitudes and arises as a result of two processes: secondary particle cascade development and neutron leakage/attenuation through the atmosphere, which is referred to as the cascade maximum. At depths in the atmosphere exceeding about 150 g/cm^2, the total neutron flux shows an exponential decrease with depth characterized with an effective attenuation length [2.18]. In the vicinity of the air/ground interface, the equilibrium of neutron fields is disturbed due to the albedo neutron effect backscattered from the terrestrial surface, which generates an increase of thermal neutron flux.

2.2 Neutron Detection Method

There is only one paper in which the cosmic-ray neutron energy spectrum at sea level was measured in the energy region of 0.05 to 2.0 MeV using hydrogen-recoil proportional counters. The experiment was conducted from June to October, 1966 at the Argonne National Laboratory [2.19]. Afterwards, various conventional and newly-developed detectors have been used for neutron measurements in the terrestrial field and aboard airplanes. In this chapter, two types of spectrometers, a multi-moderator spectrometer, the so-called Bonner ball, and an organic liquid scintillation spectrometer are described. One type of dose meter, the dose equivalent counter, the so-called rem counter, is also described along with the Phoswich-type detector for use in a mixed radiation field.

2.2.1 *Multi-moderator spectrometer (Bonner ball, Bonner sphere)*

The multi-moderator spectrometer, Bonner ball, has been widely used for neutron spectrometry in nuclear facilities and in the natural environment due to its simplicity and the usability over a wide range of neutron energies spanning from thermal to MeV. However, the energy resolution is poor and an initial guess spectrum is required. Uwamino and Nakamura developed the Bonner ball more than 20 years ago [2.20]. It has a high neutron sensitivity and uses only five polyethylene moderators. This type of Bonner ball is now widely used in Japan as a standard detector. The Bonner ball is composed of spherical ^{3}He proportional counters covered with polyethylene moderators. Neutrons having energies degraded in the polyethylene through the H(n,n)p elastic scattering reaction can be detected with a central ^{3}He counter through the ^{3}He(n,p) reaction. By increasing the polyethylene thickness, the sensitivity to neutrons increases with neutron energy. The central counter is 5.07 cm in diameter, and is filled with 5 atm ^{3}He gas (manufactured by LND Inc.). The thicknesses and the diameters of the polyethylene moderators are 0 cm (bare), 8.0 cm (1.5 cm thick), 11 cm (3.0 cm thick), 15 cm (5.0 cm thick), and 23 cm (9.0 cm thick), respectively. A schematic diagram of the detector system is shown in **Fig. 2.1**. The detector response functions are required to convert measured count rates to a neutron energy spectrum. The response functions in the energy range from thermal to 1 GeV were determined by Uwamino *et al.* [2.20] and by Nunomiya *et al.* [2.21] using the MCNPX (version 2.1.5) code [2.22]. **Figure 2.2** shows the response functions used in the present work [2.23]. The calculated results show good agreement with the results measured using the monoenergetic neutron field at the Fast Neutron Laboratory (FNL), Tohoku University, Japan, which is described in Chapter 4, and the thermal neutron field from a graphite pile using ^{241}Am-Be neutron sources at the Electro-Technical Laboratory (now, AIST: National Institute of Advanced Industrial Science and Technology), Japan.

Once the neutron count rates and response functions of the detectors are known, a deconvolution (unfolding) computer code SAND II [2.24] is applied to determine the neutron spectrum. The deconvolution process is not straightforward because *a priori* information in the form of an initial (default) spectrum that requires knowledge about the spectrum before the measurement must be applied in addition to the measurement and the response functions for obtaining a unique optimum solution.

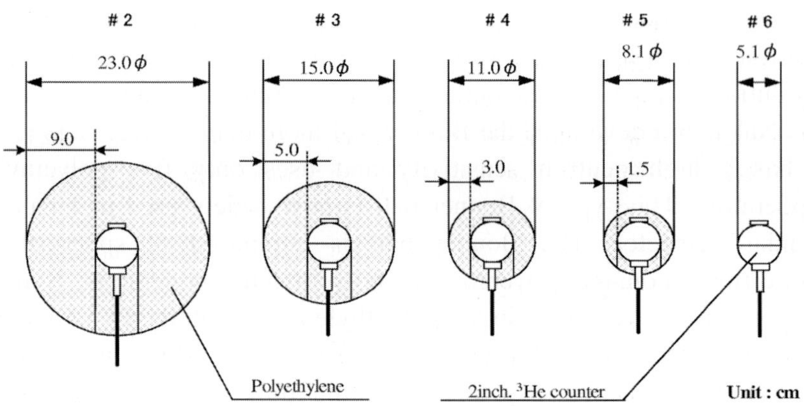

Fig. 2.1 Cross sectional view of the multi-moderator spectrometer developed by Uwamino *et al.* [2.20].

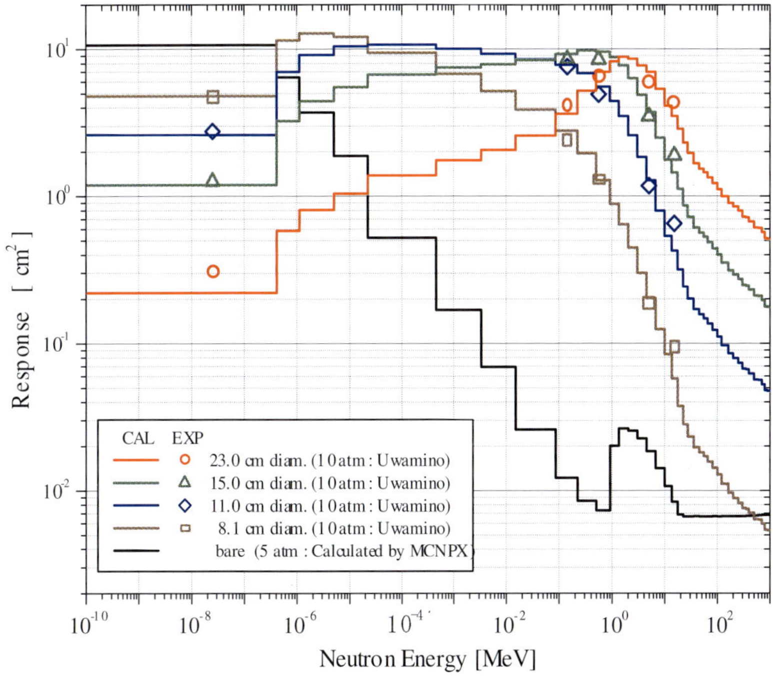

Fig. 2.2 Measured and calculated response functions of a multi-moderator spectrometer in Fig. 2.1 [2.20, 2.21, 2.23].

Goldhagen *et al.* developed the Bonner ball designed specifically for cosmic-ray neutron measurements [2.1–2.2]. Its 14 detectors are spherical 5.08 cm-diameter ^{3}He-filled proportional counters, with one unshielded, one surrounded with a layer of cadmium, and the rest surrounded with high-density polyethylene spheres having diameters ranging from 6.7 to 38 cm. Moderator diameters and masses for all the detectors are given in **Table 2.1**. To reduce the response to thermal neutrons, all but the largest moderator spheres are surrounded with thin cadmium shells. Even very large Bonner spheres with standard all-plastic moderators have responses that drop to low values as the energy increases from about 15 MeV to several hundreds of MeV (see **Fig. 2.3**, curves 8–12), so detectors 13 and 14 have the same diameter moderators (30 and 38 cm) as detectors 11 and 12, but detector 13 has a 25-kg lead shell embedded in its moderator, and 14 has an 18-kg steel shell. High-energy neutrons striking the nuclei of atoms with high atomic number cause hadronic showers with easily detected secondary neutrons, creating a rising response with increasing energy (see **Fig. 2.3**, curves 13–14).

Table 2.1 Detector moderator diameters and masses by Goldhagen *et al.* [2.1–2.2].

Detector No.	Outside Diameter (in.)	Outside Diameter (cm)	Mass of Polyethylene (kg)	Mass of Converter Shell (kg)
1 (bare)	-	-	-	-
2 (Cd)	-	-	-	-
3[a]	2.620	6.66	0.0773	-
4	3.234	8.21	0.2021	-
5[a]	3.849	9.78	0.3893	-
6	4.626	11.75	0.7306	-
7	5.626	14.29	1.3845	-
8	6.814	17.31	2.521	-
9	8.232	20.91	4.478	-
10	9.830	24.97	7.680	-
11	11.846	30.09	13.481	-
12	14.974	38.03	27.305	-
13	11.848	30.09	11.262	25.225 Pb
14	15.030	38.18	25.395	17.917 Fe

[a]Not flown because of ER-2 space constraints.

Detector responses as a function of energy (see **Fig. 2.3**) were calculated for neutrons at energies up to 150 MeV using MCNP (version 4b) [2.25] and MCNPX (version 2.1.5) [2.22] with evaluated cross sections [2.26], and for neutrons, protons and charged pions from 10 MeV to 100 GeV using MCNPX with cross sections from nuclear models. They use the MAXED unfolding code [2.27–2.28], which has very similar algorithm as the SAND II code.

Our Bonner ball consists of only five sets of moderators while Goldhagen's have 14 sets of moderators, which implies that our detectors may yield a poor energy spectrum in high energy region. Nevertheless, by using a well-selected initial guess spectrum given by Goldhagen *et al.* (described in Sec. 2.3.3), our Bonner ball can determine the cosmic-ray neutron spectrum with good accuracy, as clearly shown later in Sec. 2.4.2. Our Bonner ball is much simpler and has less space/weight compared with that by Goldhagen *et al.*, so our Bonner ball is very useful in both space and environmental neutron spectrum measurements.

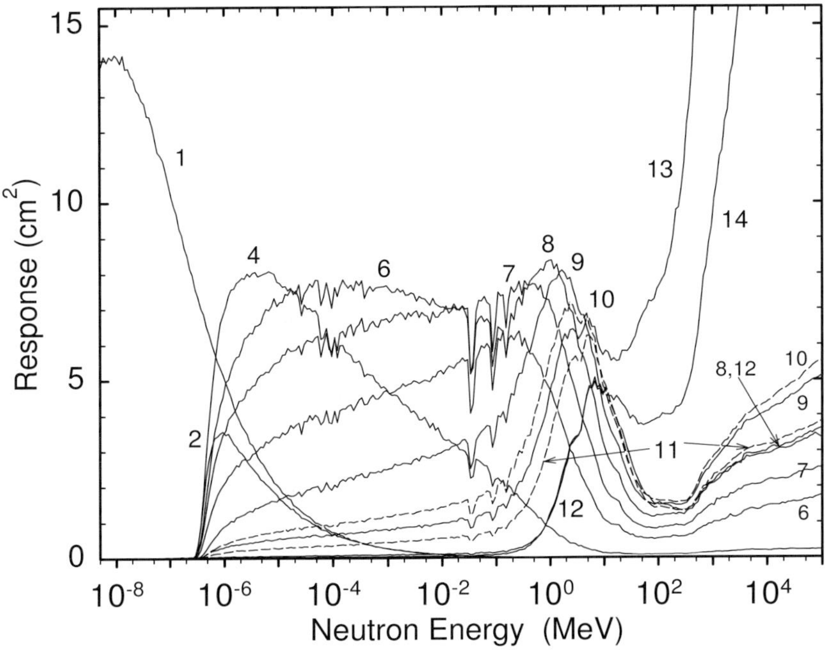

Fig. 2.3 Calculated neutron response functions for each of the detectors of the EML high-energy multisphere spectrometer. The dashed curves for detectors number 10 and 11 are only to help visually distinguish them from nearby curves [2.1–2.2].

2.2.2 *Organic liquid scintillation spectrometer*

NE213 (or BC501A) organic liquid scintillators coupled with photo-multipliers are widely used for high energy neutron spectrometry in target and shielding experiments due to the high quality of neutron and photon discrimination, along with the excellent energy resolution. In general, neutron spectra can be obtained from these detectors using an unfolding technique. Neutrons and photons incident on the organic scintillator mainly produce protons through the H(n,n)p reaction and electrons through the Compton scattering, respectively. The light output pulses from neutron-induced protons are slower than those from photon-induced electrons, which enables the pulse-shape discrimination between neutron and photon events.

The neutron and photon events are usually separated by using two-dimensional graphical plots of slow and total component pulse-heights (**Fig. 2.4**) [2.29]. When charged particles produced by neutron reactions (mainly recoil protons by the H(n,n)p elastic collision) escape from the NE213 scintillator without complete energy loss (referred to as a wall effect), the pulse shapes from high energy neutron events are close to those from gamma-ray events, and it is impossible to discriminate those events from photon events, as shown in **Fig. 2.4**. As a result, both events caused by photons and by escaping charged particles are removed. Therefore, the response functions mentioned below exclude the light outputs caused by the escaping charged particles.

The physical characteristics of the NE213 scintillator, such as the response function, detection efficiency and light output yield, are well known in detail for neutrons below 20 MeV via many studies. For higher energy neutron measurements, a large-size NE213 detector, 12.7-cm (5 inch) diam by 12.7-cm (5 inch) long, coupled with a photomultiplier (Hamamatsu R4144) has been used by our group. With regard to the response functions of the 12.7-cm diam by 12.7-cm long NE213 detector for neutrons above 20 MeV, Nakao *et al.* [2.30–2.32] measured those functions for energies up to 206 MeV using the ^{7}Li(p,n) quasi-monoenergetic and white neutron sources at the ring cyclotron facility (RRC) of the Institute of Physical and Chemical Research (RIKEN), Japan. Sasaki *et al.* [2.33] extended those functions up to 800 MeV using the white neutron source from thick carbon target bombarded by 400 MeV/nucleon carbon and 800 MeV/nucleon silicon ions at the Heavy Ion Medical Accelerator facility (HIMAC) of National Institute of Radiological Sciences (NIRS), Japan.

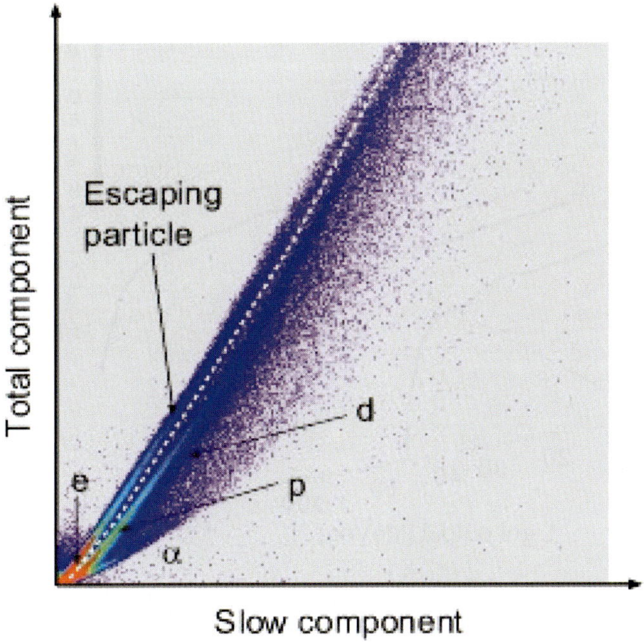

Fig. 2.4 Two-dimensional pulse height scatter plots of the slow component versus total component to identify the particles produced in the NE213 scintillator. The discrimination boundary is indicated by the broken line. The symbol e is the electron produced by photons and p, d, and, α are the proton, deuteron and alpha particle produced by neutrons, respectively [2.29].

The response functions to neutrons with energies between 50 and 800 MeV measured by Sasaki *et al.* [2.33] are illustrated in **Figs. 2.5(a)**, **(b)**. The alphabets which appear at the left shoulders of each spectrum from "A" to "O" correspond to the neutron energy boundaries listed in the figures. It has been hitherto said that the response function does not change in the region above a few hundreds of MeV where the charged particles produced by neutron reactions escape from the scintillator. Indeed, the whole shape and the upper edge of the response functions are very similar to each other beyond 200 MeV neutrons. But as shown in these graphs, a small change of the response function and a further expansion of the upper edge to higher light output can be clearly seen up to 800 MeV by choosing a wide neutron energy interval beyond 200 MeV. The event which induces such large light output might be brought

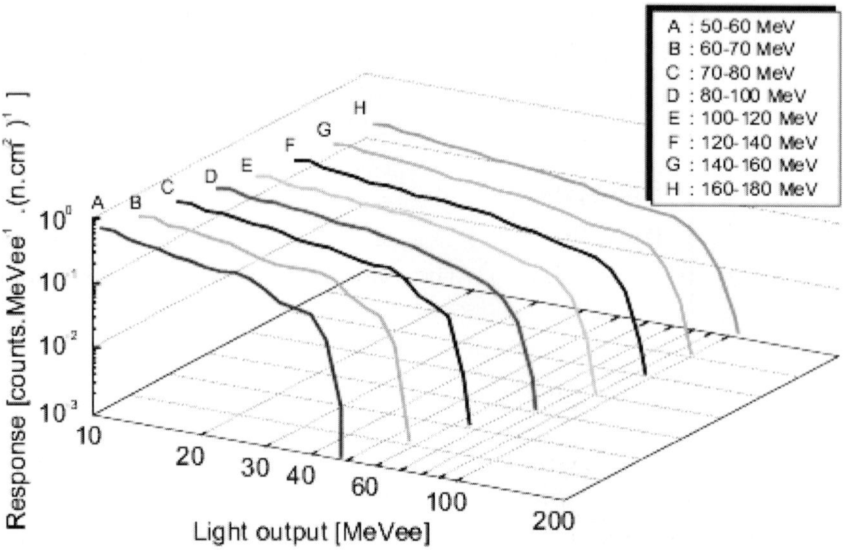

Fig. 2.5 (a) Measured response functions of the NE213 scintillator for 50–60, 60–70, 70–80, 80–100, 100–120, 120–140, 140–160 and 160–180 MeV neutrons [2.33].

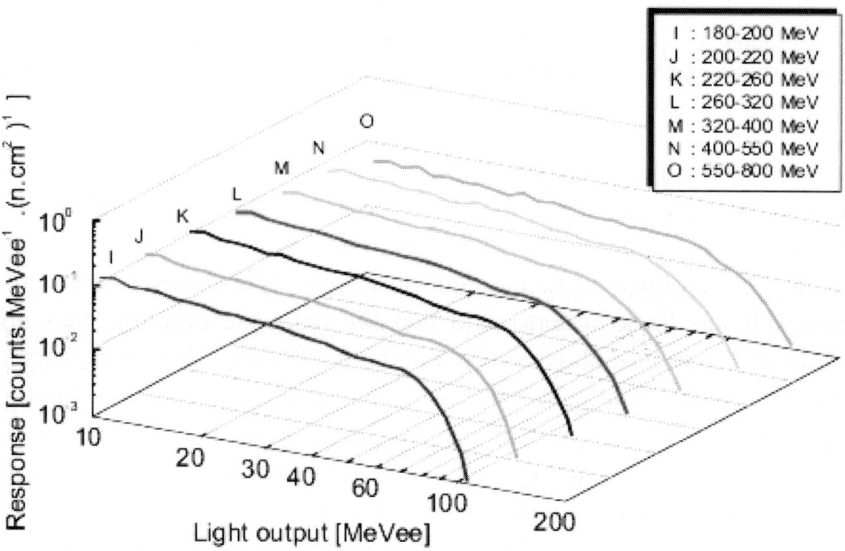

Fig. 2.5 (b) Measured response functions of the NE213 scintillator for 180–200, 200–220, 220–260, 260–320, 320–400, 400–550 and 550–800 MeV neutrons [2.33].

by multiple collisions through the C(n,x) reaction in the NE213 scintillator, and the resultant charged particles, such as Be, Li and alpha particles, have higher energy transfer from the incident neutron with increasing neutron energy.

Recently, Satoh *et al.* also measured the response functions of BC501A (similar to NE213) for neutrons up to 800 MeV at HIMAC [2.34] and compared the results with the CECIL code [2.35] and the SCINFUL-QMD code [2.36]. The latter code gave much better agreement with the measured results.

2.2.3 *Dose equivalent counter (rem counter)*

The Anderson-Braun (A-B) type rem counter, originally developed by Anderson and Braun, is still widely used. This rem counter consists of an inner thermal neutron counter, such as a BF_3 counter, ^{3}He counter, ^{6}LiI(Eu) scintillator, and so on. The incident higher energy neutrons are degraded down to thermal energy mainly through the elastic scattering with hydrogen in the polyethylene moderator. They are detected with the thermal neutron counter fixed in the center of the polyethylene. In order to realize the detector sensitivity to neutrons with energies from thermal to about 20 MeV fitted to the fluence-to-dose conversion coefficient recommended by the ICRP (International Commission on Radiological Protection), an internal thermal neutron absorber including Cd and B is inserted inside the polyethylene, and the thickness of polyethylene is adjusted to about 10 cm.

(1) High sensitivity rem counter

Two types of high sensitivity rem counters were fabricated using large volume and high pressure ^{3}He spherical proportional counters, one filled with 10 atm ^{3}He and having a 5.08 cm (2 in.) diameter, the other one filled with 5 atm ^{3}He and having a 12.7 cm (5 in.) diameter [2.37]. They are simply referred to as the 2-in.-diam ^{3}He rem counter and the 5-in.-diam ^{3}He rem counter, respectively. The design of the polyethylene moderator and the internal thermal neutron absorber was determined by the ANISN calculation [2.38]. **Figure 2.6** shows the cross-sectional view of these two rem counters. The responses of these two rem counters to neutrons of energy from thermal to 15 MeV were also measured in the neutron standard field at ETL, similarly as at FNL. The measured results are shown in **Fig. 2.7(a)** along with comparisons to the response of the first Anderson-Braun (A-B) type rem counter, the widely-used Studsvik 2202D, and the formerly-used fluence-to-dose conversion factor given by ICRP-21. Our 2-in.-diam and 5-in.-diam ^{3}He rem counters have about 10 and 70 times higher sensitivities than the Studsvik 2202D rem counter, respectively. At present, they still have highest sensitivities among the commercially available rem counters in the world. The 2-in.-diam ^{3}He portable and stationary rem counters, NSN1 and NDN1, respectively, are now commercially available from Fuji Electric Systems Co. Ltd. They are shown in **Fig. 2.8** [2.23]. Recently, the neutron energy responses of

High Sensitivity Neutron Dose Equivalent Counter

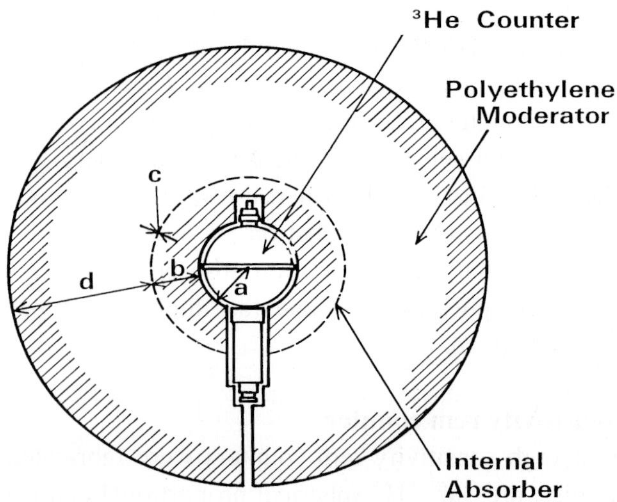

Rem Counter	a	b	c	d	weight
5.08-cm(2-in.) diam ^{3}He	5.08 cm	4.0 cm	0.114 cm	6.0 cm	8 kg
12.7 -cm(5-in.) diam ^{3}He	12.7	4.0	0.114	4.45	13

Fig. 2.6 Cross-sectional view of a high-sensitivity rem counter by Nakamura *et al.* [2.37].

(a)

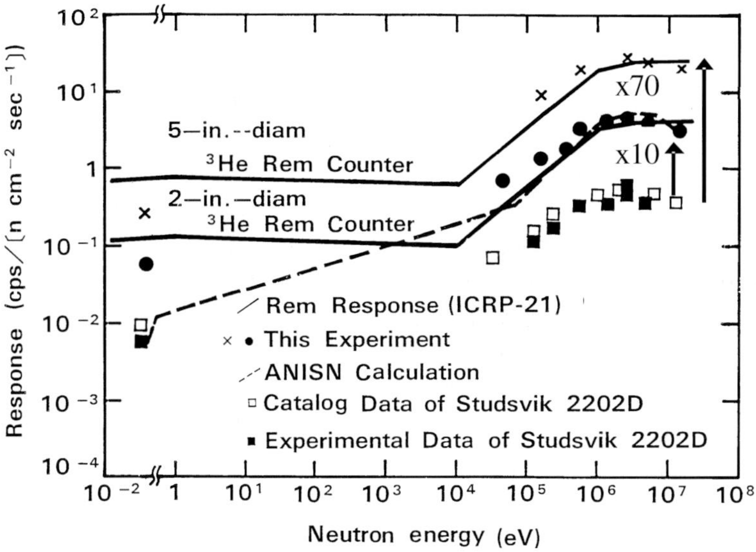

(b)

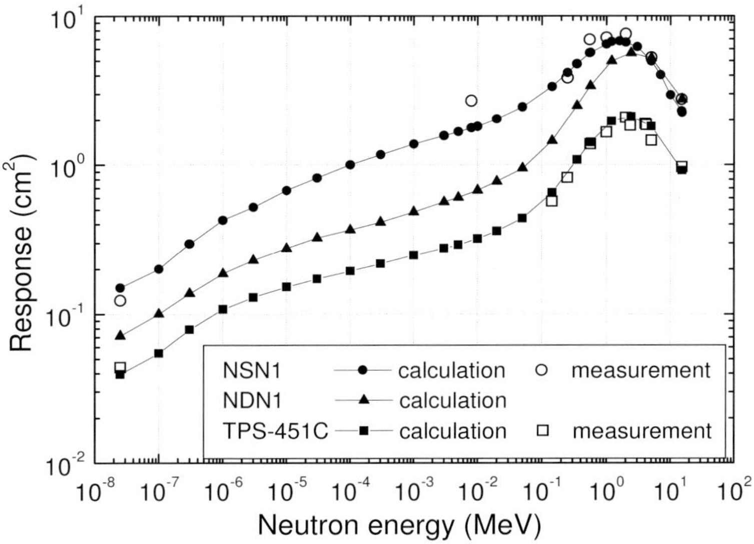

Fig. 2.7 (a) Measured response functions of two high-sensitivity rem counters using 2 and 5-inch-diam ^{3}He detectors, 2-in-diam and 5-in-diam ^{3}He rem counters, and the Studsvik 2202D rem counter. The fluence-to-dose conversion factor given by ICRP Publication 21, 1973 and the response calculated by ANISN [2.38] are also shown for comparison [2.37]. (b) Response functions of three types of Japanese rem counters calculated with MCNP-4B [2.25] by Saegusa *et al.* [2.39].

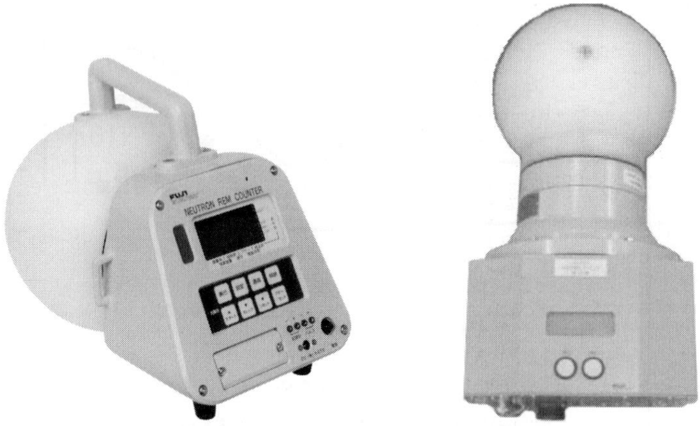

Neutron Rem Counter
NSN21001

4.5 cps/(μSv/h) ±20 %

Neutron rem area
NDN1NA13

3.0 cps/(μSv/h) ±20%

Fig. 2.8 Photograph of 2-inch-daimeter ^{3}He portable and stationary rem counters fabricated by Fuji Electric Systems Co. [2.23].

NSN1 and NDN1, together with the portable rem counter, TPS-451C (Aloka Co. Ltd.), have been measured in the monoenergetic neutron field at FNL and in the thermal, ^{252}Cf, ^{241}Am-Be neutron fields at the Facility of Radiation Standard (FRS), JAERI (Japan Atomic Energy Research Institute, now JAEA: Japan Atomic Energy Agency). The neutron energy responses were also calculated by Saegusa *et al.* [2.39] with the MCNP-4B Monte Carlo code, as seen in **Fig. 2.7(b)**. The agreement between experiment and calculation is good, except at 8 keV. They also compared the energy responses to the fluence-to-ambient dose equivalent conversion coefficient, H*(10), given by ICRP-74 [2.40]. This NSN1 rem counter has the lightest weight (about 8 kg) among of similar rem counters and is easy to handle in environmental neutron monitoring.

(2) Extended range rem counter, LINUS

Because all conventional A-B type rem counters underestimate the neutron dose above 20 MeV due to low sensitivity, Birattari *et al.*

developed a new-type rem counter, called LINUS, that extended the measurable neutron energy range above 20 MeV [2.41–2.42]. They inserted a 1-cm thick layer of lead between the inner and outer polyethylene moderator layers, just outside the borated rubber. In the central part, a spherical ^{3}He proportional counter of 3.2 cm active diameter filled with 304 kPa ^{3}He and 101 kPa Kr, is set as shown in **Fig. 2.9** [2.42]. The effect of the additional lead layer is to detect neutrons with energies greater than about 10 MeV via evaporation neutrons produced in inelastic scattering and Pb(n,2n) reactions. The evaporation neutrons can be subsequently moderated by the polyethylene and then detected by the ^{3}He counter. No significant effect is produced on neutrons with energy below about 10 MeV. Two commercial models are now available (model 6060 by Health Physics Instruments, Calif., U.S.A. and model NM500X by Muenchener Apparatebau fuer Elektronische Geraete GmbH, Munich, Germany). **Figure 2.10** gives the neutron

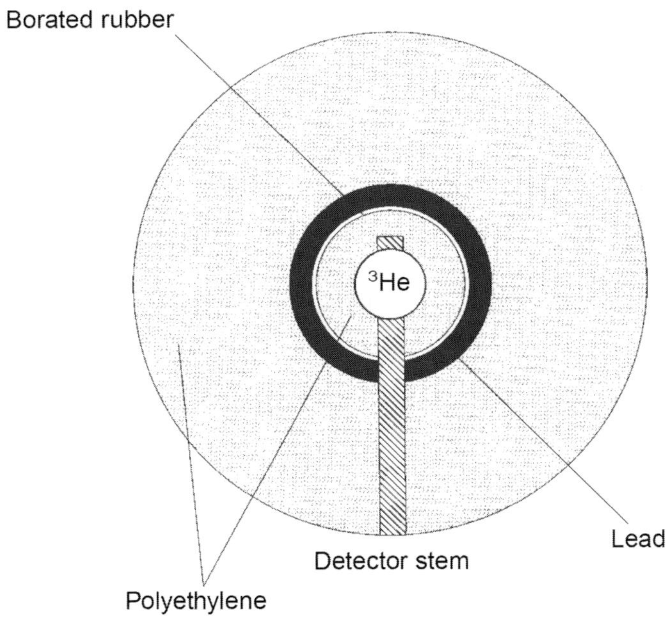

Fig. 2.9 Cross section of the spherical version of the rem counter LINUS by Birattari *et al.* [2.41, 2.42].

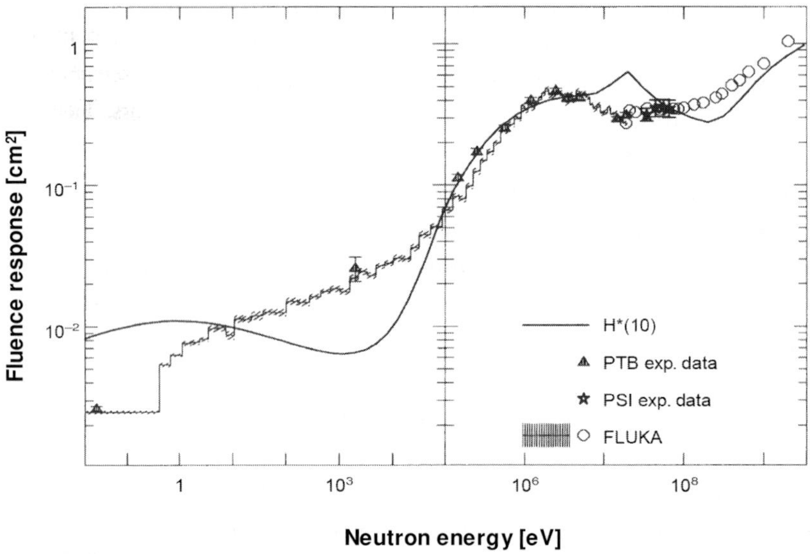

Fig. 2.10 Absolute neutron fluence response (counts per unit fluence) of the cylindrical version of the rem counter LINUS [2.41, 2.42].

energy response of LINUS measured with monoenergetic neutrons between 144 keV and 19 MeV at PTB (Physikalisch-Technische Bundesanstalt), and calculated with the FLUKA code [2.43] up to about 1 GeV, by comparing with the fluence-to-ambient dose equivalent H*(10) conversion coefficient. At 14 MeV, the response of LINUS is about 40% larger than that of the conventional A-B rem counter, 55% larger at 19 MeV and a factor 4 higher at 60 MeV, which gives much better agreement with H*(10) for neutron energies above 20 MeV.

2.2.4 *Phoswich-type detector*

(1) NE115 and NE213 Phoswich detector

In the Phoswich detector developed by Takada *et al.* [2.44–2.45], two scintillators of NE115 and NE213 are optically coupled to a single photomultiplier tube (Hamamatsu H1949). A cross-sectional view is shown in **Fig. 2.11**. The liquid scintillator is surrounded by a thin slow plastic scintillator (NE115, 5 mm thick) with a low sensitivity to neutral particles. The inner detector is a liquid organic scintillator (NE213, 58.5 mm diam and 58 mm length, which corresponds to the range of a 70 MeV proton) contained in a glass cell and has much greater sensitivity to neutrons. Charged particles are detected by both scintillators. The light output pulse in the NE115 from charged particles has a long characteristic time constant of about 225 ns, whereas the light output pulse in the NE213 from recoil protons produced by energetic neutrons has a time constant of about 30 ns. The time constant from Compton electrons produced by gamma rays is about 3.7 ns. These differences in the light-decay time constant make it possible to separate pulses of the three different particle species, as schematically drawn in **Fig. 2.12**.

Protons, neutrons and gamma rays are detected separately by use of a pulse-shape discrimination technique based on the standard CAMAC charge integration ADCs (Analog-to-Digital Converter). The charge integration of the signal is carried out during the time period specified by a gate pulse (total-gate, slow-gate). The total and slow components can be analyzed by a total gate pulse adjusted at the peak of the signal and by a wider delayed slow gate set at the long tail of the signal, respectively. A two-dimensional spectrum of the light output of the slow component versus the total component measured with the Phoswich detector in a

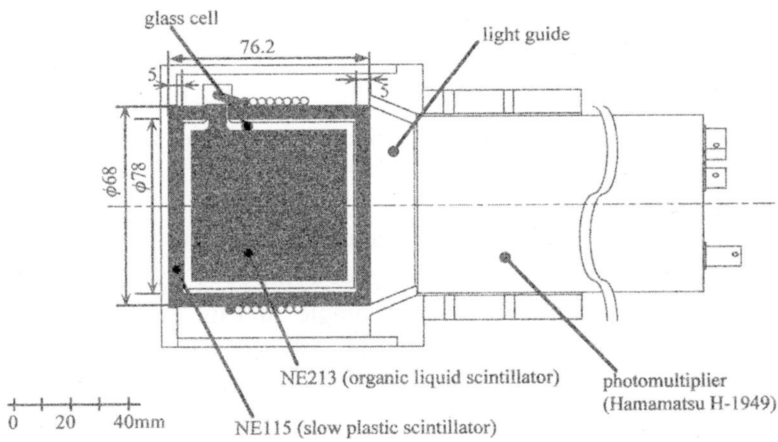

Fig. 2.11 Diagram of the Phoswich detector by Takada *et al.* [2.44, 2.45].

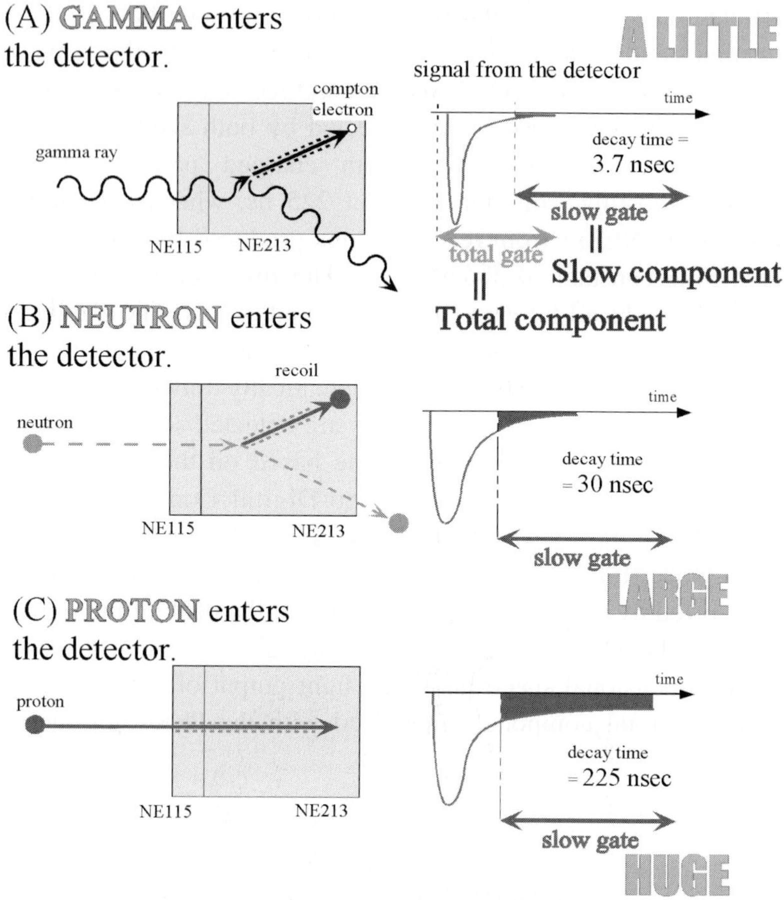

Fig. 2.12 Schematic model of particle identification principle of the Phoswich detector [2.44, 2.45].

proton and neutron mixed field, which was produced by bombarding 70 MeV protons on a 2 mm thick Be target at the AVF cyclotron facility of NIRS, Japan, is presented as an example in **Fig. 2.13** [2.45]. In this figure, there are seven distinct groups, labeled "A" through "G." The component "A" comes from the electrons scattered by gamma rays. The four components "B" through "E" containing small fractions of slow components are due to neutrons, where the components "B" and "C" correspond to recoiled protons produced by elastic collisions of neutrons with hydrogen in the liquid scintillator. The component "D" represents deuterons attributed to the $^{12}C(n,d)$ reaction, and the component "E" represents alpha particles produced in the $^{12}C(n,\alpha)$ and $^{12}C(n,n'3\alpha)$ reactions. Some fraction of component "B" is due to protons that lost some energy in the NE213 and then escaped from the NE213. The component "C" corresponds to protons that deposited all of their energy in the NE213. The components "F" and "G" doubtless contain external protons reaching the NE213 through the NE115 scintillator and the glass cell. The events where the external protons stopped in the NE115 are not seen in this figure because of the detection logic.

All events in the two-dimensional plot of **Fig. 2.13** were projected onto the total component axis to obtain individual spectra of gamma-ray, neutron and proton events, as shown in **Fig. 2.14**. We can see in these spectra that the discrimination between gamma rays, neutrons and protons is excellent, and that the proton peaks are clearly identifiable.

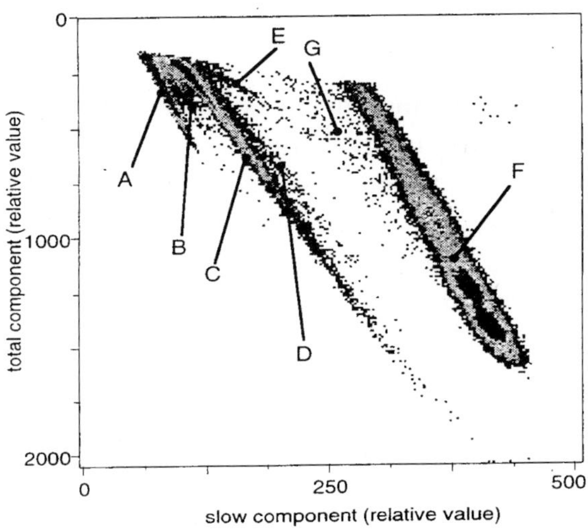

Fig. 2.13 A two-dimensional pulse height spectrum of the slow component versus the total component measured with the Phoswich detector at the NIRS. The component "A" shows gamma-ray events, the components "B" through " E" show neutron events and the components "F" and "G" show proton events [2.44, 2.45]. (© 1998 IEEE)

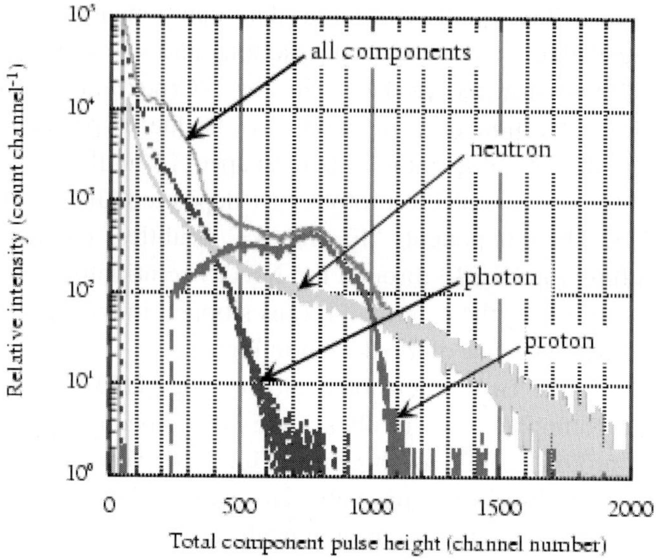

Fig. 2.14 The light output spectra projected on the total component axis obtained by discriminating between gamma-ray, neutron and proton events in Fig. 2.13 [2.44, 2.45].

(2) DARWIN (Dose monitoring system Applicable to various Radiations with WIde energy raNges)

A new, innovative radiation dose monitor, designated as DARWIN (Dose monitoring system Applicable to various Radiations with WIde energy raNges), has been developed by Sato *et al.* for monitoring doses in workspaces and the surrounding environments of high energy accelerator facilities [2.46]. DARWIN is composed of a Phoswitch-type scintillation detector that consists of liquid organic scintillator BC501A coupled with ZnS(Ag) scintillation sheets doped with ^{6}Li, and a data acquisition system based on a Digital-Storage-Oscilloscope. Scintillations from the detector induced by thermal and fast neutrons, photons and muons were discriminated by analyzing their waveforms, and their light outputs were directly converted into the corresponding doses by applying the G-function method. Dose from neutrons below 1 keV is evaluated from the number of scintillations from the ZnS(Ag) sheets, while those from neutrons above 1 MeV, as well as photons and muons, are estimated from the light output of scintillations from

BC501A. The schematic view of DARWIN is illustrated in **Fig. 2.15**. The detector consists of a cylindrical glass cell with a 12.4 cm diameter and a 12.7 cm length containing BC501A scintillator surrounded by ^{6}Li-doped ZnS(Ag) sheets and an aluminum cover. Characteristics of DARWIN were studied by both calculation and experiment.

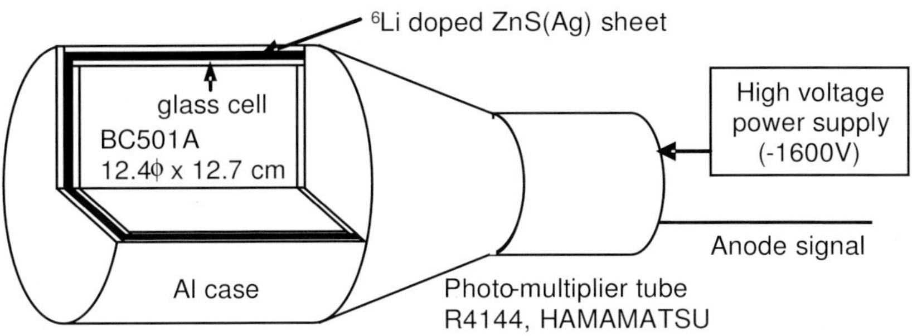

Fig. 2.15 Schematic view of DARWIN by Satoh *et al.* [2.46].

They adopted a simple pulse shape discrimination (PSD) technique called the gate integration method [2.47], which discriminates scintillations by means of a 2-dimensional plot of the fast and slow components. An example is shown in **Fig. 2.16** for the continuous energy neutron field using a ^{252}Cf source. **Figure 2.17** gives a graphical presentation of the measured response functions in the continuous energy neutron field produced by a ^{252}Cf source. The calculated result for neutron scintillation using a combination of SCINFUL-QMD [2.36] and PHITS [2.48] codes is also plotted in the figure. An excellent agreement is observed between the measured and calculated data. Doses due to each scintillation type were obtained from the response functions by applying the corresponding G-functions.

The calculated results indicate that DARWIN gives reasonable estimations of doses in most radiation fields. It was found from the experiment that DARWIN has an excellent property of measuring doses from all particles that significantly contribute to the doses in surrounding environments of accelerator facilities – neutrons, photons and muons with wide energy ranges. The experimental results also suggested that

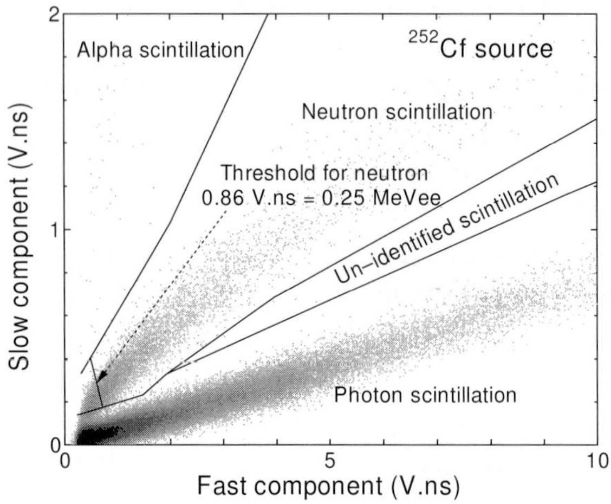

Fig. 2.16 Two-dimensional plot of the fast and slow components for the continuous energy neutron field produced by a ^{252}Cf source [2.46].

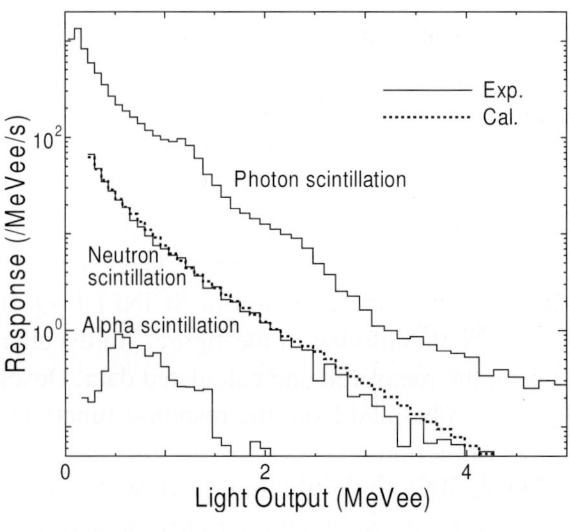

Fig. 2.17 Measured and calculated responses in the continuous energy neutron field produced by a ^{252}Cf source [2.46].

DARWIN enables us to monitor small fluctuation of neutron dose rates near background-level, owing to its high sensitivity.

The G-function method [2.49–2.50] enables us to measure the dose in real-time. This is due to the assumption adopted in the method that dose can be uniquely determined from the light output of each scintillation L from DARWIN by correlating the two quantities directly with a function named the *G-function*. Under this assumption, the total dose D due to the scintillations with a light output distribution $N(L)$ can be obtained by

$$D = \int_{L_{min}}^{L_{max}} N(L) \cdot G(L) dL, \tag{2.1}$$

where $G(L)$ denotes the G-function, and L_{min} and L_{max} are respectively the minimum and maximum light outputs of the scintillations.

The characteristics of the neutron detectors discussed here so far are summarized in **Table 2.2** for comparison.

2.3 Experimental Procedure

2.3.1 *Sequential neutron measurements on the ground at sea level*

The long-term monitoring of neutron fluence has been done using large-scale and highly-sensitive neutron monitors at several places in the world. Here, the sequential measurements of neutron energy spectrum and dose equivalent are described at four places, two in Japan, one in U.S.A. and one in Germany. The former two measurements in Japan have been done over two years, while the latter two measurements were done in less than one year. In all measurements, the atmospheric pressure was recorded at each site so that the neutron measurements could be corrected for the effect of pressure (atmospheric depth).

(1) Measurement in Sendai, Japan

In this study, a sequential measurement was carried out from April 2001 to March 2003 in Sendai, Japan during the solar maximum period [2.51–2.52] in order to investigate the long-term changes of cosmic-ray neutron flux and dose rate on the ground using three types of neutron detectors, Bonner ball, high-sensitivity rem counter and NE213 organic liquid scintillator. For the long-term sequential measurements of cosmic

Table 2.2 Characteristics of neutron detectors to be used in terrestrial neutron measurements.

Type of detector	Quantities to be measured	Characteristics
Multi-moderator detector (Bonner ball)	energy spectrum and ambient dose equivalent	gross spectrum from thermal to 1 GeV poor energy resolution strongly dependent on initial guess spectrum very easy-handling but a set of detectors
Organic liquid scintillator	same as above	1MeV to several hundreds MeV good energy resolution neutron-photon discrimination is required photomultiplier stability is essential
Phoswich type detector	same as above	1MeV to several hundreds MeV good energy resolution neutron-photon-charged particle discrimination usable in charged particle mixed field in space
Dose equivalent counter (rem counter)	ambient dose equivalent	thermal to 20 MeV for usual type 1 GeV for extended range type very-easy handling with only one detector

ray neutrons, an experimental cabin was set up on the ground at the Kawauchi campus of Tohoku University in Sendai, Japan. The cabin is 564 cm × 230 cm × 262 cm and is air-conditioned, as seen in **Fig. 2.18**. The measuring position is situated at latitude 38°15' north and longitude 140°52' east and an altitude of about 70 m. The geomagnetic latitude in Sendai is about 29°N, and the cut-off rigidity is 10.43 GV.

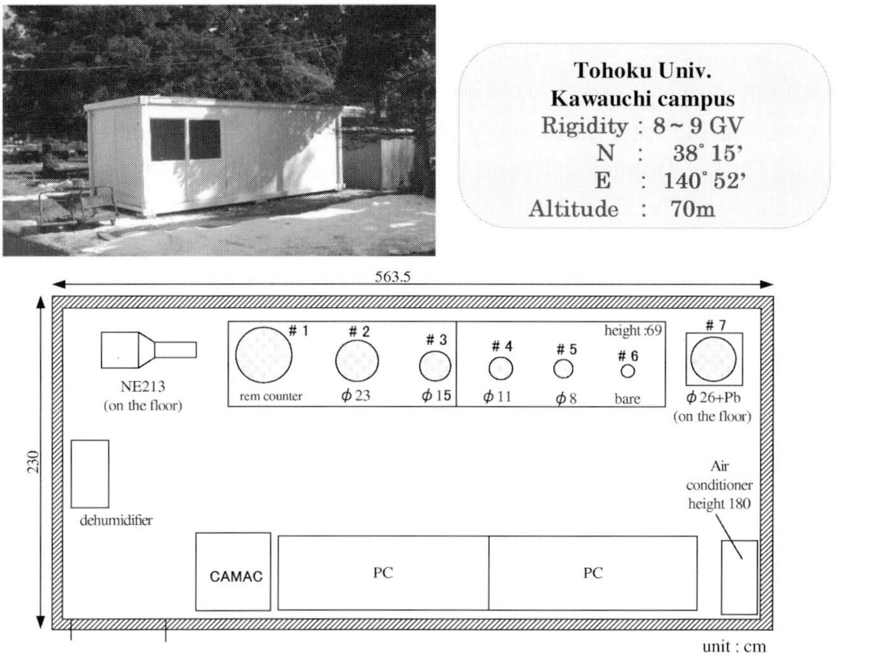

Fig. 2.18 Photograph of the experimental cabin at Tohoku University, Sendai, Japan and the arrangement of the detectors in the cabin for cosmic-ray measurement [2.51, 2.52].

The Bonner ball is composed of a 5.08-cm diam spherical ^{3}He counter filled with 5 atm or 10 atm ^{3}He gas, which is placed at the center of a set of polyethylene moderators of different diameters (0 (Bare), 8.1, 11.0, 15.0 and 23.0 cm), as given in **Fig. 2.1** [2.20]. The high-sensitivity rem counter, the so-called 5-in-diam ^{3}He rem counter, consists of a 12.7-cm-diam, 5 atm ^{3}He gas-filled spherical counter covered with a polyethylene moderator and inner absorber [2.37] for measuring the neutron ambient dose equivalent rate, as seen in **Fig. 2.6**. The 12.7 cm diam by 12.7 cm long NE213 organic liquid scintillator is also used to measure the neutron energy spectrum in the MeV region, while the Bonner ball gives the energy spectrum over a wide energy region down to thermal energy. However, the Bonner ball has poor energy resolution, especially in high-energy region above about 10 MeV. The NE213 detector has sensitivity not only for neutrons, but also for gamma rays and muons. The neutron events were separated from gamma-ray and muon events using a pulse-shape discrimination technique [2.30] which uses the total and slow gates, as described in Sec. 2.2.2. All detectors were placed at the same height (about 1 m) above ground.

The measuring interval is every 30 minutes for the rem counter, every 12 hours for the Bonner ball, and a much longer accumulation period of several days for NE213. From the total counts above the pulse-height threshold measured with the Bonner ball, the neutron energy spectrum was obtained by unfolding with the SAND-II code [2.24]. The response functions of the five detectors are calculated with the MCNPX Monte Carlo code [2.22] for neutron energies up to 1 GeV [2.21]. In this unfolding we used the typical cosmic-ray neutron spectrum by Goldhagen *et al.* as the initial guess spectrum, as described in Sec. 2.3.3 [2.2]. The neutron dose rate was then estimated by multiplying the neutron energy spectrum with the fluence-to-ambient dose equivalent conversion factor given by ICRP 74 [2.40]. The total counts of the rem counter were also converted into ambient dose rate by applying the conversion factor of 18.5 cps/(μSv h^{-1}), which was recalibrated at FRS, JAERI in Japan, using a ^{252}Cf neutron source for this measurement. The neutron energy spectrum in the MeV energy region was also obtained by unfolding the pulse height distribution of NE213 with the FORIST code [2.53] using the response functions in the energy range of 5 MeV to 800 MeV given by Sasaki *et al.* [2.33].

The data with the Bonner ball and the rem counter were taken sequentially from April 1, 2001 to March 2003, but the Bonner ball data from January to March in 2003 could not be obtained because of detector leakage trouble. The NE213 data were taken in three periods from November 7, 2001 to December 30, 2001, from August 29, 2002 to October 6, 2002, and November 11, 2002 to December 27, 2002.

(2) Measurement in Chiba, Japan

Following this study, a group in the Japan Chemical Analysis Center measured cosmic-ray neutrons with the same type of the Bonner ball and the 5-in.-diam ^{3}He rem counter (model NDN4NA11 fabricated by Fuji Electric Systems Co. Ltd.), as used in (1) Sendai. The measurements have been taken sequentially from March 2002 up to the present at a site in Chiba, Japan [2.54]. The geomagnetic latitude in Chiba is about

26.6°N. All of the results of the cosmic-ray intensity (neutron dose rate and neutron count rate by each ^{3}He proportional counter of the Bonner ball) were averaged over the accumulation of counts measured for 24 hours in order to decrease the counting error. All of the measurement apparatus are placed in a well air-conditioned room in a one-story concrete building with galvanized iron roofing (See **Fig. 2.19**). Before the beginning of the sequential cosmic-ray monitoring, a comparison of the cosmic-ray intensity between the inside and outside of the building was carried out using a survey-meter type neutron rem counter. This preliminary experiment revealed that there was no difference between the results measured inside and outside of the building. The rem counter was calibrated using a ^{252}Cf source at FRS, JAERI to be 15.2 cps/(μSv h^{-1}).

Fig. 2.19 Experimental arrangement for the Bonner ball measurement in the field at Chiba, Japan [2.54].

(3) Measurement in Braunschweig, Germany

NEMUS [2.55] is an extended Bonner ball, fabricated by PTB (Physikalisch-Technische Bundesanstalt). It consists of a ^{3}He-filled spherical proportional counter, 12 polyethylene spheres with diameters from 7.62 cm (3 in.) to 45.72 cm (18 in.), and four polyethylene spheres with copper or lead inlets, similar to the Bonner ball used by Goldhagen *et al.* that is described in **Table 2.1**. The response functions were calculated up to 10 GeV using the MCNP and MCNPX codes and adjusted to response values obtained in several calibration measurements. A long-time measurement of the natural neutron background at the PTB site (geomagnetic latitude of 48.2°N, cutoff rigidity of 2.86 GV, approx. 75 m above sea level) was done in December 1999 [2.10]. The solar activity during this period was about 89% of the cosmic ray intensity of 1954. In a wooden cottage of 3 × 3 m^2, which was temperature-controlled between $6\,°C$ and $10\,°C$, eight spheres of the NEMUS set were placed on tripods about 130 cm above ground so that the distances between them were at least 80 cm. The total measuring time was 48 days.

The neutron energy spectra were obtained by unfolding of measured count rates with two unfolding codes, MAXED [2.27–2.28] and GRAVEL [2.56]. The two codes require the initial guess spectrum, which was estimated from a calculation using the MARS code for the geographical position of the PTB [2.4].

(4) Measurement at several sites, U.S.A.

Gordon *et al.* [2.57] used an extended Bonner ball consisting of 14 detector sets developed by Goldhagen *et al.* [2.1–2.2], as shown in **Table 2.1**. A group of three detectors were deployed in an aluminum suitcase together with the preamplifiers. Two other suitcases, containing three detectors each, were part of the spectrometer. The five largest detectors were individually housed in aluminum cylindrical containers. The three suitcases and the five largest detectors were placed on top of lightweight aluminum tables, about 1 m above the ground. When used outdoors, the suitcases and electrical connections were covered with plastic bags to prevent damage from rain, and the tables were tied down to prevent movement caused by high winds. **Figure 2.20** shows the spectrometer on the roof of the IBM T. J. Watson Research Center in Yorktown Heights, N.Y.

Fig. 2.20 Fourteen-element Bonner sphere spectrometer located on the roof of the IBM T.J. Watson Research Center. The suitcases and connector ends of the large detectors were covered in plastic bags to prevent water damage [2.57]. (© 2004 IEEE)

Measurements were made outdoors at Fremont Pass, CO; Mount Washington, NH; Yorktown Heights, NY; and Houston, TX. Indoor sites included Leadville, CO, and computer labs at IBM sites in Yorktown Heights, NY, and Burlington, VT. These measurements were made between September 2002 and June 2003. The Leadville measurement could be included with the outdoor measurements because the roof of the one-story building used there was relatively thin, made primarily of wood, and did not significantly change the shape of the measured neutron spectrum. **Table 2.3** shows data relating to each of the measurement locations, including altitude, atmospheric depth, cutoff rigidity, and so on.

The response of each detector to neutrons in the energy range 10^{-10} MeV to 100 GeV was calculated using the MCNPX code (versions 2.5.d and 2.5.e), as shown in **Fig. 2.3**. The unfolding code MAXED was used to get the energy spectra, together with the initial guess calculated by

Roesler *et al.* [2.58] using the FLUKA code for locations at sea level and 811 g/cm^2 atmospheric depth.

Table 2.3 Data relating to the measurement locations by Gordon *et al.* [2.57].

Site	Alt. [m]	Atmosph. Depth [g/cm^2]	R_c [GV]	F_{BSYD} (11/02)	Solar modulation factor	F_{alt}
Fremont Pass, CO	3450	682.6	2.94	0.874	1.005	14.74
Leadville, CO	3150	719.7[a]	2.97	0.880	1.010	11.09
Mt. Wash. NH	1905	826.6	1.58	0.970	0.990	4.89
Yorktown Hts. NY	167	1011.3	2.00	0.997	1.000	1.19
Houston, TX	14	1028.0	4.68	0.896	1.008	1.04

[a]Includes 8.75 g/cm^2 for the roof in Leadville.

2.3.2 *Neutron measurements aboard an airplane and at mountain level*

Several measurements of neutron energy spectra and dose equivalent at mountain levels and aboard an airplane are selectively described here in order to investigate the altitude variation of cosmic-ray neutrons.

(1) Neutron measurements at mountain level

Kowatari *et al.* investigated the altitude variation of the cosmic-ray neutron component in low geomagnetic latitude [2.59]. To assess the natural background neutron dose rate in Japan which locates at a lower geomagnetic latitude, the cosmic-ray neutron energy spectrum and the ambient dose equivalent were evaluated by measuring the neutron energy spectrum at various altitudes in the Mt. Fuji area (26°N geomagnetic latitude, altitude: 40 m – 2400 m) using the Bonner ball previously described. In parallel with the neutron energy spectrum measurement, the measurement of the neutron ambient dose equivalent was also done by using a commercially available neutron rem counter in the same area (altitude: 40 m – 3776 m). The neutron rem counter used for the measurement was a portable survey-meter type (Type NSN 21011) equipped with 5.07-cm diam spherical ^{3}He proportional counter, manufactured by Fuji Electric Systems Co. Ltd. An additional polyethylene moderator (2 cm thickness) was placed on the rem counter in order to improve the neutron sensitivity in the high energy region. The neutron ambient dose equivalent rate was measured using 9 sets of rem counters in one measurement, since the cosmic-ray induced neutron dose rate in the lower geomagnetic region is quite low [2.52, 2.54, 2.60]. From the counting rates obtained by 9 neutron rem counters, the ambient dose equivalent rate for neutrons was obtained by multiplying the counting rate to the ambient dose equivalent rate (H*(10)) conversion coefficient given by ICRP 74 [2.40] and the calculated values by Sannikov and Savitskaya [2.61] in the energy range of 200 MeV to 1 GeV. The counting rate to ambient dose equivalent rate conversion coefficient of each rem counter was estimated to be 2.91 (cps/(μSv·h^{-1})) from results of a calibration with a ^{252}Cf source at FRS, JAERI, averaged over 9 sets of rem counters.

A series of measurements were performed in an air-conditioned freight car from September 2, 2002 to September 9, 2002 (See **Fig. 2.21**). At the high altitude locations above 2,800 m, only two rem counters were used for the neutron measurement because the area was inaccessible by car. More than 5 hours were required for neutron measurements at one position in order to reduce the counting errors of each Bonner ball and rem counter down to less than 3%. The temperature in the cargo container was kept within 23 ± 2°C. The centers of each ^{3}He counter of the Bonner ball and of each rem counter were fixed at about 2 m and 1 m heights above the ground level, respectively. An open space was selected for the measurement in the Mt. Fuji area in order to suppress the influence from surrounding objects. Throughout the neutron measurements, the temperature, atmospheric pressure and the geographical position of the measuring point were also measured. The conversion of geographical position into geomagnetic coordinate was carried out on the basis of the International Geomagnetic Reference Field (IGRF) 2000.

Schraube *et al.* [2.8] measured the neutron spectrum on the roof of an atmospheric observatory positioned at the summit of the Zugspitze, the highest mountain in Germany (2963 m). Its geographical position is 47°25'N and 11°E (geomagnetic latitude of 42.3°N, cutoff rigidity of 4.29 GV). The Bonner ball used by Schraube et al. consisted of a set of 14 polyethylene spheres with a ^{3}He counter of 32 mm inner diameter (172 kPa ^{3}He gas and 100 kPa Kr gas). The diameters of polyethylene spheres were 2.5, 3, 3.5, 4, 5, 6, 7, 8, 9, 10, 11, 12, and 15 in. In addition, the 9 in. diameter sphere had an internal lead layer. The Bonner ball was placed in the open air in a thin aluminum container whose temperature could be kept at 20 ± 3°C. Total experiment time was 144 h. The counting time for each sphere was chosen in such a way that the uncertainty due to counting statistics was slightly less than 2.5%. This experiment is still going on at present by the GSF - National Research Center for Environment and Health, Institute of Radiation Protection. **Figure 2.22** shows the photographs of the experimental arrangement using the Bonner ball at Zugspitze [2.62].

Florek *et al.* [2.7] did a similar experiment using the Bonner ball 0.7 m above the ground at High Tatras (2632 m) in August 1983 and August 1992. During those times, dry and stable weather was maintained in Central Europe and the 11-year solar cycle was near its maximum (~90%).

Fig. 2.21 Photographs of field measurements in air-conditioned freight cars [2.54, 2.59].

Fig. 2.22 Photographs of the experimental room and the arrangement of the Bonner ball at Zugspitze, Germany [2.62].

(2) Neutron measurements aboard an airplane

Nakamura *et al.* [2.17] measured the altitude variation of the cosmic-ray neutron energy spectrum and dose equivalent aboard an airplane by using the Bonner ball and the 5-in.-diam ^{3}He rem counter described in Secs. 2.2.1 and 2.2.3. The data were obtained on a February 27, 1985 (solar minimum period) flight of a DC-8 aircraft of Japan Air Lines chartered by the Institute of Physical and Chemical Research (RIKEN). The flight started at 10:06 a.m. from Narita Airport, proceeded west, keeping the geomagnetic latitude at approximately 24°N, and returned at about 12:00 a.m. to Narita Airport. During a total flight time of 2.2 h, the aircraft was flown for 60 min along the 24.1 ± 0.5°N parallel of geomagnetic latitude at an altitude of 4880 m, and for a period of 20 min along the 24.2 ± 0.4°N parallel at 11280 m for the cosmic-ray neutron spectral measurements. The flight path is plotted in **Fig. 2.23**.

To help reduce the significant uncertainties in calculations of such exposures, the Atmospheric Ionizing Radiation (AIR) Project, an international collaboration of 15 laboratories organized by the NASA (National Aeronautics and Space Administration) Langley Research

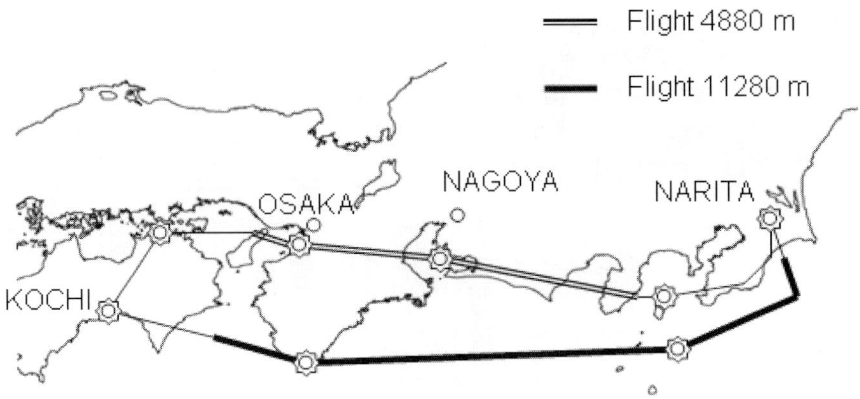

Fig. 2.23 Path of the DC-8 flight chartered by the Institute of Physical and Chemical Research [2.17].

Center, made simultaneous radiation measurements with 14 instruments on a NASA ER-2 high-altitude aircraft [2.1–2.2]. Of the many AIR measurements, Goldhagen *et al.* discussed neutron spectrometry using the extended Bonner ball consisting of 14 detector sets developed by them, as described in **Table 2.1**.

The AIR ER-2 flights were scheduled for June 1997, a time of maximum galactic cosmic radiation (solar minimum), and were designed to cover as wide a range of latitude and altitude as possible within the operational capabilities of the ER-2. There were five measurement flights. Their four paths (there were two identical South flights) are shown on a map in **Fig. 2.24**. All flights originated at the NASA Ames Research Center (37.4°N, 122°W) in California and started with a northward climb. The East flight provided a long period of nearly constant radiation level by flying at constant geomagnetic cutoff. The rest of the flights measured the effects of latitude and of altitude at low geomagnetic cutoffs. The East and the two North flights included an altitude dip while flying magnetic west at constant geomagnetic cutoff, allowing measurements at altitudes from 21.3 km (70,000 ft) down to 16 km (52,500 ft). **Figure 2.25** shows graphs of altitude as a function of time after takeoff for three of the flights. Overall, the flights covered

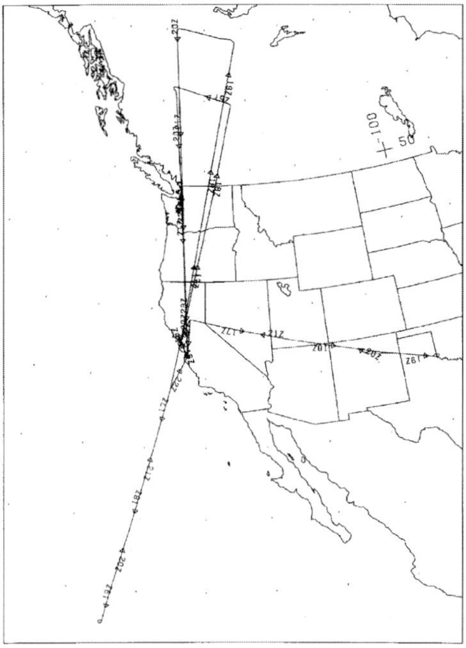

Fig. 2.24 Flight paths of the AIR measurement ER-2 flights [2.1, 2.2].

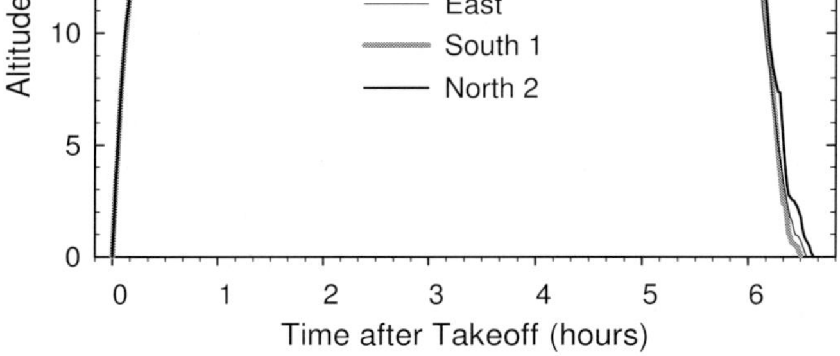

Fig. 2.25 Altitude as a function of time during three of the ER-2 flights [2.1, 2.2].

latitudes from 18° to 60°N, corresponding to geomagnetic vertical cutoff rigidities from 12 GV to 0.4 GV, and atmospheric depths from 50 to 110 g/cm^2, not including initial climb and final descent, which yield some data at lower altitudes for one latitude. Neutron spectrometry measurements were also made on the ground.

Nagaoka *et al.* measured the altitude variation of neutron ambient dose equivalent rate using a balloon [2.63]. A balloon was launched from sea level by JAXA on August 25, 2004 from the Sanriku Balloon Center, Iwate, Japan, and rose to an altitude of 25 km. The geomagnetic latitude at the Sanriku Balloon Center is 30.2°N and the cutoff rigidity is 9.2 GV. **Figure 2.26 (a)** shows a photograph of the balloon experiment, and **Fig. 2.26 (b)** gives the flight route of the balloon. A portable survey-meter type (Type NSN 21011) neutron rem counter and an identical counter covered with additional polyethylene moderator (2 cm thickness) were installed into a 2-mm-thick aluminum pressure vessel. The pressure vessel was used because the atmospheric pressure decreased to less than 30 hPa. The balloon was collected on the next day after 1 day of flight. The measurement was taken during a period of average solar activity.

(3) Neutron measurements in space

The same type of Bonner ball shown in **Fig. 2.1** was fabricated by the National Space Development Agency (NASDA, now JAXA: Japan Aerospace Exploration Agency) for use in space. The Bonner ball detector was mounted in the SPACEHAB module located in the space shuttle cargo bay. **Figure 2.27** shows the Bonner ball on board the space shuttle and space station. The neutron energy spectrum in the shuttle was measured from January 24 to 28, 1998, in S/MM-08 (STS-89) [2.64]. Measured data and orbital information were transferred through a downlink to the NASA Johnson Space Center, Mission Control Center and transmitted for analysis to Tsukuba Space Center, NASDA. This Bonner ball was also used in the US module of ISS (International Space Station, altitude 400 km) by NASA to measure neutron dose from March 23 to July 6, 2001 [2.65].

(a)

(b)

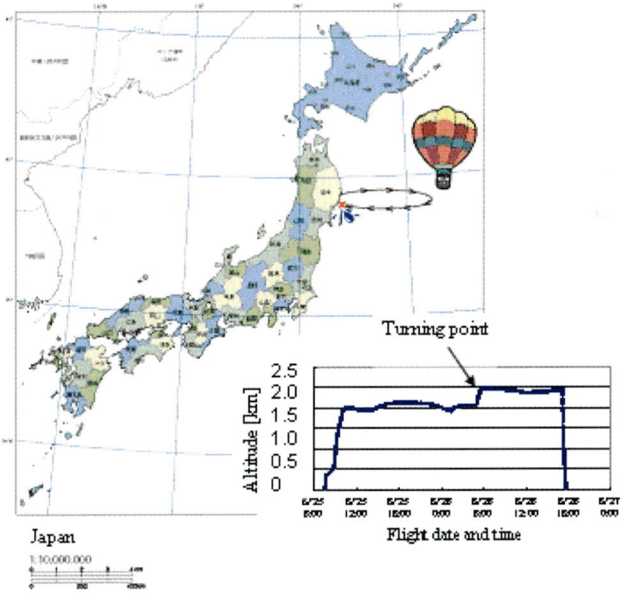

Fig. 2.26 (a) Photograph of the balloon experiment at Sanriku, Japan on August 25, 2004 [2.63]. (b) Flight route of the balloon experiment by JAXA, Japan Aerospace Exploration Agency [2.63].

Fig. 2.27 Photographs of the Bonner ball on board the shuttle in the SPACEHUB module and the US module of ISS, International Space Station [2.64, 2.65].

2.3.3 *Data analysis*

Once the neutron response function for each of the detectors, $R(E,P)$, is known, the neutron energy spectrum, $\phi(E)$ can be obtained from the measured counts, $C(P)$ as follows,

$$C(P) = \int R(E,P) \cdot \phi(E) dE\,, \tag{2.2}$$

where E is neutron energy and P is output pulse height. By solving this integral equation through a deconvolution, or unfolding, the neutron energy spectrum can be obtained with a computer program. Various unfolding programs have ever been developed for this purpose.

For a Bonner ball spectrometer, the number of $C(P)$ data is always much less than the number of $\phi(E)$ values, and as such many reasonable solutions may exist. To select the optimum solution of $\phi(E)$, an initial, or default, spectrum that represents prior knowledge about the shape of the spectrum is always required. From the ensemble of possible spectra that fit the experimental data within uncertainties, the optimum spectrum is selected which is closest to the default spectrum. The unfolding codes SAND-2, BUNKI, BON, MAXED, GRAVEL, *etc.* are used. Among them, SAND-2 [2.24] using the successive iteration method and MAXED [2.27–2.28] using the maximum entropy method are most widely used. These codes are constrained to use just the shape of the default spectrum and not the amplitude (total flux). Because the choice of default spectrum can influence the final measured spectrum, the most appropriate default spectrum must be selected. Since these codes apply no smoothing, they preserve any structure in the default spectrum that is finer than the resolution of the spectrometer.

As an initial (default) spectrum for unfolding, Goldhagen *et al.* examined available calculated cosmic-ray neutron spectra. **Figure 2.28** shows graphs of three calculated cosmic-ray neutron spectra for different locations in the atmosphere made by different groups [2.3–2.4, 2.66] using different Monte Carlo radiation transport codes. Neutron flux, sometimes called fluence rate, is plotted on the vertical axis in units of (neutrons cm^{-2} s^{-1}) per lethargy (the natural logarithm of energy) versus neutron energy in MeV, with a logarithmic scale on the horizontal axis. (Flux per lethargy, the so-called lethargy flux, is equivalent to $E(d\phi/dE)$, the so-called energy flux, where E is particle energy and ϕ is flux.) The calculated spectra all have a large "evaporation" peak centered at 1 or 2 MeV, a smaller peak at about 100 MeV, and detectable numbers of neutrons up to about 10 GeV (10^4 MeV). The spectrum of Roesler *et al.* [2.3], calculated for 200 g/cm^2 at a 4.3 GV cutoff, shows fine structure from nuclear resonances in nitrogen and oxygen in the atmosphere. It also has a larger ratio of high-energy neutrons to evaporation neutrons than the calculations by Kurochkin *et al.* [2.4] (200 g/cm^2, 2.9 GV cutoff) and by Armstrong *et al.* [2.66] (50 g/cm^2, 4.6 GV cutoff). Only the Armstrong calculation included thermal energies. They compared the calculated spectra with their measured spectra on board the aircraft and at sea level [2.1–2.2]. Except for the widths of their bins, the three unfolded spectra are practically identical, and the measured spectrum is generally similar to the calculated spectra.

For the organic liquid scintillator, the revised FERDO [2.67] and the FORIST [2.53] unfolding codes are widely used without the need of any initial guess spectrum. These codes include the window function which represents the Gaussian broadening of the output pulses, reflecting the detector energy resolution, $\sigma(E)$. The revised FERDO code uses the functional value of $\sigma(E)$, and the FORIST code can change $\sigma(E)$ values arbitrarily.

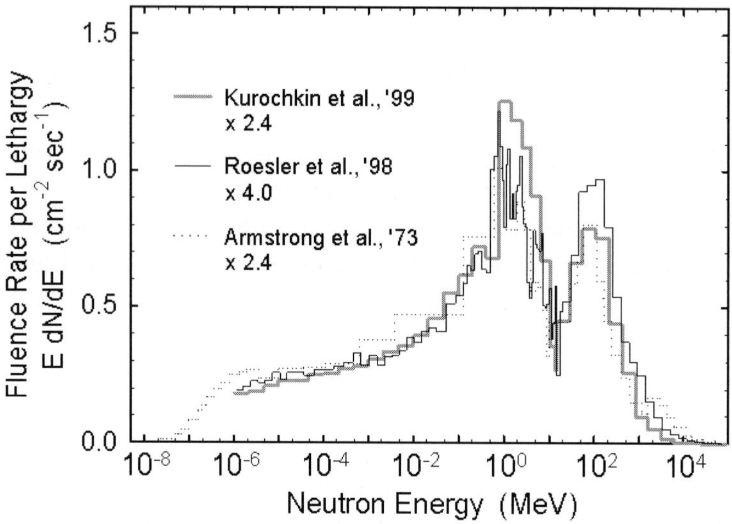

Fig. 2.28 Three calculated cosmic-ray neutron spectra [2.3, 2.4, 2.66] used as default spectra in unfolding the measured spectra. The calculated spectra are shown scaled to give best fit to the data that were used to unfold the spectra in Fig. 2.31.

After obtaining the neutron energy spectrum, the ambient dose equivalent (H*(10)) was obtained by multiplying the fluence-to-ambient dose equivalent (H*(10)) conversion coefficients given by ICRP 74 and the calculated values by Sannikov and Savitskaya [2.61] for the energy range of 200 MeV to 1 GeV.

The G-function described in Sec. 2.2.4 (2) is introduced as follows. The measured pulse height distribution, *P(L)*, is given by

$$P(L) = \int R(E,L) \cdot \phi(E) dE, \tag{2.3}$$

where L is light output, *R(E, L)* is the response function per neutron and $\phi(E)$ is the energy spectrum (n/cm^2).

The ambient dose equivalent, H*(10) (Sv) is given by

$$H^*(10) = \int \phi(E) \cdot h(E) dE, \tag{2.4}$$

where *h(E)* is the fluence-to-ambient-dose-equivalent conversion coefficient (in Sv cm^2) given by ICRP74 [2.40]. Then, the G function is introduced,

$$h(E) = \int R(E,L) \cdot G(L) dL . \tag{2.5}$$

By combining Eq. (2.3) and Eq. (2.4), we can get

$$H = \int P(L) \cdot G(L) dL . \tag{2.6}$$

The G-function can be calculated by solving Eq. (2.5) using the response function of the detector *R(E,L)* (where *E* represents the energy of incident particles), and the fluence-to-dose conversion coefficients *h(E)*, along with employing the unfolding code SAND-II [2.24].

2.4 Results and Discussions

2.4.1 *Atmospheric pressure effect*

The effect of atmospheric pressure on the neutron dose rate during the period of March 9 to October 4, 2001 is exemplified in **Fig. 2.29** [2.51–2.52]. The dose rate clearly indicates an inverse effect to atmospheric pressure and can be fitted to the following exponential attenuation curve:

$$y = 2.2 \ exp\ (-0.0065\ x), \tag{2.7}$$

where y is ambient dose rate in μSv/h and x is atmospheric pressure in hPa. Eq. (2.7) is generally written as

$$y = A \ exp(\ -\ \mu x\ ), \tag{2.8}$$

where y is the measured total counts of the Bonner ball or the rem counter. All measured total counts, y, at any atmospheric pressure are extrapolated to the value, y', at 1013 hPa as

$$y' = y \ exp[\ -\ \mu(1013\ -\ x)]. \tag{2.9}$$

The μ values obtained by least mean square fitting for two periods from June 10, 2001 to Dec. 31, 2001 in Sendai and from Jan. 1, 2002 to March 31, 2003 in Chiba [2.54] are tabulated in **Table 2.4 (a)**.

Schraube *et al.* [2.8] also gave a similar exponential attenuation effect of atmospheric pressure measured at the summit of the Zugspitze, as shown in **Table 2.4 (b)**. The μ values in **Table 2.4 (b)** are 20 to 50% larger than those in **Table 2.4 (a)** in Japan. This may reflect the effect of

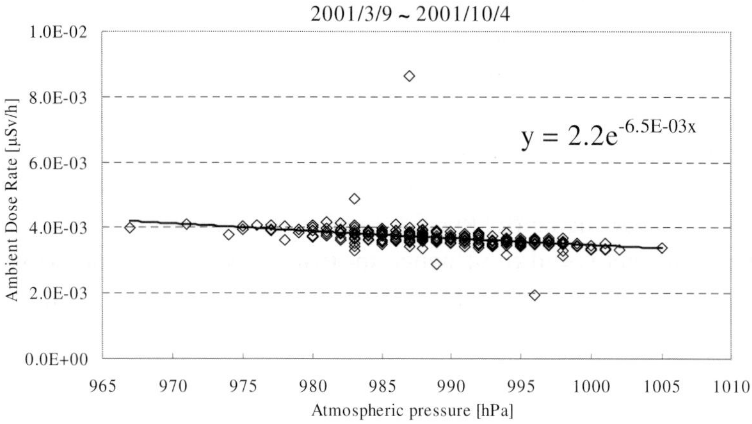

Fig. 2.29 Correlation of ambient dose equivalent rates with atmospheric pressure, as measured with a rem counter during the period from March 9, 2001 to October 4, 2001 [2.51, 2.52].

Table 2.4 Atmospheric pressure correction factor, μ in Eq. (2.9) (cited from Refs. [2.8], [2.52], [2.54]).

(a) Low geomagnetic latitude in Japan

	Observation period	Rem Counter $\times 10^{-4}$	Bonner Ball $\times 10^{-4}$				
Moderator diameter (cm)		29.6	23	15	11	8.1	5.1
μ	2001,6.10-2001,12.31	45	30	55	46	45	48
	2002,1.1-2003,3.31	47	46	46	43	47	67

(b) High geomagnetic latitude in Europe

	Data Source	5"(12.5 cm) Bonner Ball
μ	Zugspitze (meas.)	82 ± 16
	Kiel-monitor (meas.)	72.1 ± 0.1
	FLUKA (calc.)	60 ± 4

geomagnetic latitude, since the change of cosmic-ray neutron intensity due to atmospheric pressure is smaller at low geomagnetic latitude than at high geomagnetic latitude.

2.4.2 *Neutron energy spectra*

(1) Energy spectra on the ground at different elevations

Figure 2.30 shows the neutron spectrum measured outdoors on a concrete roof in Yorktown Heights, New York by Gordon *et al.* [2.57]. Neutron energy, E, times the differential neutron flux, $d\phi/dE$, (referred to as the energy flux $E\ d\phi/dE$), is plotted on the vertical axis versus neutron energy, E (in MeV), using a logarithmic scale on the horizontal axis. The spectrum has three broad peaks: a high-energy peak centered at about 100 MeV and extending up to about 10 GeV, a nuclear evaporation peak centered at around 1 or 2 MeV, and a thermal peak due to neutrons slowing down by scattering with surrounding materials. Between the evaporation peak and the thermal peak, there is a plateau region where $d\phi/dE$ is approximately proportional to $1/E$. The evaporation peak has fine structure from nuclear resonances in nitrogen and oxygen in the atmosphere and in materials in the concrete roof. Most of this structure is finer than the resolution of the Bonner spectrometer, and it appears in the measured spectrum because it is present in the calculated spectrum [2.58] that was used for the initial guess spectrum in the unfolding. The measured spectra at the other 4 locations given in **Table 2.3** have also similar shapes, as shown in **Fig. 2.31**.

The representation of the neutron spectrum in **Fig. 2.30** is standard in the field of radiation protection, but not in the literature of SEUs or cosmic-ray physics, where it is simply plotted as $d\phi/dE$. But neutrons typically require a log-log plot covering many orders of magnitude on both axes. In the energy range where neutron flux has 1/E spectrum, $E\ d\phi/dE$ is nearly constant and is mathematically identical to $d\phi/d(ln(E))$, referred to as lethargy flux, as described in Sec. 2.3.3.

Wiegel *et al.* also gave a similar spectrum having three peaks with a flat region in the intermediate energy range at PTB, Germany, as did Gordon *et al.* [2–10].

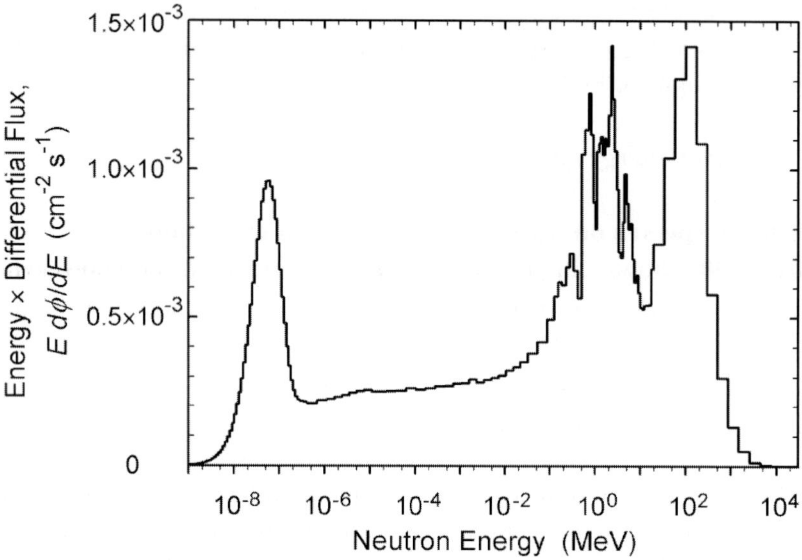

Fig. 2.30 Neutron spectrum measured on the roof of the IBM T.J. Watson Research Center in York Town Heights, N.Y. [2.57]. (© 2004 IEEE)

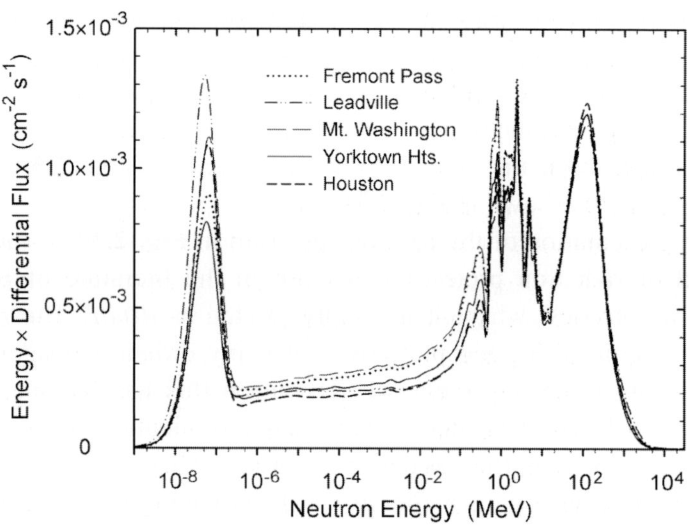

Fig. 2.31 Measured neutron spectra for all five sites indicated in the plot. Each spectrum has been scaled to sea level at the cutoff of New York City and solar modulation for November 2002, as described in the text [2.57]. (© 2004 IEEE)

Figure 2.32 shows the neutron energy spectra measured by Nakamura *et al.* at Tohoku University, Sendai on three different days (May 12, July 12, September 6 in 2002) [2.51–2.52]. Each data set was obtained for 12 hours using the Bonner ball. Three peak components can also be seen in **Fig. 2.32**. The first highest peak is a cascade peak component, which appears around 100 MeV. This component fluctuates about 30% from day to day and has a clear tendency to increase on the day of relatively higher solar activity, which in turn corresponds to a higher neutron flux. The second middle peak is an evaporation peak component, which appears around a few MeV. The third lowest peak is a thermal neutron peak component, which appears below about 1 eV. The peak values of the second and, in particular, the third components keep almost constant values of about 7.5×10^{-4} (n cm^{-2} sec^{-1} lethargy^{-1}) and about 1.8×10^{-4} (n cm^{-2} sec^{-1} lethargy^{-1}), respectively. **Figure 2.33** shows the comparison of neutron energy spectra obtained during the daytime of December 14 without snowfall and of December 16 with a 20 to 30 cm-thick snowfall. The evaporation peak and the cascade peak decrease about 25% due to snowfall, while on the other hand the thermal peak does not change at all. The thermal peaks in **Figs. 2.30** and **2.31** given by Gordon *et al.* are sharper than those in **Figs. 2.32** and **2.33**. This difference is due to the fact that the thermal energy regions in **Figs. 2.32** and **2.33** are expressed in only one energy group having a wide lethargy bin, whereas there are several lethargy bins divided according to a Maxwellian distribution in **Figs. 2.30** and **2.31**.

Figure 2.34 shows the three energy spectra obtained for long accumulation periods of pulse counting with a NE213 detector [2.52]. These three data are for three periods of high count rate, low count rate, and average (medium) count rate from November to December 2001. The NE213 detector could not obtain data for neutrons of energy below about 7 MeV because of the difficulty of photon and muon discrimination, as described before. The unfolded spectra give two peaks: an upper edge of the peak below 10 MeV corresponding to evaporation neutrons, and a big peak around 80 MeV due to cascade neutrons. Similarly, as in **Fig. 2.32**, the high-energy peak increases with increasing the neutron flux, although the increasing rate is smaller than that in **Fig. 2.32** due to the longer accumulation periods of pulse counting

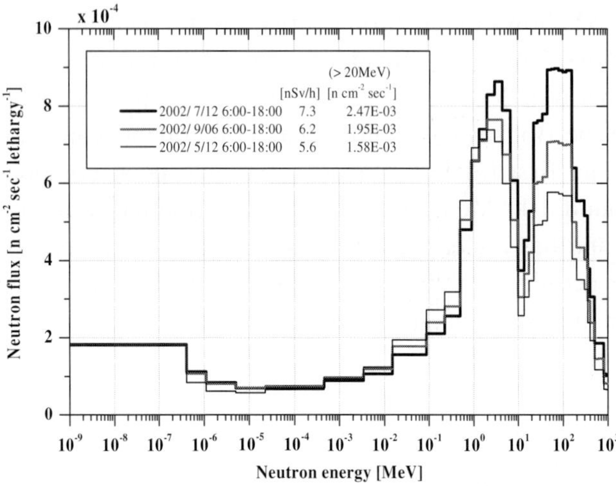

Fig. 2.32 Comparison of neutron energy spectra on three different days, July 12 (high flux), September 6 (medium flux) and May 12 (low flux) in 2002, measured by the Bonner ball [2.51, 2.52].

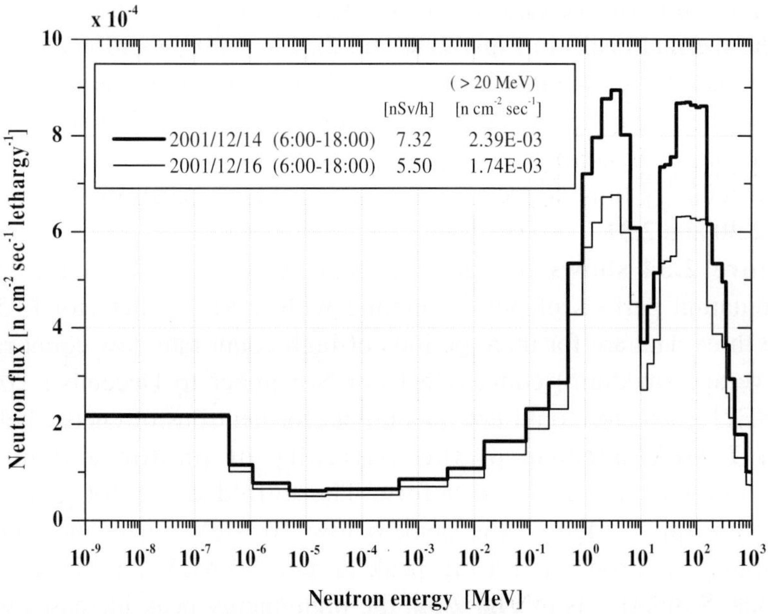

Fig. 2.33 Comparison of neutron energy spectra measured by the Bonner ball on the ground with (Dec. 16, 2001) and without (Dec. 14, 2001) snowfall [2.51, 2.52].

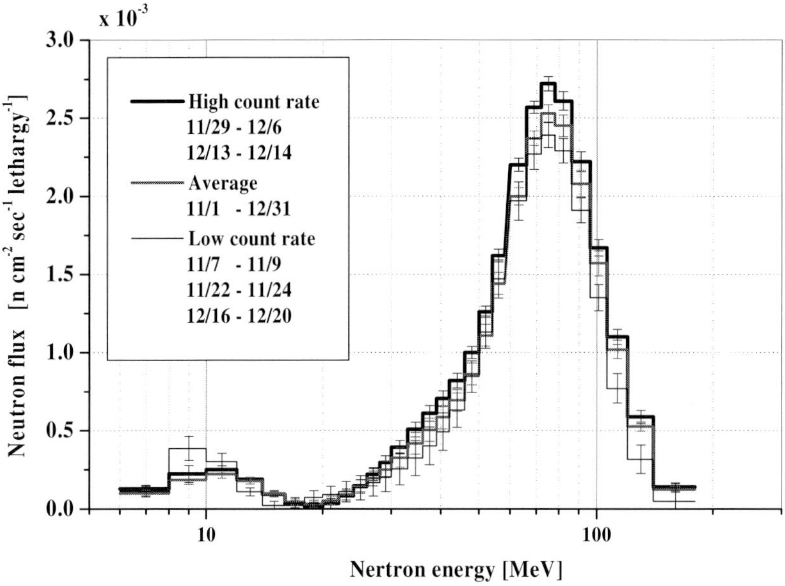

Fig. 2.34 Comparison of neutron energy spectra measured with NE213 on days of high, medium (average) and high count rates in 2001 [2.51, 2.52]

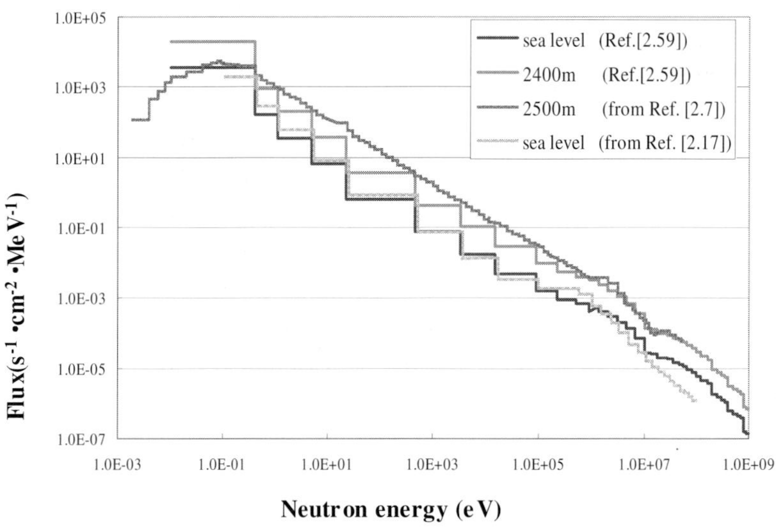

Fig. 2.35 Neutron energy spectra measured by the Bonner ball (Sea level and 2400 m height) in the Mt. Fuji area [2.59] compared with other measurements reported in Refs. [2.7] and [2.17].

with NE213. The peak around 80 MeV given by NE213 is much narrower than the peak around 100 MeV given by the Bonner ball. This can be explained by the fact that the NE213 has much better energy resolution than the Bonner ball, especially in the high energy region. At the same time, however, there may be some possibility that the light output pulses of NE213 were saturated in high energy region, which could lead to a narrower peak in the high-energy region.

Figure 2.35 shows the neutron energy spectra, $\phi(E)$, measured by a Bonner ball (at sea level and 2400 m height) in the Mt. Fuji area along with other measured results for comparison [2.59]. At sea level, the neutron energy spectrum obtained by Kowatari *et al.* [2.59] is in good agreement with the spectrum obtained by Nakamura *et al.* [2.17], except for the high energy region (above 10 MeV). The neutron energy spectrum obtained at 2400 m altitude is also in rather good agreement with that obtained at 2632 m in High Atlas at geomagnetic latitude 50°N by Florek *et al.* [2.7], particularly in the region from 1 MeV up to several tens MeV. On the other hand, in the lower energy region (below several hundreds keV), the spectrum obtained by Florek *et al.* is higher than that by Kowatari *et al.* This discrepancy is mainly due to the difference of the geographical coordinates (geomagnetic latitude and altitude) of the measuring points. In the thermal energy region, the discrepancy is partly due to the fact that the number of energy bins used in the spectrum unfolding process was different.

Figure 2.36 shows the neutron flux per lethargy measured at each altitude in the Mt. Fuji area by Kowatari *et al.* [2.59]. Three major peaks (the peak in the thermal neutron region, the evaporation peak around 1 MeV, and the peak around 100 MeV caused by nuclear spallation reaction) are clearly seen. **Figure 2.37** shows the neutron spectra normalized to the highest values around the evaporation peaks in each whole spectrum. It can be seen that the neutron energy spectra measured at each altitude are almost the same in their shapes. This is because an equilibrium between generation and attenuation of neutrons is achieved [2.68] in this region of altitudes. As the measuring point approaches the ground surface, the shape of the neutron energy spectrum changes. This reflects the disturbance of equilibrium caused by the ground surface effect [2.68], as can be seen in the next section.

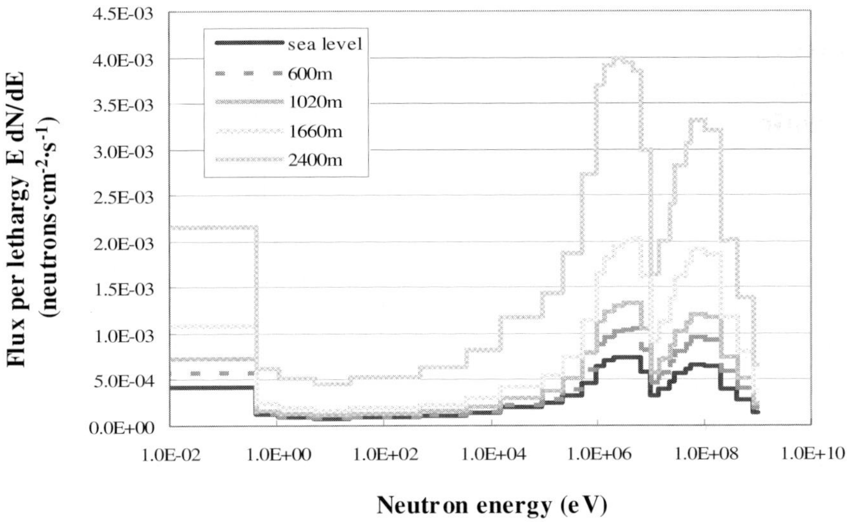

Fig. 2.36 Measured neutron spectra at various altitudes around Mt. Fuji [2.59].

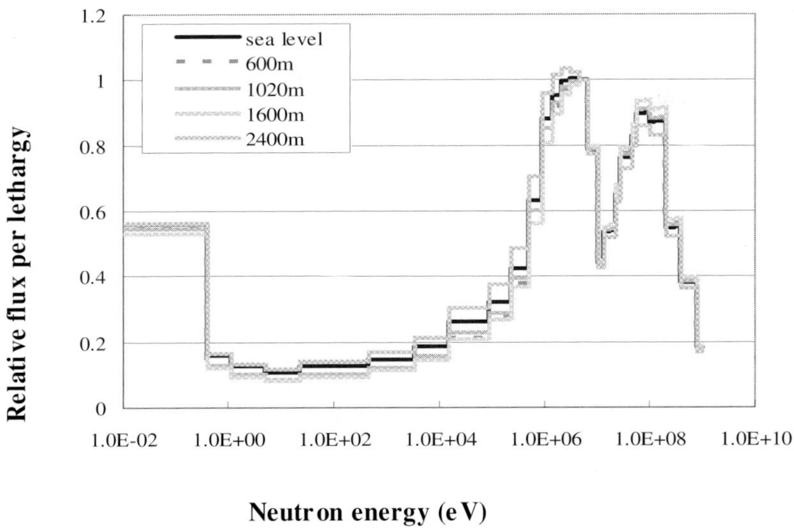

Fig. 2.37 Measured neutron spectra at various altitudes around Mt. Fuji (normalized according to the highest count) [2.59].

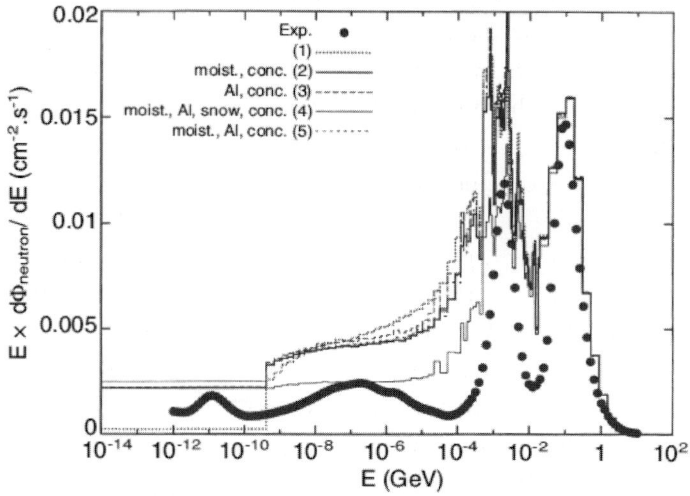

Fig. 2.38 Measured (symbols drawn at the geometric average of lower and upper bin edges) and calculated (histograms) ground-level neutron spectra for the experiment at Schneefernerhaus [2.58].

Schraube *et al.* also gave a similar neutron energy spectrum at the summit of Zugspitze (4.29 GV) [2.8]. At the same location, Mares *et al.* continued the measurements shown in **Fig. 2.22** [2.62]. **Figure 2.38** shows the measured spectrum [2.58]. In that spectrum, the thermal peak is not so remarkable, and a small broad peak can be seen around several hundreds eV.

(2) Energy spectra in air and space

Figure 2.39 shows cosmic ray neutron spectra unfolded from measurements aboard an ER-2 high-altitude airplane by Goldhagen *et al.* [2.1–2.2]. The measured spectrum is generally similar to the calculated spectra in **Fig. 2.28** that is used as default spectra in unfolding, but it has a taller evaporation-neutron peak and is lower from 10^{-4} to 0.2 MeV. The measured spectrum has a smaller 100-MeV peak than the Roesler spectrum [2.3], a wider valley between the two peaks than the Kurochkin spectrum [2.4], and almost no thermal neutrons. The data unfolded in **Fig. 2.39** were taken on the shorter North flight near the northern extreme of its outbound leg (54°N, 117°W, cutoff = 0.8 GV) at an atmospheric depth of 56 g cm^{-2} (20 km (65,600 ft) altitude). Over 2×10^5 neutron counts were recorded in 10 minutes. **Figure 2.39** also shows the same measured spectrum together with a spectrum measured for 24 minutes near the southern extreme of the South 1 flight (18.7°N, 127°W, 12 GV cutoff) at almost the same atmospheric depth (53.5 g/cm^2, 20.3 km). The total neutron fluence rate (flux) at the northern location was 8 times the fluence rate (flux) at the southern location; the southern spectrum is shown multiplied by 8. The spectra have almost the same shape, just like the spectra at various altitudes around Mt. Fuji area in **Fig. 2.37**. What difference there is may be due to the lower statistics of the southern measurement and the relatively large uncertainty of the correction for charged hadrons when applied to the two locations with such different geomagnetic cutoffs.

Figure 2.40 shows cosmic-ray neutron spectra measured at three different altitudes on the ER-2 and on the ground. The locations, atmospheric depths, and altitudes of the measurements are given in **Table 2.5**. The measurement at 56 g/cm^2 (dotted line) is the same northern spectrum shown in **Fig. 2.39**. At 101 g/cm^2, the atmospheric depth is 1.8 times greater, but the total neutron fluence rate (flux) decreases there by less than 3%. The small difference between the spectral shapes at 56 g/cm^2 and 101 g/cm^2 is large enough to affect the ratio of dosimetric quantities to the total neutron fluence. The difference is probably real, because the two measurements were taken on the same flight with nearly the same total counts and count rate at nearly the same cutoff. The difference between the spectrum measured at 201 g/cm^2

Table 2.5 Neutron integral quantities measured at various locations by Goldhagen *et al.* [2.2].

Geographic location	Cutoff [GV]	Atmospheric depth [g/cm^2]	Altitude [km]	Altitude [ft]	Neutron flux [(cm^{-2}s^{-1})]	Effective dose rate [μSvh^{-1}]	H*(10) Rate [μSvh^{-1}]
19N°,127W°	12	53.5	20.3	66500	1.28	0.91	1.06
54N°,117W°	0.8	56	20.0	65600	10.2	6.9	8.5
56N°,121W°	0.7	101	16.2	53300	10.0	6.2	7.8
38N°,122W°	4.5	201	11.9	39000	3.4	2.1	2.7
37N°,76W°	2.7	1030	0	0	0.0122	0.0083	0.0093

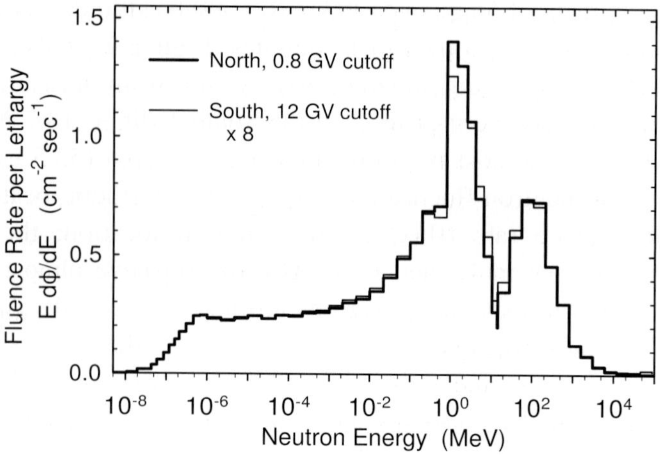

Fig. 2.39 Cosmic-ray neutron spectra measured at high altitude and high northern latitude (56 g/cm^2 atmospheric depth, 20 km altitude; 54°N, 117°W, 0.8 GV cutoff rigidity) and at the south end of the South-1 flight (19°N, 127°W, 12 GV cutoff; 54 g/cm^2 atmospheric depth). The south spectrum is shown multiplied by 8 [2.1, 2.2].

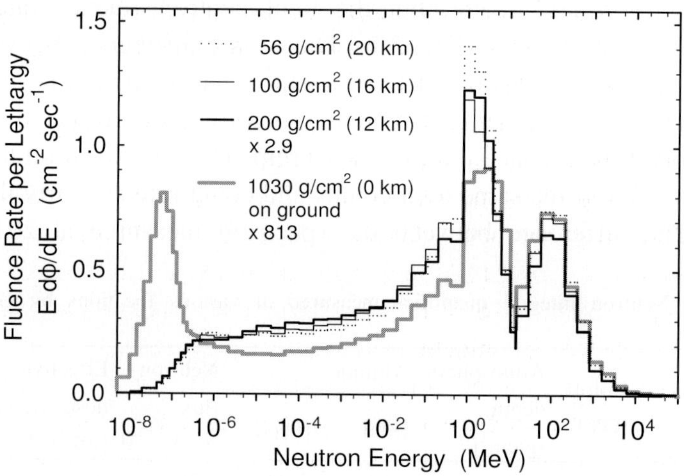

Fig. 2.40 Comparison of cosmic-ray neutron spectra measured at different atmospheric depths (altitudes) during the ER-2 flights and on the ground at sea level [2.1, 2.2].

(11.9 km, 39,000 ft) and the spectra at higher altitudes may be within the uncertainties. The measurement at 201 g/cm^2 was taken by combining 2 minutes of data from each of the 4 flights as the ER-2 rapidly climbed through normal commercial aviation altitudes shortly after takeoff (see **Fig. 2.25**). About 54,000 neutron counts were recorded in those 8 minutes. The cosmic-ray neutron spectrum measured on the ground shows a distinctly different shape, because soil reflects neutrons differently than air does. As expected, a significant number of thermal neutrons is produced in the ground. Normalized to the same total fluence rate (flux) as the in-flight measurements, the ground spectrum is lower from 10^{-6} to 2 MeV and enhanced or the same at higher energies. The data for the ground measurement shown here were collected for a week and had about 2.1×10^5 neutron counts.

Figure 2.41 gives the cosmic-ray neutron energy spectra in units of lethargy at different atmospheric depths (altitudes) obtained by Goldhagen *et al.* [2.2] in June, 1997 (solar minimum period) and by Nakamura *et al.* [2.17] on February 27, 1985 (solar minimum period), together with the average value at sea level obtained in the study conducted on September 6, 2002 (solar maximum period) by Nakamura *et al.* [2.51–2.52]. Goldhagen *et al.* measured the neutron energy spectra and doses using a Bonner ball aboard an airplane over the U.S.A. Nakamura *et al.* also measured the neutron spectra and doses using a Bonner ball aboard an airplane over Japan. The spectra by Nakamura *et al.* are revised from the original ones by re-unfolding with the initial guess spectrum given by Goldhagen *et al.* [2.1–2.2]. The three spectra at high altitudes (20 km, 11.28 km and 4.88 km) having a similar cutoff rigidity of 12 GV are very close together, although Goldhagen's spectrum is a little bit higher than Nakamura's spectrum in the energy region below the evaporation peak around 1 MeV. They are also similar in the MeV region to the spectrum measured in the study [2.51–2.52] conducted at sea level at a cutoff rigidity of 10 GV, with two eminent peaks around 1 MeV and 100 MeV. However, in the energy region below 1 MeV, the spectrum becomes softer with increasing atmospheric depth, and the thermal energy peak can be found at sea level, which indicates neutron thermalization through the atmosphere and backscattering from the earth, as described before. In spite of the large

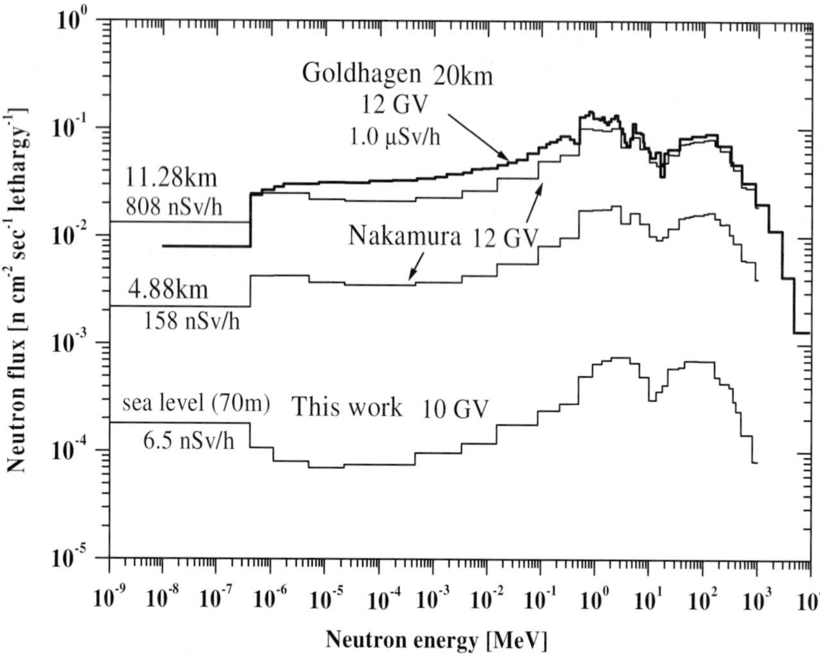

Fig. 2.41 Cosmic-ray neutron energy spectra in units of lethargy at different atmospheric depths (altitudes) around the high cutoff rigidity of 10 to 12 GV obtained by Goldhagen *et al.* [2.1, 2.2] in June, 1997 (solar minimum period) and Nakamura *et al.* [2.17] on February 27, 1985 (solar minimum period), together with the average value at sea level obtained in the present study on September 6, 2002 (solar maximum period) [2.52].

time intervals between these neutron measurements, the energy spectra have not changed very much. This spectral equilibrium might be due to poor energy resolution and the strong dependence of the initial-spectrum guess of the Bonner ball, especially in high energy region. Therefore, it is recommended to measure the neutron spectrum with some other detector having better energy resolution, such as an organic scintillator. It is also strongly recommended to do long-term cosmic-ray neutron measurements at various areas in the world.

Bonner ball measurements were performed in space from January 24 to 28, 1998 on S/MM-08 (STS-89) in the SPACEHUB module in the space shuttle [2.64], and from March 23 to November 14, 2001 in the US module of the ISS (International Space Station) [2.65]. **Figure 2.42** shows the neutron energy spectra observed at three geomagnetic regions: the Polar region, the SAA (South Atlantic Anomaly) region, and the Equatorial region. The spectra were determined by unfolding with the SAND-2 code using a 1/E initial guess spectrum. The three spectra are very close each other, but the absolute values are different. The total neutron fluxes ranged from 7.64 to 112.95 cm^{-2} s^{-1} for the SAA region and 0.35 to 7.97 cm^{-2} s^{-1} for the other two regions. The average ambient dose equivalent rates were 45.8 μSv/h for the SAA region and 1.0 to 5.4 μSv/h for the other regions.

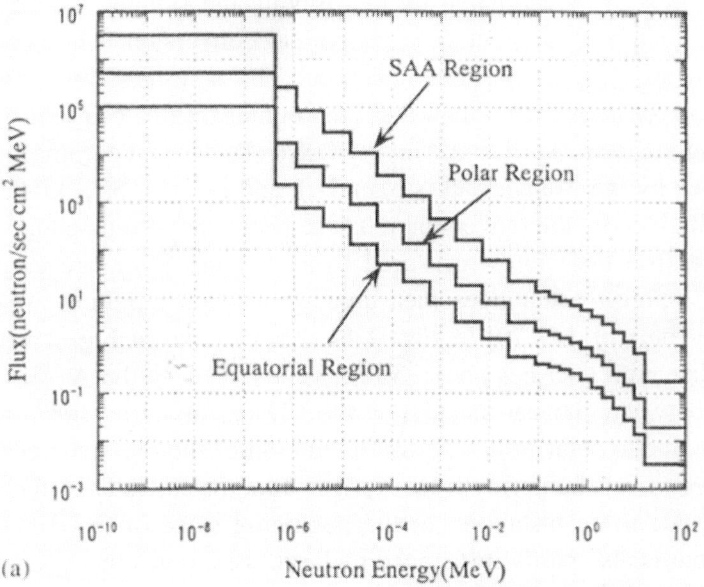

Fig. 2.42 Neutron energy spectra in the space shuttle STS-89 measured with a Bonner ball [2.64].

2.4.3 *Time-sequential results of neutron ambient dose equivalent rates*

Figure 2.43 shows the time-sequential data of neutron ambient dose equivalent rates obtained by using a rem counter at Tohoku University, Sendai by Nakamura *et al.* [2.52]. The dose rate remains almost constant at about 4.0 nSv/h, with a variation of about 5% over the entire time period between April 2001 and March 2003. However, there are some remarkable variations at points A and B, especially on April 16, which will be discussed later. At point A, the neutron ambient dose steeply decreased by about 30%, and this phenomenon remained for a few days. During those particular days, it snowed on the Kawauchi campus, and about 20 cm of snow piled up on the hutch. This steep decrease in ambient dose equivalent was caused by neutron attenuation through the hydrogen in the snow. At point B, large solar flares were observed by the ACE satellite [2.69] (See **Fig. 2.44**) and the neutron dose rates decreased by about 10%. This phenomenon is called the Forbush decrease. In **Fig. 2.43**, the dose rates estimated by multiplying the Bonner ball spectrum with the fluence-to-ambient dose equivalent conversion factor given by ICRP 74 are also given for comparison. These results show a higher fluctuation (of about 15%) than the rem counter results, especially in the beginning stage in April and May of 2001. The average value of dose rates given by the Bonner ball is 6.5 nSv/h and is about 60% higher than 4.0 nSv/h given by the rem counter. These results are discussed later.

The neutron dose rates measured by Nakamura *et al.* are compared with the neutron count rates measured by neutron monitors at several places in the world [2.70] in **Fig. 2.44**. Also shown in **Fig. 2.44** is the proton flux with energies above 30 MeV observed by the ACE satellite located between the earth and the sun. The proton flux occasionally increases due to solar flares. A big fluctuation of the rem counter counts can be seen over the period from April to June in 2001. This fluctuation is partly due to an instability in the beginning stage of the data taking system, but is also partly influenced by large solar flares that occurred on April 3 and 16. This fluctuation can also clearly be seen at high latitude locations, the South Pole, Oulu and Newark. It is already known that the secondary cosmic-ray intensity in the atmosphere surrounding the earth decreases, in some cases, during a couple of days after a large solar flare due to a big geomagnetic disturbance, called the Forbush decrease (FD). On the other hand, an increase is seen in some cases just after a big solar flare emits a large amount of high energy protons that reach the earth with energies above several hundreds of MeV, called a ground-level enhancement (GLE). The big fluctuation on April 16 is the successive effect of FD and GLE.

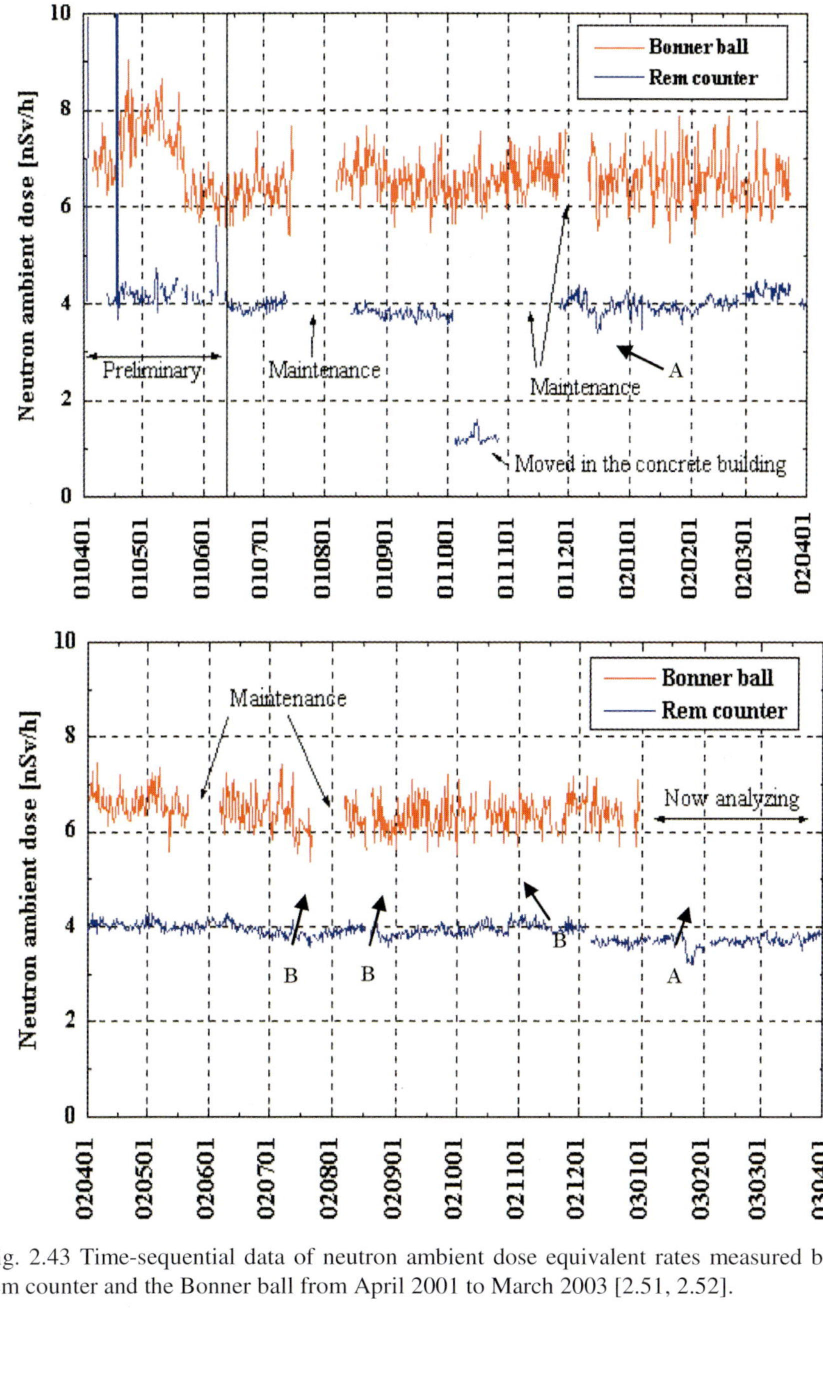

Fig. 2.43 Time-sequential data of neutron ambient dose equivalent rates measured by a rem counter and the Bonner ball from April 2001 to March 2003 [2.51, 2.52].

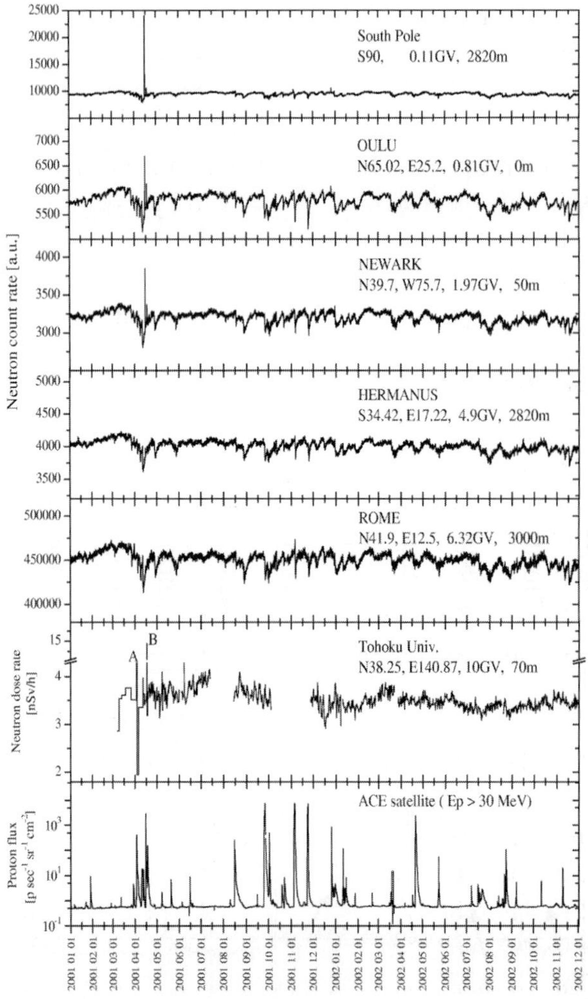

South Pole

http://www.bartol.udel.edu/~neutronm/

Oulu University (Finland)

http://cosmicrays.oulu.fi/

Hermanus (South Africa)

http://www.puk.ac.za/physics/data/

NEWWARK(USA)

http://www.bartol.udel.edu/~neutronm/welcome.html

Roma (Italy)

http://www.fis.uniroma3.it/~svirco/index.html

ACE satellite

http://helios.gsfc.nasa.gov/ace/ace.html

http://sec.noaa.gov/ace/ACErtsw_home.html

Fig. 2.44 Comparison of neutron dose rates and neutron count rates measured on the ground at various positions [2.70] and high-energy proton flux above 30 MeV measured by the ACE satellite [2.69]. The graph is cited from Ref. [2.52].

Sequential measurements using a similar rem counter have been done at the Japan Chemical Analysis Center (JCAC), Chiba from March 2002 up to the present [2.54]. **Figure 2.45** shows the neutron ambient dose equivalent rates averaged over each day in units of nSv/h measured with a 5-in.-diam ^{3}He rem counter. The dose rates are normalized to 1 atm (1013 hPa) to correct for differences in atmospheric pressure. The normalized dose rates indicate a big Forbush decrease from the end of October to the beginning of November, 2003, and a small Forbush decrease at the beginning of September, 2005, which reflect the large solar flares from October 28 to November 4, 2003, and from September 7 to 9, 2005. **Figure 2.46** gives the relative values of neutron dose rates together with the neutron intensities measured with monitors at the Bartol Research Institute at Newark, Fort Smith and the South Pole [2.69]. These values are normalized to the results on July, 2004, where the solar activity indicated by the heliocentric potential was average relative to the activity over the past 50 years. The long-term tendencies of the neutron intensities at four sites from 27°N to 90°N are very similar, although the Forbush decrease becomes larger with increasing geomagnetic latitude.

2.4.4 *Average values of neutron flux and ambient dose equivalent*

The average values of neutron flux and ambient dose equivalent measured from April 2001 to December 2002 at Tohoku University, Sendai, are summarized in **Table 2.6** [2.52]. It is found that the neutron flux with energies above 20 MeV (0.00207 cm^{-2} s^{-1}) is about 30% of total flux (0.00747 cm^{-2} s^{-1}). The neutron ambient dose equivalent rate estimated from the Bonner ball measurement is about 6.5 nSv/h. This value is about 60% larger than 4.0 nSv/h given by the rem counter measurement. Similar values of neutron ambient dose equivalent rates of 6.5 and 4.0 nSv/h, respectively, have also been observed during the sequential measurements from March 2002 up to the present at Chiba, Japan. The difference between two detectors mainly comes from the fact that the rem counter has a low sensitivity to neutrons above 20 MeV with a large deviation from a curve of the fluence-to-ambient-dose-equivalent

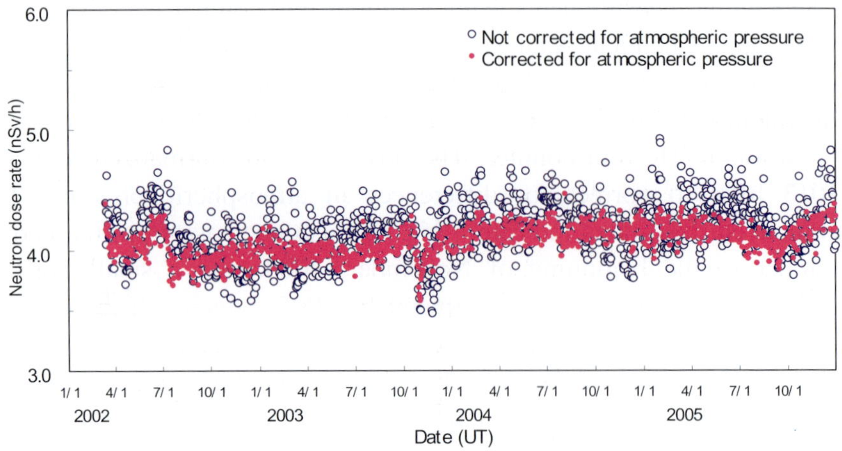

Fig. 2.45 Time-sequential data of neutron ambient dose equivalent rates measured by a rem counter from March, 2002 to December, 2005 in Chiba, Japan [2.54].

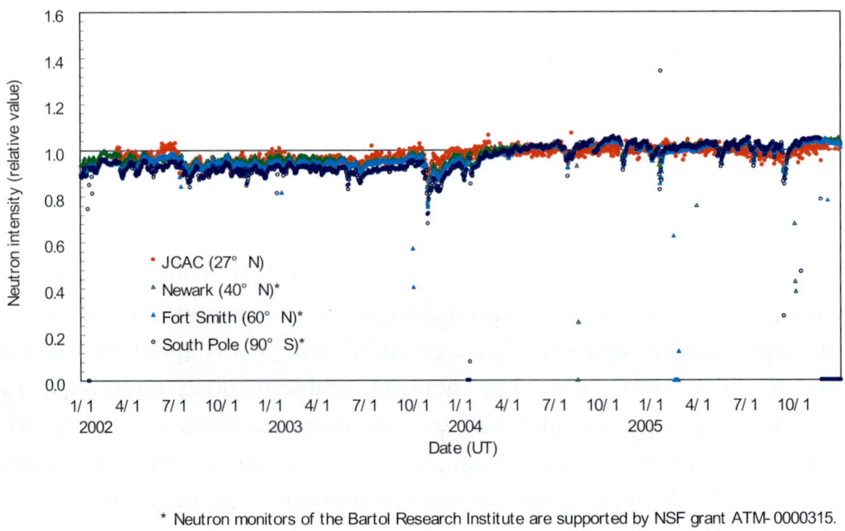

Fig. 2.46 Comparison of relative values of neutron dose rates measured at JCAC, Chiba [2.54] and neutron monitor intensities measured at Newark, Fort Smith and South Pole.

Table 2.6 Average values of neutron flux and ambient dose equivalent measured at Sendai, Japan (Geomagnetic latitude 29°N, Cutoff rigidity 10.43GV) from 2001.4.1 to 2002.12.31, compared with other measurement by Kowatari *et al.* [2.59].

(a) 2001.4.1 - 2002.12.31 in Sendai at 29°N by Nakamura *et al.* [2.52].

	Neutron flux (x 10^{-3} n cm^{-2} sec^{-1})			Neutron ambient dose equivalent (nSv/h)	
	Total	> 20 MeV		Total	< 20 MeV
Bonner ball	7.47 ± 0.52	2.07 ± 0.14	Bonner ball	6.5 ± 0.5	3.5 ± 0.3
NE213		1.78*	Rem counter	4.0 ± 0.2	

*NE213 data were taken in three periods from Nov. 7, 2001 to Dec. 30, 2001, from Aug. 29, 2002 to Oct. 6, 2002, and Nov. 11, 2002 to Dec. 27, 2002.

(b) 2002.3 to 2003.12 in Chiba at 26.6°N by Kowatari *et al.* [2.59].

	Neutron flux (x 10^{-3} n cm^{-2} sec^{-1})			Neutron ambient dose equivalent (nSv/h)	
	Total	> 20 MeV		Total	< 20 MeV
Bonner ball	7.69	1.95	Bonner ball	6.47	2.68
			Rem counter	4.0 ± 0.1	

conversion factor, while on the other hand, the Bonner ball can give the neutron spectra up to 1 GeV by using an initial guess spectrum extending up to 1 GeV, although the response function decreases for high-energy neutrons. The ambient dose equivalent obtained by the Bonner ball becomes 3.5 nSv/h for the neutron energy range below 20 MeV, which is close to the rem counter value. The neutron dose obtained by the rem counter, therefore, gives an underestimation to that obtained by the Bonner ball when the neutron field has a component with energies above 20 MeV. The average values of total neutron flux and neutron flux above 20 MeV obtained by the Bonner ball are about 7.5 × 10^{-3} n cm^{-2} sec^{-1} and about 2.1 × 10^{-3} n cm^{-2} sec^{-1}, respectively. The latter value is very close to the average value of 1.8×10^{-3} n cm^{-2} sec^{-1} obtained by NE213. In **Table 2.6**, our results measured at Sendai (geomagnetic latitude 29°N) are compared with the averaged results measured shortly afterwards from February 2002 to December 2003 with the same-type Bonner ball and rem counter at Chiba (geomagnetic latitude 26.6°N), Japan [2.54]. Both results are quite similar to each other, which verifies the good reproducibility and reliability of these two long-term sequential measurements.

From the measurement by Gordon *et al.* [2.57] shown in **Fig. 2.30**, it is apparent that about 30% of the neutron flux in the measured spectrum is at energies above 10 MeV, which is close to the value in **Table 2.6** given by Nakamura *et al.* [2.52]. The total neutron flux at Yorktown Heights, New York at 50.7°N geomagnetic latitude (cutoff rigidity of 1.86 GV) was 0.0134 cm^{-2} s^{-1}, which is roughly a factor of 1.8 higher than 0.00747 cm^{-2} s^{-1} at Sendai at geomagnetic latitude 29°N (10.43 GV).

Wiegel *et al.* [2.10] reported a total neutron flux of 0.012 cm^{-2} s^{-1} and an ambient dose equivalent rate of 8.5 nSv/h at PTB at 48.2°N geomagnetic latitude (2.86 GV), which are about 60% and 30% higher than those at Sendai at 29°N geomagnetic latitude (10.43 GV), respectively.

The total flux, ambient dose equivalent rate H*(10) and effective dose rate of the cosmic-ray induced neutrons have been determined from the neutron energy spectra obtained at Mt. Fuji area and at sea level at Chiba by Kowatari *et al.* [2.59]. The results are shown in **Table 2.7**. All values vary with the altitude at the measurement location. The cosmic-ray neutron flux obtained at sea level (40 m) is estimated to be about 0.0075 cm^{-2}·s^{-1}, and the ambient dose equivalent rate (H*(10)) there is about 6.4 nSv/h, which are close to the values obtained at Sendai. These values are both lower than other results measured at higher geomagnetic latitude, as described above. As shown in **Table 2.7**, the ambient dose equivalent rates H*(10) are slightly lower than the effective dose rates H_E (AP geometry) obtained at all altitudes. However, the effective dose rate H_E (ISO geometry) at each altitude is estimated to be lower than H_E (AP geometry) and the ambient dose equivalent rate. The similar tendency is pointed out by Goldhagen *et al.* [2.2]. It might be caused by the fact that the fluence-to-dose conversion factors for ISO geometry are lower than those for AP geometry and the ambient dose equivalent H*(10), especially in the low energy region (below 1 MeV).

The total fluence, effective dose and ambient dose equivalent rates measured by Goldhagen *et al.* [2.2] are also given in **Table 2.5**. They used the neutron fluence to effective dose (isotropic irradiation) and H*(10) conversion factors from ICRU report 57 [2.71] for neutron energies up to 20 and 201 MeV, respectively, and from Ferrari *et al.* [2.72] above 200 MeV. Ambient dose equivalent was always greater than effective dose. For the high-altitude spectra, neutrons with energies higher than 10 MeV made up 24% of the total flux and contributed 38–39% of H*(10) and 68–70% of the effective dose rates. Those percentages are very close to the results measured in Japan (See **Table 2.6**).

Table 2.7 Neutron flux and dose rate at the indicated altitudes in the Mt. Fuji area by Bonner Ball by Kowatari *et al.* [2.59].

Altitude [m]	Atmospheric depth [g/cm^2]	Neutron flux [#/(cm^{2}s)]	Effective dose rate (AP geometry) [nSv/h]	Effective dose rate (ISO geometry) [nSv/h]	Ambient dose equivalent rate (H*(10)) [nSv/h]
40	1025	0.0075±0.00029	6.8±0.26	5.0±0.20	6.4±0.25
620	965	0.010±0.00039	9.6±0.36	7.2±0.27	9.0±0.3
1020	923	0.013±0.00044	12±0.4	9.1±0.31	11+0.4
1660	856	0.020±0.0014	19±1.4	14±1.0	17±1.3
2400	785	0.041±0.0014	35±1.2	26±0.9	34±1.1

2.4.5 *Variation with latitude, altitude and solar activity*

The intensity of cosmic-ray induced neutrons (and other secondary cosmic radiation) in the atmosphere varies with altitude, location in the geomagnetic field, and solar magnetic activity [2.18]. Atmospheric shielding at a given altitude is determined by the mass thickness per unit area of the air above, called atmospheric depth. The geomagnetic field deflects low-momentum primary cosmic particles back into space, lowering the neutron flux produced in the atmosphere. The minimum momentum per unit charge (magnetic rigidity) that an incident (often, vertically incident) particle can have and still reach a given location above the earth is called the geomagnetic cutoff rigidity for that point. The varying magnetic field carried outward from the sun by the solar wind plasma that permeates the solar system also reduces the cosmic-ray intensity at earth. This solar modulation has been measured for decades by a number of neutron monitors on the ground at various locations (see [2.70], for example). The cosmic-ray induced neutron flux is highest when solar activity is at a minimum (solar minimum), and lowest during solar maximum.

Table 2.8 summarizes the numerical values of neutron fluxes (total and above 20 MeV) and dose rates (ambient dose equivalent rates) at various altitudes measured aboard the airplane and at sea levels in Sendai and Chiba, all having the cutoff rigidities of 10 to 12 GV. The dose rates given by rem counters are about 60% of those by Bonner balls, as shown before.

The total neutron flux and ambient dose equivalent rate can be both approximated by an exponential function of the altitude at the measuring location Z (m), as pointed out by some references [2.7, 2.73], as follows,

$$\phi_{total} = \phi_0 \times \exp(\alpha \cdot Z)\,, \tag{2.10}$$

$$H_z = H_0 \times \exp(\alpha \cdot Z)\,. \tag{2.11}$$

In the Mt. Fuji measurements by Kowatari *et al.* [2.59], the α-values for the total neutron flux and ambient dose equivalent rate are estimated to be about 0.00071 and 0.00070 m^{-1}, respectively. The α-values for the total neutron flux and the ambient dose equivalent rate obtained from Bonner ball measurements are in good agreement with each other, although the α-value of 0.00076 m^{-1} obtained by rem counter measurements is slightly higher. The good agreement is because the neutron energy spectrum at each altitude has the same shape, as shown in **Fig. 2.37**.

Table 2.9 indicates the comparison of the α-values obtained for various geomagnetic latitudes. The value referred in UNSCEAR 2000 [2.73] is derived by converting the attenuation factor per atmospheric depth into attenuation factor per altitude in units of m^{-1}, assuming standard air. It is confirmed that the α-value decreases according to the geomagnetic latitude and that the magnitude of the attenuation of cosmic-ray neutron is influenced by the geomagnetic intensity.

Table 2.8 Neutron fluxes and dose rates measured at various altitudes around the 10 GV cutoff.

Altitude (m)	Atmospheric depth ($g\ cm^{-2}$)	Neutron flux ($n\ cm^{-2}\ s^{-1}$)		Neutron dose rate (nSv/h)		Ref.
		Total	>20MeV	Bonner ball	Rem counter	
20,300	53.5	1.24	0.28	1,000	---	Goldhagen[2.2]
11,280	221	1.04	0.24	808	515	Nakamura[2.17]
4,880	559	0.189	0.049	158	86	Nakamura[2.17]
2,400	786	0.0407	----	33.6	---	Kowatari[2.59]
1,660	856	0.0195	----	17.5	---	Kowatari[2.59]
1,020	923	0.0131	----	11.5	---	Kowatari[2.59]
620	965	0.0102	----	9.01	---	Kowatari[2.59]
70	1,024	0.00747	0.00207	6.5	4.0	Nakamura[2.52]
40	1,026	0.00769	0.00195	6.47	4.0	Kowatari[2.59]

Table 2.9 Comparison of α -values obtained at various geomagnetic latitudes [2.59].

Geomagnetic latitude (degrees)	α - value ($\times 10^{-3} \cdot m^{-1}$)
26.0	0.7
48.0	0.78 * [2.73]
49.0	0.85 [2.7]
50.0	1.04 #

*Converted from the attenuation factor per atmospheric depth, assuming standard air.
#A. Bouville, W.M. Lowder, "Human population exposure to cosmic radiation", *Radiat. Protec. Dosim.*, 24, 293–299 (1988).

Figure 2.47 shows the comparison of the altitude variation of the relative neutron dose rate using various α-values. In **Fig. 2.47**, the neutron dose rate is normalized according to the dose rate at sea level

(0 m altitude). For the estimation of the neutron dose rate at given altitude, it is indispensable to adopt reliable α-values obtained by the evaluation of the variation of dose rate with altitude.

The atmospheric depth D (g/cm^2) can be estimated as a function of the altitude X (km), as follows [2-74],

$$\begin{aligned} X &= 47.05 - 6.9\ln(D) + 0.299(\ln 0.1\,D)^2 && D < 25 \\ &= 45.5 - 6.34\ln D && 25 < D < 230 \\ &= 44.34 - 11.861\,D^{0.19} && D > 230 \end{aligned} \qquad (2.12)$$

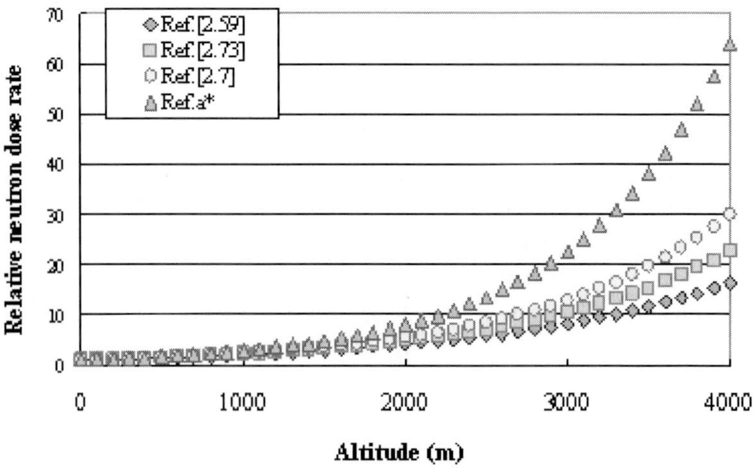

Fig. 2.47 Comparison of the altitude variation of the relative neutron dose rate using various alpha-values [2.59]. a* is referred from K.Fujimoto, K.O'brien, " Estimation of dose from cosmic rays in Japan", *Hoken Buturi*, 37[4], 325–334 (2002), [in Japanese].

Figures 2.48 and **2.49** give the altitude variation of total neutron flux, neutron flux above 20 MeV (cm^{-2} s^{-1}) and neutron ambient dose equivalent rates (nSv/h), respectively, as measured by Nakamura *et al.* [2.17], Goldhagen *et al.* [2.2] (both in an airplane), by Kowatari *et al.* [2.59], Matsumoto *et al.* [2.75] (both on the ground at the Mt. Fuji area), along with the result on the ground by Nakamura *et al.* [2.52]. Both graphs show that the cosmic-ray neutron variation with altitude exhibits an exponential attenuation, exp($-D/\lambda$), as a function of atmospheric depth,

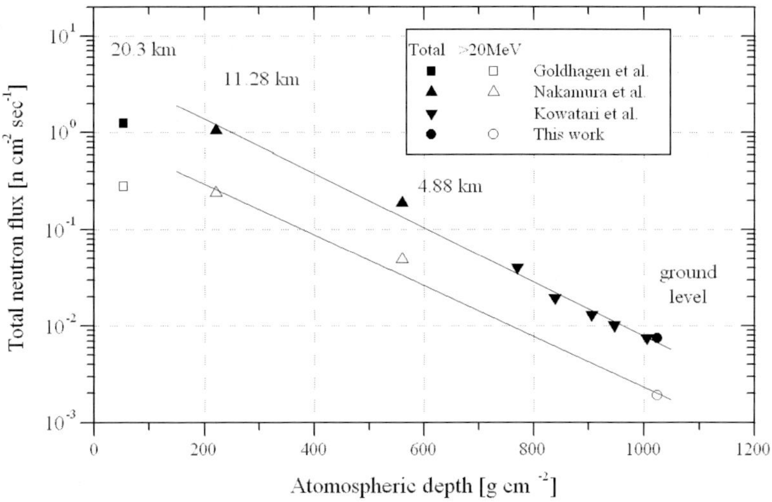

Fig. 2.48 Altitude variation of total neutron flux, neutron flux above 20 MeV (cm^{-2} s^{-1}) measured by Nakamura *et al.* [2.17], Goldhagen *et al.* [2.2] (both in an airplane), by Kowatari *et al.* [2.59] on the ground at the Mt. Fuji area, and by the work of Nakamura *et al.* [2.52] at sea level.

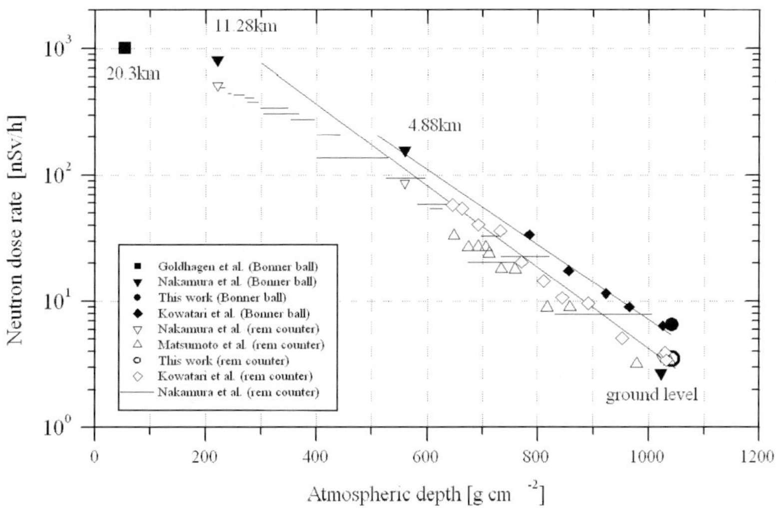

Fig. 2.49 Altitude variation of neutron ambient dose equivalent rates (nSv/h) measured by Nakamura *et al.* [2.17], Goldhagen *et al.* [2.2] (both in an airplane), by Kowatari *et al.* [2.59], Matsumoto *et al.* [2.75] (both on the ground at the Mt. Fuji area), and by the work of Nakamura *et al.* [2.52] at sea level.

D, up to about 7 km above the ground. It approaches a saturated value near 20 km, which is near the cascade maximum. From the exponential attenuation curves, the attenuation lengths, λ, are determined to be 155 g/cm^2 for total neutron flux, 165 g/cm^2 for neutron flux above 20 MeV. The attenuation length for neutron ambient dose equivalent determined by Bonner ball measurements is 144 g/cm^2. A smaller value of 135 g/cm^2 is determined by rem counter measurements, since the rem counter is mainly sensitive to neutrons below 20 MeV. These values are quite similar to the values between 140 and 165 g/cm^2 listed by Ziegler *et al.* [2.18]. The λ values have reciprocal relationship to the α values in Eqs. (2.10) and (2.11).

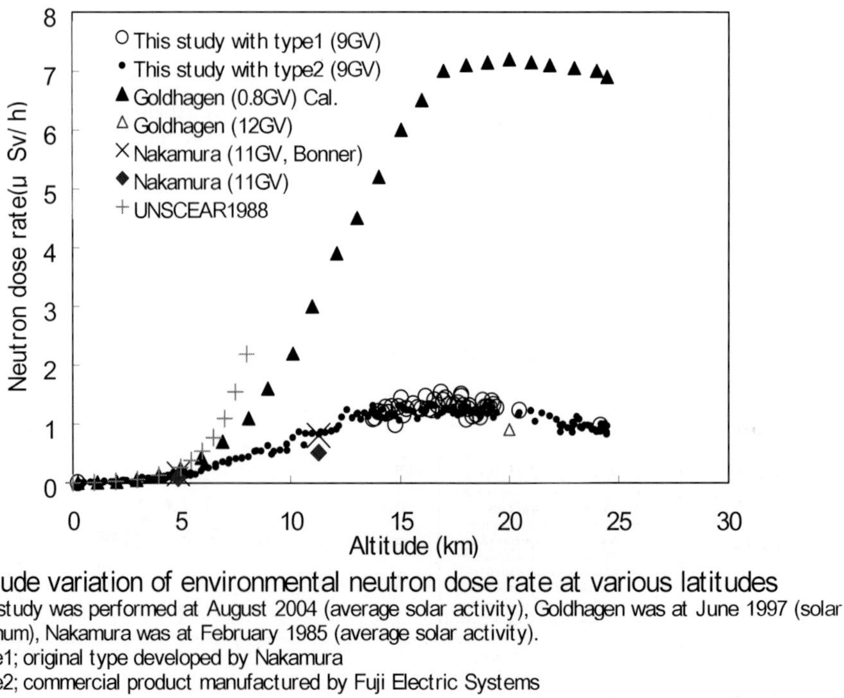

Fig. 2.50 Altitude variation of ambient neutron dose equivalent rates measured with two types of rem counters launched in a balloon up to 25 km above Japan on August 25, 2004 [2.63]. The measured results are compared with other results by Nakamura *et al.* [2.17] and Goldhagen *et al.* [2.1–2.2].

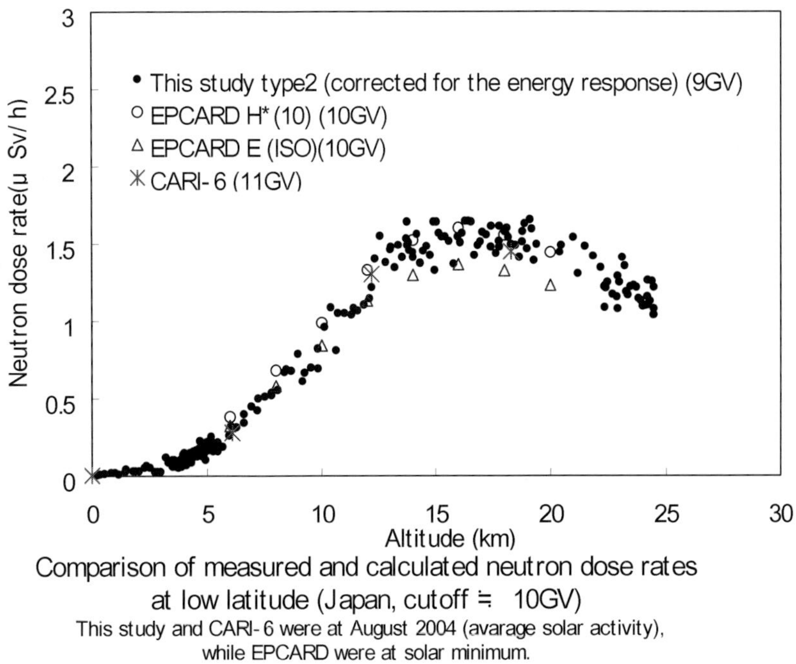

Fig. 2.51 Comparison of measured ambient neutron dose equivalent rates [2.63] with calculations by EPCARD [2.76] and CARI-6 [2.77].

The ambient dose equivalent rates measured with two types of rem counters launched in a balloon up to 25 km above sea level in Japan (about 10 GV) are shown in **Fig. 2.50** [2.63]. The results are compared with other experimental and calculated results at 0.8 GV and 12 GV.

The UNSCEAR results calculated at high latitude [2.73] and the Goldhagen results at 0.8GV [2.1] are much higher than the results at low latitude with 10 and 12 GV cutoffs. **Figure 2.51** shows the altitude variation of the measured results compared with the EPCARD [2.76] and CARI-6 [2.77] calculations. The agreement between experiment and calculation is quite good.

The variations of total neutron flux and ambient dose equivalent rate as a function of geomagnetic latitude have also been measured by many researchers. **Figures 2.52** and **2.53** give the summarized results of total

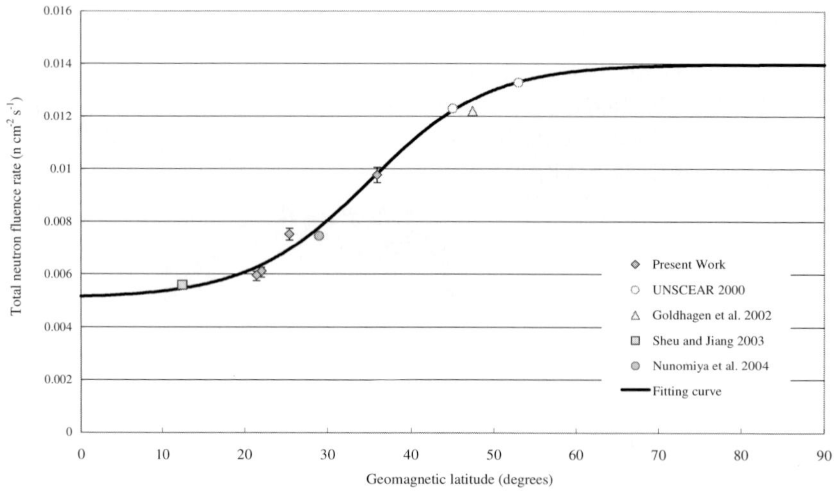

Fig. 2.52 Geomagnetic latitude variation of the cosmic-ray neutron flux (fluence rate) [2.54].

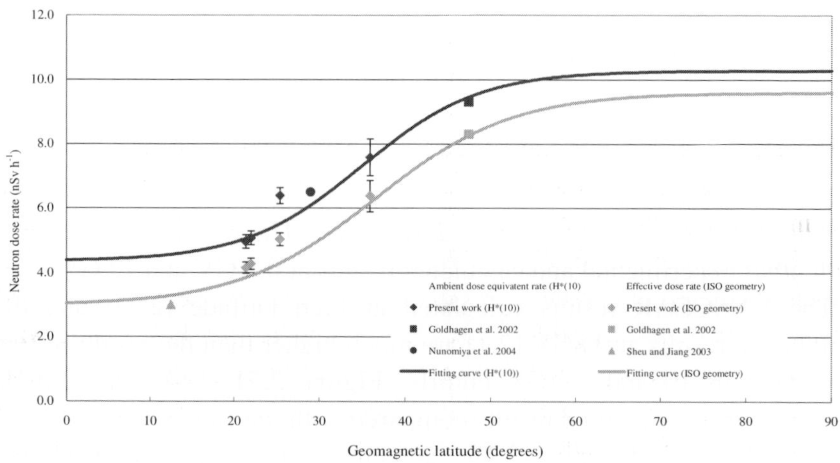

Fig. 2.53 Geomagnetic latitude variation of the cosmic-ray neutron ambient dose equivalent and effective dose rates [2.54].

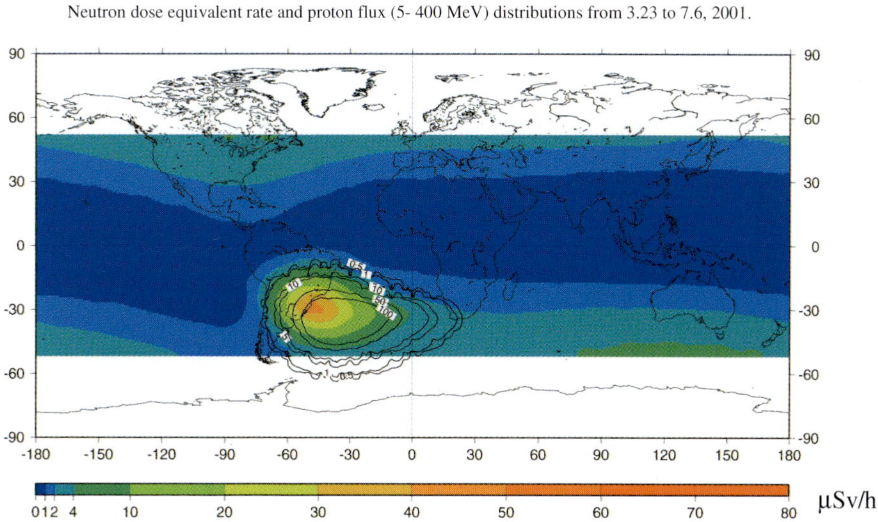

Fig. 2.54 World map of neutron ambient dose equivalent rates measured by the Bonner ball at about 400 km from March 23 to July 6, 2001, overlapped with proton flux distribution of energies from 5 to 400 MeV in solid curves [2.65].

neutron flux and ambient dose equivalent rate, respectively [2.54]. The neutron flux and dose rate increase with the latitude from the equator to the pole, which is a result of the geomagnetic effect.

Figure 2.54 shows the neutron ambient dose equivalent rate distribution in μSv/h on a world map. The ambient dose equivalent rate was measured in the US module of the ISS about 400 km above the earth from March 23 to July 6, 2001 [2.65]. The neutron dose rate distribution is overlapped with the proton flux distribution for energies from 5 to 400 MeV. The neutron and proton flux distributions are well correlated with each other, giving high intensities at the SAA region. This indicates that neutrons are produced mainly from protons through nuclear reactions in the air. **Figure 2.55** indicates the neutron ambient dose-equivalent rate on April 16, 2001, when a large solar flare occurred. The Bonner ball clearly detected the increase of neutron dose rates due to that flare.

As already described in Sec. 2.4.3, solar activity also has an effect on neutron flux and dose rate on the ground. **Figure 2.56** gives the

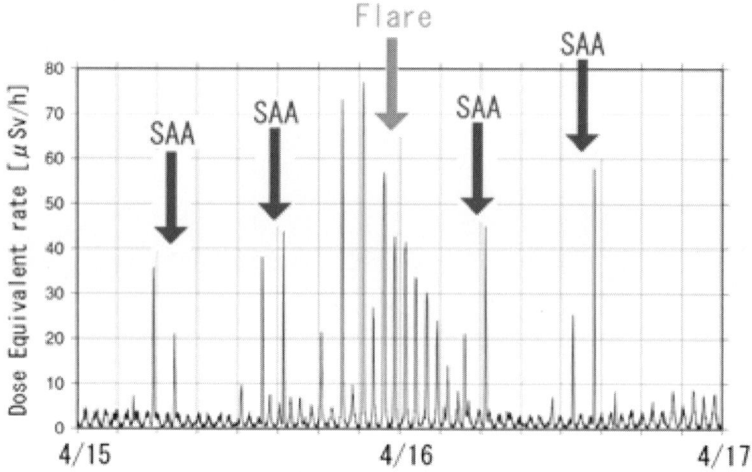

Fig. 2.55 Neutron ambient dose equivalent rate measured by a Bonner ball in the US module of the ISS around April 16, 2001 [2.64]. A steep increase of dose rate is found due to a large solar flare on April 16, along with large increases due to passage through the SAA.

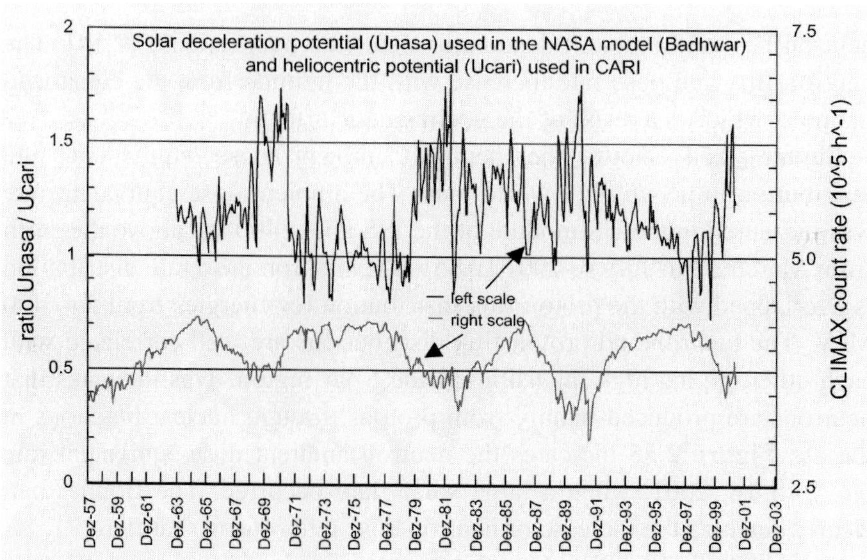

Fig. 2.56 Correlation of the ratio of solar deceleration potential used in the NASA model and heliocentric potential used in the CARI code to the CLIMAX neutron monitor [2.62].

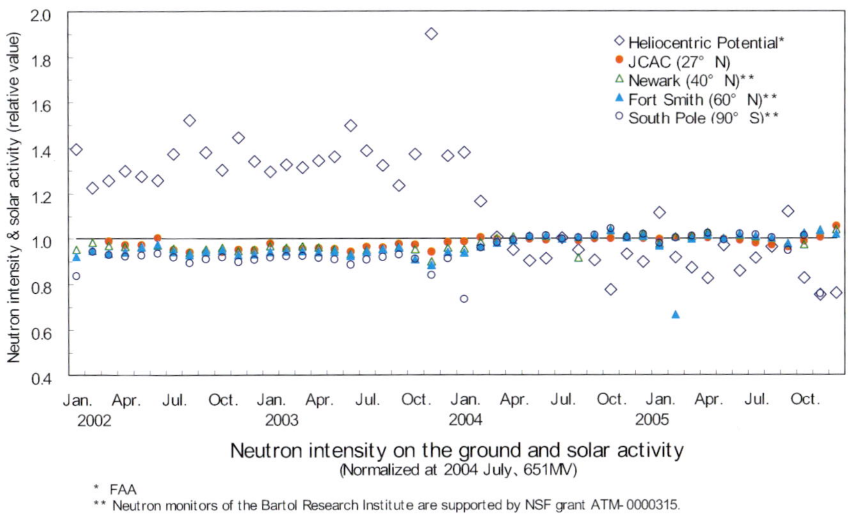

Fig. 2.57 Comparison of relative values of measured neutron intensities at various locations on the ground, JCAC in Chiba [2.54], Newark, Fort Smith, and the South Pole, with the heliocentric potential.

correlation of the ratio of solar deceleration potential used in the NASA model and heliocentric potential used in the CARI code to the CLIMAX neutron monitor. An inverse correlation is clearly observed between the two. Similarly, as in **Fig. 2.46**, **Fig. 2.57** also gives the relative values of neutron dose rates measured on the ground at Chiba together with the neutron intensities measured with the monitors of the Bartol Research Institute at Newark, Fort Smith and the South Pole, but with the heliocentric potential for comparison. The inverse correlation between heliocentric potential and neutron intensity is clearly seen from 2002 towards 2005, which means that the shielding effect on primary cosmic radiation entering into the earth becomes stronger with increasing the solar activity.

2.4.6 *Calculation of the cosmic-ray neutron spectrum*

A number of studies have been devoted to the estimation of the neutron spectra induced by cosmic rays. For example, Ziegler [2.18] proposed an empirical formula for predicting the high energy neutron spectra in order to estimate SERs on the ground. O'Brien *et al.* [2.78] developed a deterministic code LUIN based on an analytical two-component solution of the Boltzmann transport equation. The code is capable of estimating the altitude dependences of aircrew doses and integrated neutron fluxes precisely, and is adopted in the route dose calculation code CARI-6 [2.77]. Several authors [2.3, 2.79–2.80] adopted a Monte Carlo particle transport code, FLUKA [2.43], for the calculation of the cosmic-ray propagation in the atmosphere. Their calculation methods were successful in reproducing the neutron spectra measured at high altitudes, and a similar model was employed in the European Program Package for the Calculation of Aviation Route Doses EPCARD [2.76]. The PCAIRE code [2.81] was also developed on the basis of semi-empirical formula.

Gordon *et al.* [2.57] developed a semi-empirical analytical formula for estimating the neutron energy spectrum by considering that the outdoor ground-level neutron spectrum above about 5 MeV does not change significantly with altitude, cutoff rigidity, or solar modulation. To account for the effects of these variables on the total flux, the neutron energy spectrum at any location outdoors can be expressed as follows:

$$d\phi/dE = d\phi_0(E)/dE \; F_{alt}(d) \; F_{BSYD}(R_c, d, I), \qquad (2.13)$$

where $d\phi_0(E)/dE$ is the energy spectrum at a reference location (e.g., New York City at sea level and mid-value solar modulation), d is the atmospheric depth, R_c is the vertical geomagnetic cutoff rigidity, I is the relative count rate of a neutron monitor measuring solar modulation, $F_{alt}(d)$ is a function describing the dependence on altitude (i.e., on atmospheric depth) and $F_{BSYD}(R_c, d, I)$ is a function describing the dependence on geomagnetic location and solar modulation (and also depth). Atmospheric depth is given by $d = h/g$, where h is the barometric pressure and g is the acceleration of gravity. Vertical cutoff depends primarily on the horizontal component of the earth's magnetic field. It is near zero at the poles and has a maximum of 15 to 17 GV at the equator. Consequently, the cosmic-ray induced neutron flux is higher at the poles and lower at the equator. Values of R_c for cosmic rays reaching the atmosphere have been calculated by Shea and Smart [2.82] for a grid of locations covering the globe and updated for the Federal Aviation Administration [2.57].

In Eq. (2.13), the main altitude dependence is an exponential attenuation, as described in Sec. 2.4.4,

$$F_{alt}(d) = exp[(d_{SL} - d)/L_n], \tag{2.14}$$

where d_{SL} (= 1033.7 g/cm^2) is the atmospheric depth at sea level, and L_n (= 131.3 g/cm^2) is the effective attenuation length in the atmosphere for neutrons above 10 MeV in New York, as determined by a fit to the data measured by Gordon *et al.* F_{alt} varied by almost a factor of 15 from sea level to the 3450 m altitude at the measurement site in Climax, CO,—a far larger variation than the global variation with cutoff, which is about a factor of 2 from equator to pole at sea level, and 3 at the highest inhabited altitudes. Solar modulation is smaller still, about a 25% decrease from the maximum to minimum recorded monthly-averaged rates at polar locations near sea level (McMurdo and Oulu neutron monitors [2.70, 2.83]), 7% at the equator, and 30% to 12% for polar and equatorial sites at high elevations (Climax and Huancayo-Haleakala monitors [2.84]).

The function F_{BSYD} in Eq. (2.13) is a depth-dependent Dorman function developed by Dorman and Yanke and parameterized by Belov, Struminsky, and Yanke (BSY) to describe the location dependence of ground neutron monitor rates in terms of cutoff and barometric pressure [2.85–2.86]. It has the general form

$$F_{BSYD}(R_c, d, I) = N[1 - exp(- \alpha/R_c^{\ k})], \tag{2.15}$$

where N is a normalization factor, and α and k depend on pressure (depth) and solar modulation. Values of these parameters for solar minimum and maximum (S_{min} and S_{max}) are given in Ref. [2.57].

Figure 2.58 shows a graph of $E\ d\phi/dE$ vs neutron energy for the evaporation and high-energy regions of the reference spectrum and the analytic model. The data are shown as a histogram and the analytic model is the solid smooth curve. The fit is very good above 10 MeV and reasonably good in the evaporation region down to about 0.4 MeV. **Figure 2.59** shows the same data as **Fig. 2.58**, but presented as the more familiar differential flux (neutrons). **Figure 2.59** also shows the spectrum from Appendix E of JEDEC Standard JESD89 [2.87]. The JEDEC model

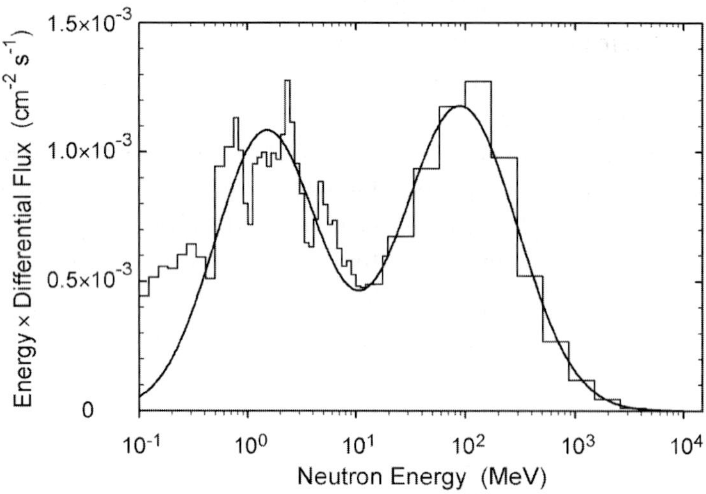

Fig. 2.58 Upper-energy portion of the reference neutron spectrum at New York City, sea level, and mid-level solar modulation (histogram), the analytic fit (solid smooth curve) [2.57]. (© 2004 IEEE)

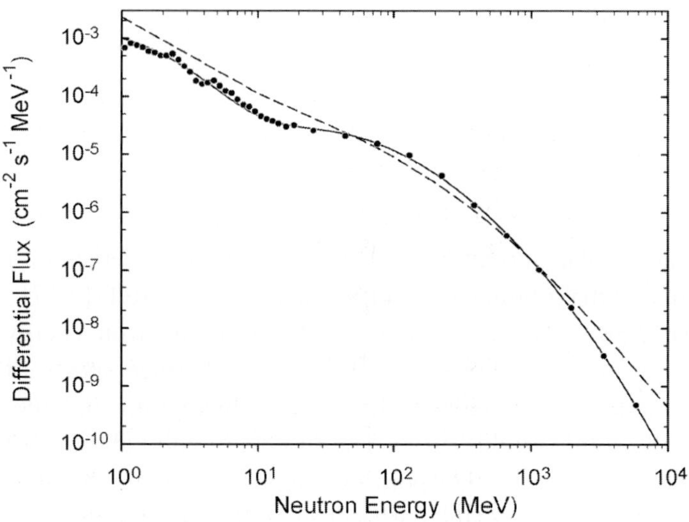

Fig. 2.59 Differential flux, d ϕ = dE, (cm^{-2} s^{-1} MeV^{-1}) of cosmic-ray induced neutrons as a function of neutron energy [2.57]. The data points are their reference spectrum from the measurements, the solid curve is their analytic model, and the dashed curve is the JEDEC model [2.87]. (© 2004 IEEE)

underestimates the reference measured flux integrated from 50 MeV to 1 GeV and overestimates it from 5 to 50 MeV and again from 1 to 10 GeV by factors of 1.7 and 1.3, respectively. The analytic model reproduces the measured spectrum better than the JEDEC model.

The cosmic-ray neutron spectra depend not only on the atmospheric depth, cut off rigidity and solar modulation (abbreviated to "global conditions"), but they also depend, in an intricate manner, on the structure of the aircraft [2.88] and the water density around the point of interest [2.4] (abbreviated to "local geometries"). Because of this, none of the existing models are able to reproduce the measured neutron spectra at any location and time with satisfactory accuracy.

With this in mind, Sato and Niita [2.89] have calculated the cosmic-ray neutron spectra by performing a Monte Carlo simulation of particle transport in the atmosphere based on the Particle and Heavy Ion Transport code System PHITS [2.48], utilizing the latest version of the nuclear data library JENDL-High-Energy File (JENDL/HE) [2.90]. In the simulation, cosmic rays were incident on the earth at an altitude of 86 km, up to which the US-Standard-Atmosphere-1976 has atmospheric data. Proton, alpha and heavy ions with charges up to 28 (Ni) were considered as the source particles, although the contributions of heavy ions to the cosmic-ray neutron spectra are generally small.

The spectra of the incident particles were estimated by the CREME96 code [2.91]. In the code, the effect of the geomagnetic field on the spectra was considered by specifying the McIlwain L, where the vertical cut-off rigidity r_c is related to this parameter by the simple formula,

$$r_c = 14.5 / L^2 . \tag{2.16}$$

It should be noted that the output fluxes of CREME96 are the averaged values for all directions, after accounting for the solid earth's shadowing effect. Hence, the values from CREME96 should be converted into downward fluxes by considering the an-isotropy of the radiation field in order to be used in the PHITS simulations. The atmosphere was divided into 28 concentric spherical shells. The densities and temperatures of each shell were determined by referring to the US-Standard-Atmosphere-1976. The atmosphere was assumed to be composed of 75.4% nitrogen, 23.3% oxygen and 1.3% argon by mass above an altitude of 2 km, and an additional 0.06% hydrogen by mass below this altitude, due to the existence of water vapor. The earth was represented as a sphere with the radius of 6378.14 km, and its composition was assumed to be 59.2% oxygen, 28.0% silicon, 10.6% aluminum and 2.2% hydrogen by mass. This constitution corresponds to 60% SiO_2, 20% Al_2O_3 and 20% H_2O by mass.

Figure 2.60 shows the comparisons of the calculated neutron spectra with the corresponding experimental data obtained by Goldhagen et al. [2.92] and Nakamura *et al.* [2.52]. Note that all the neutron spectra are expressed in units of neutrons per cm^2 sec lethargy. The results at altitudes below 2 m (~1035 g/cm^2) on the ground level are depicted in the figure in order to consider the effect of the earth's albedo neutrons precisely. The spectra predicted by the analytical functions given below are also plotted in the figures. It is evident from the figure that the simulations reproduce the experimental data very well over a variety of global conditions.

Based on this PHITS calculation, the neutron spectrum in the semi-infinite atmosphere, ϕ_{Inf}, was expressed by the product of the basic spectrum, $\phi_{B,}$ and the integrated neutron flux below 15 MeV, $\phi_{L,}$ as follows:

$$\phi_{\mathrm{Inf}}(s, r_c, d, E) = \phi_{\mathrm{B}}(s, r_c, d, E)\Phi_{\mathrm{L}}(s, r_c, d), \tag{2.17}$$

where s, r_c, d and E denote the solar modulation potential, vertical cut-off rigidity, atmospheric depth and neutron energy, respectively. The advantage of introducing ϕ_{B} is that the spectrum below 15 MeV is almost independent of the global conditions for altitudes under 20 km. The analytical functions for predicting the neutron spectra are therefore applicable to altitudes under 20 km.

For the purpose of reproducing the neutron spectra at ground level, $\phi_{G,}$ by an analytical function, the following equation was proposed as

$$\phi_{\mathrm{G}}(s,r_{\mathrm{c}},d,E,w)=\Phi_{\mathrm{L}}(s,r_{\mathrm{c}},d)\left[\phi_{\mathrm{B}}(s,r_{\mathrm{c}},d,E)f_{\mathrm{G}}(E,w)+\phi_{\mathrm{T}}(E,w)\right] \tag{2.18}$$

where w is the weight fraction of water, $f_G(E,w)$ denotes the functions representing the disturbance of the spectrum due to the local geometry effect, and $\phi_T(E,w)$ expresses the thermal neutron peak. For neutron energies above 1 eV, ϕ_T becomes negligible, and $f_G(E,w)$ corresponds to the ratio of neutron flux with the local geometry effect to neutron flux in the semi-infinite atmosphere, ϕ_G/ϕ_{Inf}. The reason for assuming f_G and ϕ_T to be independent of the global conditions is that the local geometry effect has a dominant influence on the low energy region, where the shape of the basic spectrum ϕ_B is independent of the conditions.

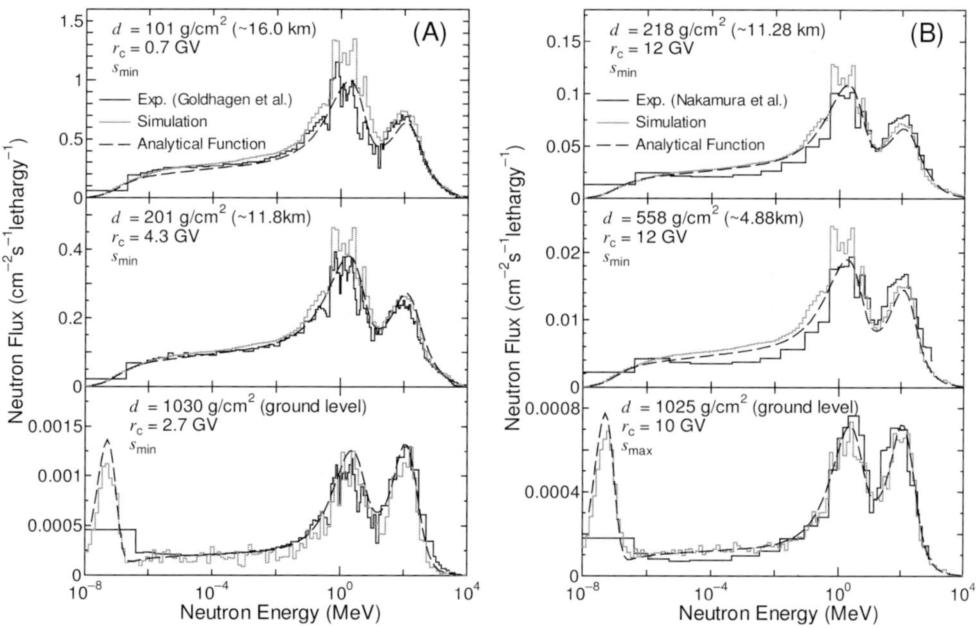

Fig. 2.60 Calculated and experimental neutron spectra in the atmosphere for various global conditions. The panels (A) and (B) show the comparison with the data measured by Goldhagen *et al.* [2.92] and Nakamura *et al.* [2.52], respectively. The values of d and r_c are the atmospheric depth and the cut-off rigidity, respectively, while s_{min} and s_{max} indicate the solar minimum and maximum, respectively.

The accuracy of the analytical functions was also verified by comparing the calculation results with the experimental data by Kowatari *et al.* [2.59] that was described in Sec. 2.4.2, as shown in **Fig. 2.61**. It can be clearly seen from the graph that the analytical function can reproduce the experimental data very well. They have developed software named EXPACS, which can calculate cosmic-ray neutron spectra at any location using the above analytical functions. The software is available on-line from their web site [2.93].

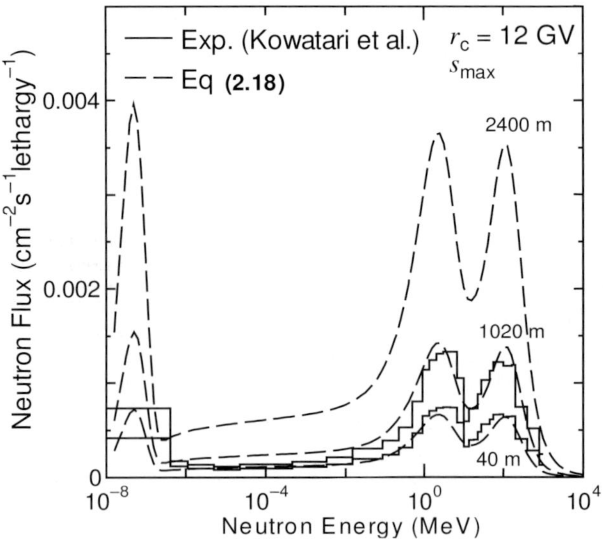

Fig. 2.61 Neutron spectra calculated with Eq. (2.18), in comparison with the measured data in the Mt. Fuji area at altitudes of 40, 1020 and 2400 m [2.59], where the corresponding atmospheric depths are approximately 1025, 923 and 785 g/cm^2, respectively.

2.5 Concluding Remarks

The terrestrial neutron energy spectrum and ambient dose equivalent rate were reviewed in this chapter.

The shapes of neutron energy spectra on the ground for different geomagnetic altitudes and latitudes are similar and typically have three prominent broad peaks. There is a peak around 100 MeV caused by cascade neutron production, a peak around 1 MeV caused by evaporation process, and a peak in thermal energy region due to neutron thermalization through the atmosphere and backscattering from the earth. In the air aboard an airplane and in space, however, thermal neutron peaks can not be found in the spectra.

The absolute values of the neutron flux and ambient dose equivalent are strongly dependent on the altitude and latitude. **Figure 2.62** shows the comparison of neutron fluxes at low (Narita and Sendai, Japan) and high (New York, U.S.A.) geomagnetic latitudes, together with a calculation, which clearly indicates a significant difference between Japan and U.S.A. This means that accurate neutron flux and dose rate measurements are required at the site of any soft error investigation.

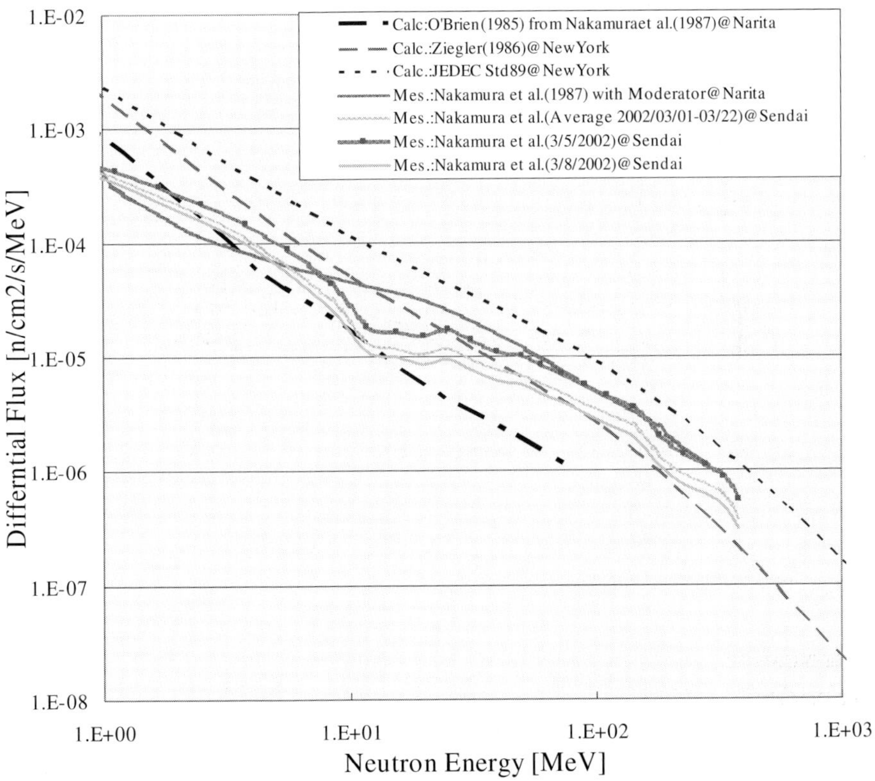

Fig. 2.62 Comparison of measured neutron fluxes (energy spectra) at low (Narita and Sendai, Japan) and high (New York, U.S.A.) geomagnetic latitudes with the indicated calculations.

Chapter 3

Irradiation Testing in the Terrestrial Field

Hideaki Kameyama and Yasuo Yahagi

Methods and equipment to measure neutron soft-errors in a terrestrial field are described. The chi-square method used to estimate the soft-error rate (SER) is described. Correction methods used to normalize SER rates are also introduced.

3.1 What Does Real-Time SER Mean?

Modern life without semiconductor devices is hard to imagine; almost everyday they are used in such applications as home electronics and industry. Many applications in industry require highly reliable instruments that are constructed with robust semiconductor devices that have a high level of error-mitigation against the impact of noise. In general, however, it is too expensive to manufacture commercial based goods with these types of semiconductor devices. In this chapter, a renewed understanding of the soft error rate (SER) is presented that has a direct effects on the reliability of semiconductor devices used in commercial and industrial applications

Soft error due to noise and other effects can lead to mechanical failure of semiconductor devices. One method used to mitigate the effects of the SER is to employ an error detection/correction code (EDC/ECC) [3.1] into the devices or the system so as to keep the system at a specific safety level. Therefore, it is necessary, in this context, to know the SER of the system in advance of their commercial stage. The difficulty of soft error evaluation at the system level is that it strongly depends on the characteristics of the semiconductor devices. For example, devices with general purpose-use memory devices like SRAMs (Static Random Access Memories) and DRAMs (Dynamic Random Access Memories) fundamentally function through simple read/write operations. The SER in this case depends on the frequency of access to those devices, which in turn is a function of the application of the device. Therefore, one practical way to measure SER is by using memory devices because of their simple write/read operations. On the other hand, the SER test has some difficulties for logic devices such as the FPGA (Field Programmable Gate Array) device or SOC (System On Chip), because it is difficult in these devices to properly identify the physical location of the failure on the chip. Also, it takes a long time to decode error logs even when the logical failure path can be found. Another difficulty is that the SER of an FPGA strongly depends on the application because the FPGA is composed of several types of components, such as a MPU (Main Processing Unit) or random access memories (RAMs).

The terrestrial neutron flux with energies above 1 MeV, which is a concern for SEU, is very low.

As a result, the reliability of the calculated value of the SER is poor. Increasing the number of chips is one method to improve statistical reliability, which is the case for server systems that contain thousands of chips. However, Real-Time SER (RTSER) testing requires too much time and is expensive, due to low flux of the terrestrial neutrons, as mentioned above. Because of this, RTSER testing is impractical for the development and marketing of semiconductor devices. It is, however, very important because the test is carried out in conditions very close to the actual operation of device. In order to obtain the SER of devices efficiently, a lot of accelerated testing is carried out using particle accelerators. The results of accelerator-based testing will be described in detail in Chapters 4 and 5. What should be stressed here is that RTSER is carried out in nearly actual operating conditions of the device, including the natural radiation environment present during operations. As such, the test results are useful to the system design. At the same time, it is important to acquire actual data of RTSER in order to verify both the simulation model (including the device model) and the method of accelerated testing. Estimation of the SER of complex devices can be carried out just by simulation in the near future without experimental testing in the field.

After the methodology for reliability is discussed in Sec. 3.2, examples of real-time testing, a system for the RTSER, and remarks in carrying it out are described in the subsequent sections.

3.2 Statistics and FIT Estimation Methodology

From the point of view of the reliability of the failure rate estimation, the FIT (Failure in Time) rate and ppm (parts per million) are well-known estimation methodologies using the failure rate(λ), time(h), and confidence level based on a χ^2 distribution for FIT. In general, it is also well known to use a 60% confidence level for FIT estimations in the case of commercial applications and a 90% confidence level for high - reliability uses such as military applications, space applications, power plant systems and large-scale computer systems. The confidence level (= C.L.) indicates the certainty of the occurrence probability ranges depending on the probability distribution.

In general, a C.L. of 90% or 60% is used. Below is Eq. (3.1), which is used to calculate the SER FIT rate.

$$\mathrm{FIT} = \frac{\chi^2(1-\mathrm{C.L.}, 2r+2) \times r \times 10^9}{2 \times n \times h} \tag{3.1}$$

$\chi^2(1 - \mathrm{C.L.}, 2r + 2)/2$ = one sided chi-square function
r = number of failures
1 − C.L. = hazard ratio = ε
$2r + 2$ = Degree of freedom
n = total number of devices
h = operation hours

The explanations of confidence level and the detailed statistical calculation methodology of the SER FIT rate are discussed in the next section.

3.2.1 *Confidence level*

When testing SER at accelerator facilities, the total number of particles incident on the DUT (device under test) must be considered in order to establish, with a high degree of statistical confidence, that all of the sensitive volume has been irradiated uniformly. After a given particle fluence, each memory cell either passes or fails. The probability of failure for each memory cell is simply calculated by Eq. (3.2)

$$\lambda = \frac{r}{n} \tag{3.2}$$

where λ is a probability of failure, n is the total number of memory cells, and r is the total number of failures. In order to get the estimated probability of failure λ within $\pm\,\pi\%$ at a given C.L. ($= 1 - \varepsilon$), the following condition must be satisfied, according to the definition of the confidence interval:

$$\Theta^{-1}\left(\frac{\varepsilon}{2}\right) \bullet \frac{1}{\sqrt{r}} \bullet \sqrt{\frac{n-r}{n-1}} \le \pi\% \tag{3.3}$$

where ε is hazard ratio, $\Theta^{-1}(\varepsilon/2)$ is the inverse cumulative standard normal distribution function, r is the total observed errors and n is the total number of components.

The relation between confidence level and the number of samples is shown below in Eq. (3.4), where λ is the probability of failure and n is the number of samples.

$$n = \frac{log(1 - \text{C.L.})}{log(1 - \lambda)} \tag{3.4}$$

The relationship between each confidence level and the number of failures is illustrated in **Fig. 3.1**. The number for each confidence level is calculated by a χ^2 function described as a one sided chi-square function. When the number of failures, r is less than "1", the value of the χ^2 function increases with r, but beyond 1 it decreases with r. At r = 1, the χ^2 function gives the maximum peak value. In general, the χ^2 function converges to 1 as the number of failures increases.

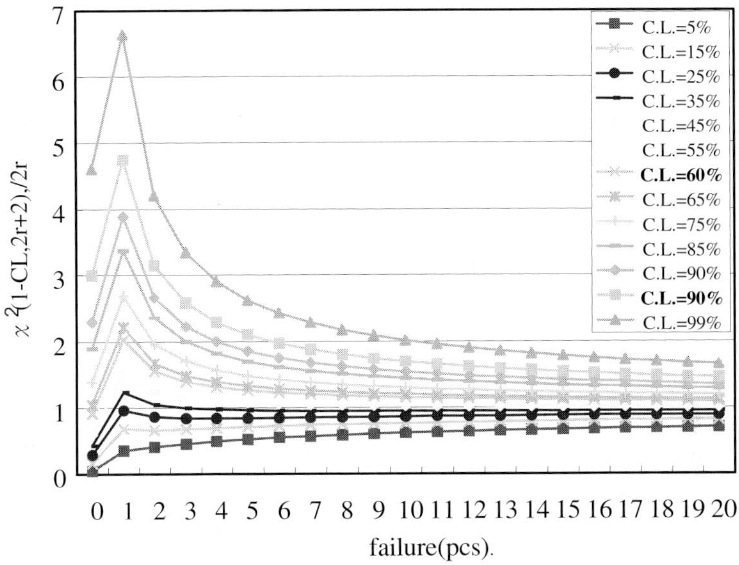

Fig. 3.1 Relationship between the $\chi 2$ function and the number of failures.

3.2.2 *SER FIT rate calculation (example)*

A FIT calculation is exemplified as follows: When the number of failures is 3 out of 50,000 pcs. for SRAM devices after 1000hrs of testing, the FIT rate is 83.4 FIT. This value was obtained by using Eq. (3.5), along with the value of the one-sided chi-square function in **Table 3.1**, using C.L. of 60%,

$$\mathrm{FIT} = \frac{1.39 \times 3\mathrm{pcs.} \times 10^9}{50{,}000\mathrm{pcs.} \times 1{,}000\mathrm{h}} = 83.4\mathrm{FIT} \tag{3.5}$$

When the number of failures is 0 out of 30,000 pcs. for SRAM devices after 3000hrs of testing, the FIT rate is 30.7 FIT

$$\mathrm{FIT} = \frac{0.92 \times 10^9}{30{,}000\mathrm{pcs.} \times 3{,}000\mathrm{h}} = 30.7\mathrm{FIT} \tag{3.6}$$

Table 3.1 χ^2 table @ C.L. = 60%.

C.L (%)	No. of failure(n)							
	0	1	2	3	4	5	6	7
15	0.16	0.68	0.67	0.68	0.70	0.71	0.72	0.74
20	0.22	0.82	0.77	0.77	0.77	0.78	0.79	0.80
25	0.29	0.96	0.86	0.85	0.84	0.84	0.85	0.85
30	0.36	1.10	0.96	0.92	0.91	0.90	0.90	0.90
35	0.43	1.24	1.05	1.00	0.97	0.96	0.95	0.95
40	0.51	1.38	1.14	1.07	1.04	1.02	1.01	1.00
45	0.60	1.52	1.24	1.15	1.10	1.08	1.06	1.05
50	0.69	1.68	1.34	1.22	1.17	1.13	1.11	1.10
55	0.80	1.84	1.44	1.31	1.24	1.19	1.17	1.15
60	0.92	2.02	1.55	1.39	1.31	1.26	1.22	1.20
65	1.05	2.22	1.67	1.48	1.39	1.33	1.29	1.25
70	1.20	2.44	1.81	1.59	1.47	1.40	1.35	1.32
75	1.39	2.69	1.96	1.70	1.57	1.48	1.43	1.38
80	1.61	2.99	2.14	1.84	1.68	1.58	1.51	1.46
85	1.90	3.37	2.36	2.00	1.82	1.70	1.62	1.56
90	2.30	3.89	2.66	2.23	2.00	1.85	1.76	1.68
95	3.00	4.74	3.15	2.58	2.29	2.10	1.97	1.88
99	4.61	6.64	4.20	3.35	2.90	2.62	2.43	2.29

3.3 Overview of the Real-Time SER Evaluation System for Memory Devices

An overview of memory devices is described here, and then examples of an overall Real-Time SER evaluation system for two types of SRAM devices will be described. One device is a low-power consumption SRAM, which retains digital data for a relatively long time. Such SRAMs are used, for example, in mobile phones. The other device is a high speed synchronous SRAM (SSRAM) which is used for cache memory in high-end server systems. Such SSRAMs are frequently accessed in high-end server systems, and their error rates must be evaluated in advance of the system design because of the high-reliability requirements of the server system.

3.3.1 *Overview of the memory devices*

There are roughly two types of memory devices [3.3]; one with volatile memories, such as SRAMs and DRAMs, and another with non-volatile memories, such as flash memories and EEPROMs (electrically erasable programmable read only memories). Stored data in volatile memories are lost when power is lost. On the other hand, non-volatile memories can retain stored data even when power is switched off. From the point of view of operational speed, the writing speed of flash memories is much slower than that of SRAM; on the order of 10^{-6} sec and 10^{-8}–10^{-9} sec, respectively [3.3]. Because of the advantage of operational speed, SRAM is used for cache memory of the CPU (Central Processing Unit) or MPU (Main Processing Unit). The SER evaluation of SRAM is, therefore, essential for the system's design.

The operations of SRAM, DRAM and flash memory are briefly reviewed here.

(1) SRAM

The equivalent circuit for a SRAM memory cell is shown in **Fig. 3.2**. The memory cell of a SRAM basically consists of one flip-flop circuit and two switching transistors. Before 800-nm process technology, it was normal to employ a traditional SRAM cell consisting of two driver/transfer nMOS and two load resistors utilized with poly-silicon material, as shown in **Fig. 3.2(A)**. After 800-nm to 350-nm process technologies were employed, SRAM cells consisted of two driver/transfer nMOS and TFTs (called Thin Film Transistors), instead of load resistors, as shown in **Fig. 3.2(B)**.

The next generation SRAMs are used with low-voltage applications such as mobile phones, and with high-speed applications, such as server applications or telecommunications. The SRAM cell structure itself has been shifted to a full CMOS SRAM cell structure [3.4] called the 6-transistor SRAM cell structure, as shown in **Fig. 3.2(C)** [3.5]. One reason for the rise of 6-transistor SRAM cell relative to poly-silicon load resistors or thin film transistors (TFTs) [3.6] is due to the fact that traditional SRAMs have relatively much higher leakage and standby currents, as well as lower stability than 6-transistor SRAM cells.

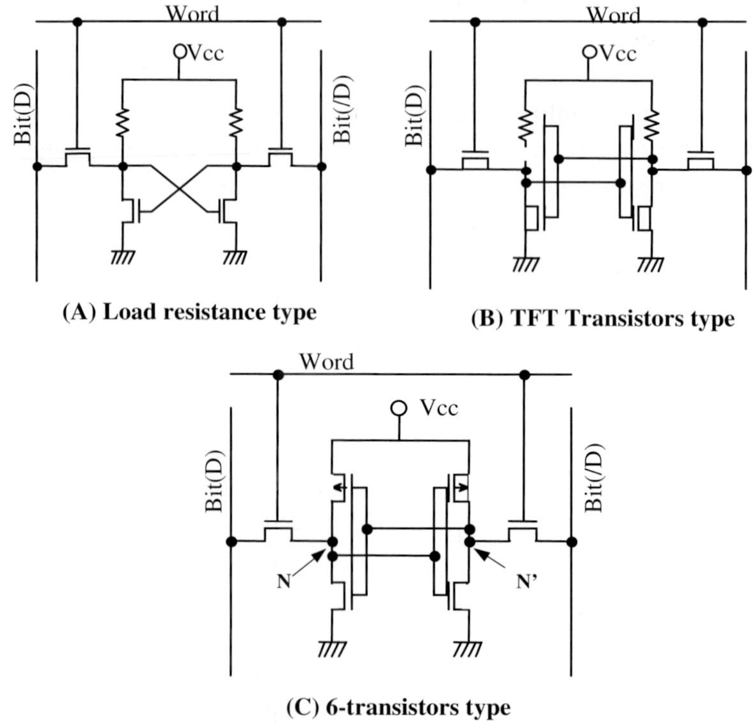

Fig. 3.2 Three SRAM-type cells.

The advantages of 6-transister SRAM cells have led them to become more popular than other traditional SRAMs. In each memory cell structure, there are metal lines shared with other memory cells in each row and column. The lines in a row are called the word lines, and the lines in a column are called the bit lines, along which the data are transmitted in and out of the memory cells. Addressing a specific memory cell from the memory array matrix can be achieved through selecting the relevant word and a bit line. When the access transistors are switched on applying a pulse signal along the word line, a datum is transferred between a pair of bit lines (D, D') and flip-flops. In writing data into a memory cell, one of the data lines (for example, bit line D) is given a higher voltage "H" (logic high), while another bit line D' is given a lower voltage "L" (logic low), and as a result the storage nodes

N and N' are in the state of "H" and "L", respectively. This results in a binary expression for the information (**Table 3.2**): a logic 1 or a logic 0. In reading the stored data, the voltage levels applied to the data lines are read by the sense amplifier and then transferred to the data I/O pin of the memory chip. The data are then read corresponding to the logic described in **Table 3.2**.

Table 3.2 Binary states of SRAM.

Node	Voltage	
N	H	L
N'	L	H
State	1	0

(2) DRAM

Figure 3.3(A) shows the equivalent circuit for a DRAM memory cell called a 1T-1C cell. It means that the cell is composed of one T (transistor) that functions as a switch, and one C (capacitor) that stores charge of information. Since it is possible to manufacture the cell area of the DRAM to be small, DRAM is suitable for large-scale integration of memory devices and has an advantage of smaller cost per bit. When the higher voltage "H" (logic high) is applied to the word line and the transistor is ON, the signal due to the stored charge in the capacitor appears on the bit line. Leakage current, however, exists in the thin film in the capacitor. As a result, the stored charge in the capacitor decreases as time passes, and information is lost. In order to compensate the leakage current, charge is injected into the capacitor, referred to as a "refresh operation." This property is particular to DRAMs, and such RAMs are therefore referred to as "dynamic" rather than "static". The interval of the refresh operation is approximately 10^{-4} sec.

(3) Flash memory

Flash memory, unlike SRAM or DRAM, is essentially one of the types of EEPROMs categorized as a non-volatile memory device using a floating gate transistor (**Fig. 3.3(B)**), and the stored data are retained even if the power supply is lost. It requires no refresh operation during data retention. The Fowler-Nordheim (FN) tunneling effect is used as the mechanism for programming/reading data, where hot electrons play an important role. The word lines and bit line are highly biased either to inject electrons into the floating gate or to discharge electrons from the floating gate. Due to this memory cell structure and electron conduction mechanism, flash memory is well known as a device providing excellent radiation tolerance [3.7].

(4) MONOS EEPROMs

MONOS (Metal-Oxide-Nitride-Oxide-Semiconductor) type EEPROMs (**Fig. 3.3(C)**) are also categorized as non-volatile memory devices using a floating gate transistor. The stored data are retained either in operation mode or non-operation modes, such as stand-by mode or retention mode. The FN tunneling effect or channel-hot-electron effect is used as a mechanism of programming/reading data via the assistance of the select gate, where both hot electrons and hot holes play an important role. Word lines and bit lines are highly biased (plus or minus) to inject electrons or holes into the floating gate or to discharge the electrons or holes from the floating gate.

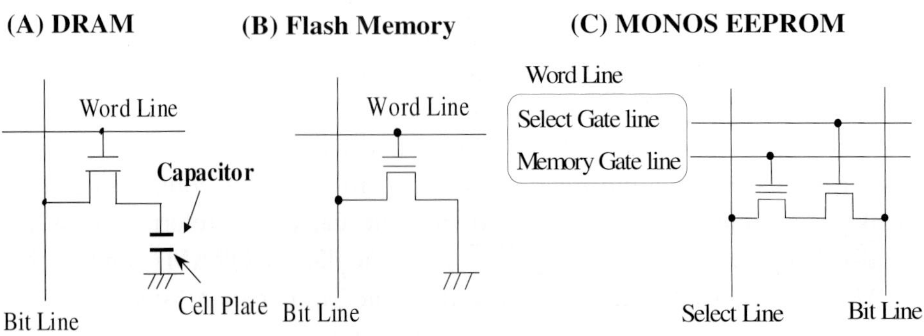

Fig. 3.3 Three types of memory devices: DRAM (A) and Flash Memory (B).

MONOS EEPROMs are also well known as excellent radiation tolerant devices [3.8, 3.9], and as a result these kinds of MONOS cells have often been used for aerospace or military applications [3.10, 3.11].

3.3.2 *General description of a Real-Time SER evaluation system*

One of the systems illustrated in **Fig. 3.4(A)** is designed for Real-Time SER evaluation of fast synchronous 16 Mbit SRAMs (called as SSRAMs) fabricated by a 130-nm process technology node. This system is able to carry 1024 SSRAM devices, with 64 devices set on each circuit board. The system configuration illustrated in **Fig. 3.4(B)** provides 4 major parts: 16 test board units mounted with CPU, a Changeover unit, the Control PC, and power supply unit. A specific data pattern is written into the memory chips and then read out under constant applied voltage during the test period.

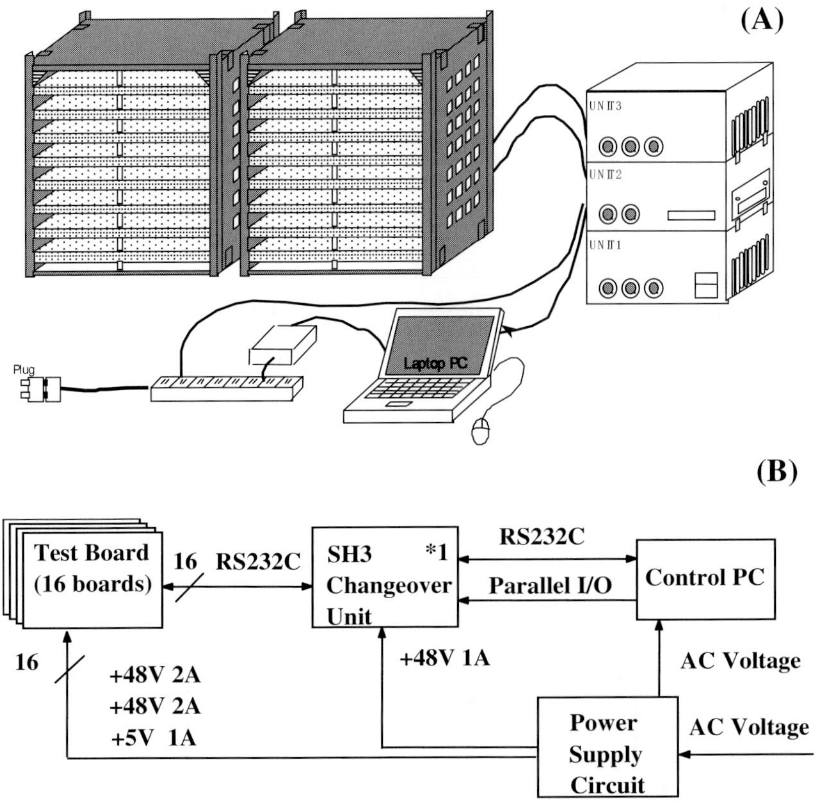

Fig. 3.4 Overview (A) and block diagram (B) of a Real-Time SER evaluation system.

A data pattern is selected from "checker board (010101…)" including "checker board compliment (101010…)", all "0", all "1" and the alternation between all "0" and all "1". Typical operation of the device during the test period is implemented with a frequency of 32 MHz, and the read time is repeated every 10 seconds with both the checkerboard pattern and reverse pattern. (**Fig. 3.5**)

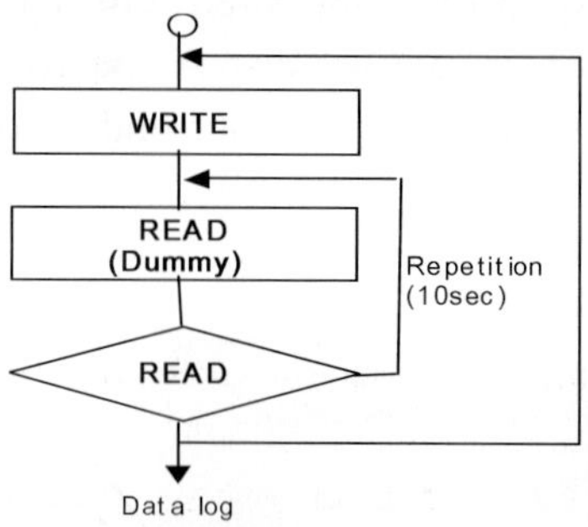

Fig. 3.5 Flow chart of the device operation.

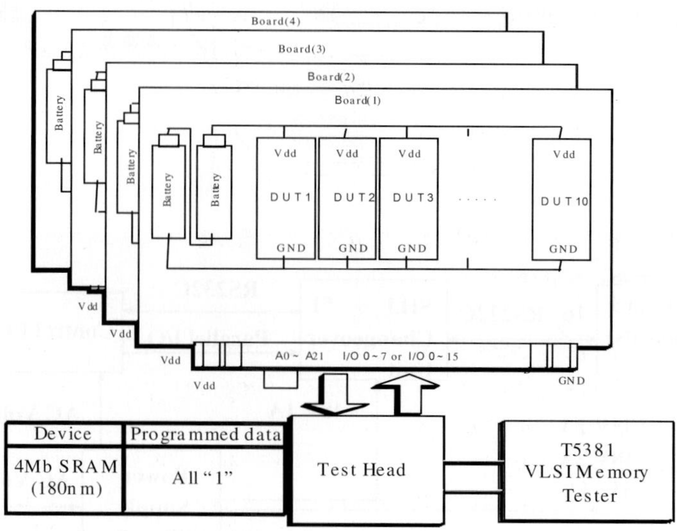

Fig. 3.6 Overview of the test procedure for low-power consumption SRAMs.

For low-power consumption SRAMs, a battery-backup-type of test board is usually employed. An overview of the test procedure is depicted in **Fig. 3.6**. 10 pieces of SRAM chips are set on a test board, and the data pattern of "all 1" is written on the chips by an LSI tester. In this case, a T5381 Advantest VLSI memory tester was used. More than 10 test boards, i.e. more than 1000 devices in total, are set at a specific location for real-time testing and are kept in data retention mode. After a certain period of testing, data on the chips are read out and checked for errors. The batteries must be fully checked and fixed before long term field-testing.

3.4 Environmental Conditions of Real-Time SER Testing

The terrestrial neutron flux varies according to the geomagnetic longitude and latitude and the altitude of the location at ground level, as described in Chapter 2. It also shows temporal changes, mainly depending on the activity of the sun. It is therefore very important to have knowledge of the geophysical properties of the Real-Time SER testing location, such as the temporal and spatial variation of the terrestrial neutron energy spectrum, flux, and dose in order to directly measure the SER of a device under actual conditions, or to estimate the SER from the results of accelerator tests (which will be discussed in Chapter 5). In this section the temporal and spatial variation of the terrestrial neutron energy spectrum and dose is discussed.

3.4.1 *Spatial and temporal variation of the terrestrial neutron energy spectrum and dose*

Primary cosmic rays consist of charged particles such as protons, with a vast range of energies up to 10^{20} eV. They are produced by stellar explosions (super novae) in the galaxy, and by particles emitted from the sun [3.12, 3.13]. They are influenced by both of heliomagnetic and geomagnetic fields when incident upon the atmosphere of the earth. Charged particles move spirally around the magnetic field, and the subsequent energy spectrum and the flux depends on the geomagnetic latitude and longitude. In the polar region, where the geomagnetic intensity (expressed as cutoff rigidity) is weak, the observed primary cosmic-ray flux is much higher than that near the equatorial region.

Terrestrial neutrons observed on the ground are produced through various interactions of primary cosmic rays with the atmosphere and therefore have a close relation with the activity of the sun. The activity of the sun varies with a cycle of about 11 years (Schwabe cycle).

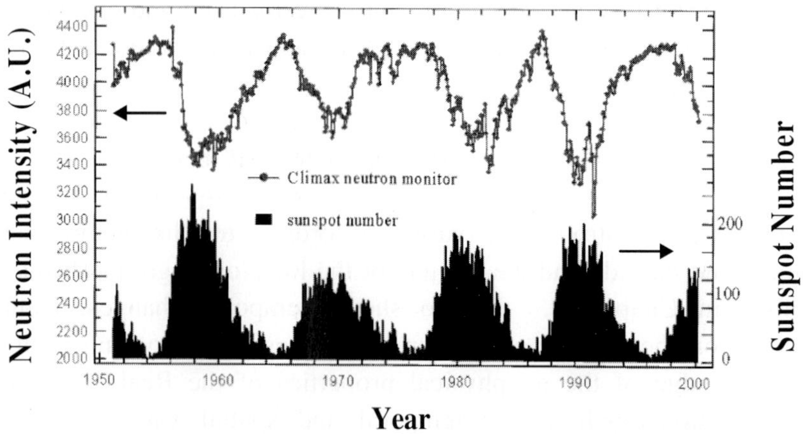

Fig. 3.7 Annual variation of the solar activity (sunspot number) and terrestrial neutron intensity observed at Climax, USA [3.12].

Figure 3.7 shows a temporal variation on neutron intensity measured at Climax, USA, and the observed number of black spots in the sun [3.14], where a negative correlation can be seen between them. This is due to the fact that the incoming flux of galactic cosmic rays with energies below 10^9 eV is reduced when stronger heliomagnetic fields are produced during periods of high solar activity [3.13, 3.14].

3.4.2 *Geomagnetic latitude, longitude and altitude of Real-Time SER tests*

As described in the previous section, neutron flux has a close relation with geomagnetic latitude, longitude and altitude at the observation location. For investigating the results of real-time testing or estimating

SER of devices from accelerator-based tests, the information on geomagnetic latitude, longitude and altitude of a specific location is essential. It is convenient to use a GPS (Global Positioning System device such as the eTrex Legend of GARMIN® to obtain information on geographic latitude, longitude, and altitude at the testing location. The measured geographic data are then converted into geomagnetic data by referring to the web page of NASA (National Aeronautics and Space Administration)/Goddard Space Flight Center [3.15] or that of CERN (Conseil Européen pour la Recherche Nucléaire) Cosmic Ray Telescope Muon Detector Group [3.16]. Detailed geomagnetic information is obtained from the NASA web site, and the CERN site gives the geomagnetic cutoff rigidity R_c which represents the shielding effect on protons. R_c has its smallest value around Greenland, near the North Pole, and its largest value around Singapore, at the equator.

The geomagnetic correction coefficient δ ($\delta = 1$ when $R_c = 0$) is calculated by the following equation Eq. (3.7), which is obtained by curve fitting the data in Ref. [3.17].

$$\delta = \begin{cases} -0.0009R_c^{\ 2} - 0.0012R_c + 1.0026 & (R_c < 1.76) \\ -0.0068R_c^{\ 2} + 0.0092R_c + 1.0007 & (1.76 \leqq R_c < 3.37) \\ 0.0006R_c^{\ 2} - 0.0486R_c + 1.1134 & (R_c \geqq 3.37) \end{cases} \quad (3.7)$$

Since the atmosphere also plays a role as a shield for cosmic-ray neutrons, when the terrestrial-neutron spectrum at a specific location is generally applied to calculate SER, a correction to the spectrum based on altitude is also needed. As already given in Sec. 2.4.5, according to Ref. [3.13], the correction factor for the altitude effect γ is given by

$$\gamma = \frac{\Phi_n}{\Phi_0} = \exp\left(-\frac{A - A_0}{L}\right), \quad (3.8)$$

where A_0 is the atmospheric pressure (g/cm^2) at the standard altitude H_0, Φ_0 the neutron flux at H_0, Φ_n the neutron flux at a specific location, A

the corresponding atmospheric pressure to Φ_n, and L the effective attenuation length in the atmosphere. For cosmic-ray neutrons, L has a value of 148 g/cm^2 in Ref. [3.2], but other several L values are also given in Sec. 2.4.5. In case of sea level, which is used as a standard altitude, Eq. (3.8) is written as

$$\gamma = \frac{\Phi_n}{\Phi_0} = \exp\left(-\frac{A-1033}{L}\right). \tag{3.9}$$

To estimate the atmospheric pressure at a given altitude, the following equation is given in Ref. [3.2]

$$A = 1033 - \left\{0.03648 \times \left(\frac{a}{0.3}\right)\right\} + \left\{4.26 \times 10^{-7} \times \left(\frac{a}{0.3}\right)^2\right\} \tag{3.10}$$

which is referred to as the IBM Model. The NASA-Langley Model [3.10] also gives the following equation,

$$A = 1033 \times \exp\left\{-0.03813 \times \left(\frac{a}{300}\right) - 0.00014 \times \left(\frac{a}{300}\right)^2 + 6.4 \times 10^{-7} \times \left(\frac{a}{300}\right)^3\right\}, \tag{3.11}$$

where the variable a in Eqs. (3.10) and (3.11) represents the altitude in meters. The atmospheric depth D (g/cm^2) can be also estimated as a function of the altitude X (km), as given in Eq. (2.12) in Sec. 2.4.5 of Ref. [3.2],

$$X = \begin{cases} 47.05 - 6.9\ln D + 0.299(\ln 0.1D)^2 & D < 25 \\ 45.5 - 6.34\ln D & 25 < D < 230. \\ 44.34 - 11.836D^{0.19} & D > 230 \end{cases} \tag{3.12}$$

Comparing the three equations, Eq. (3.10) coincides with Eqs. (3.11) and (3.12) only at relatively low altitude (< about 10km). Both Eqs. (3.11) and (3.12) give almost the same results, as shown in **Fig. 3.8**.

Figure 3.9 shows the relation between the correction factor γ and the atmospheric pressure A of Eqs. (3.10) and (3.11), with variation of the L value. Here, we have considered the several L values that have been obtained by several research groups, as stated in Sec. 2.4.5(from 131.3 to 165 g/cm^2). In case of the IBM Model Eq. (3.10), the calculated results are shown in the figure for atmospheric pressure above about 250 g/cm^2, which corresponds to an altitude of about 10 km. As shown in **Fig. 3.9**, Eqs. (3.10) and (3.11) give almost the same results for correction factors up to an altitude of about 10 km. The correction described above is taken into consideration in the estimation of SER from accelerator-based testing for devices at a specific location.

It is also included in the SER simulation code CORIMS (COsmic-Ray IMpact Simulator), which is described in detail in Chapter 6.

Since the simulator CORIMS is intended to calculate the SER of devices, particularly at the surface level of the earth, it includes Eq. (3.10), which is valid for altitudes up to about 10 km.

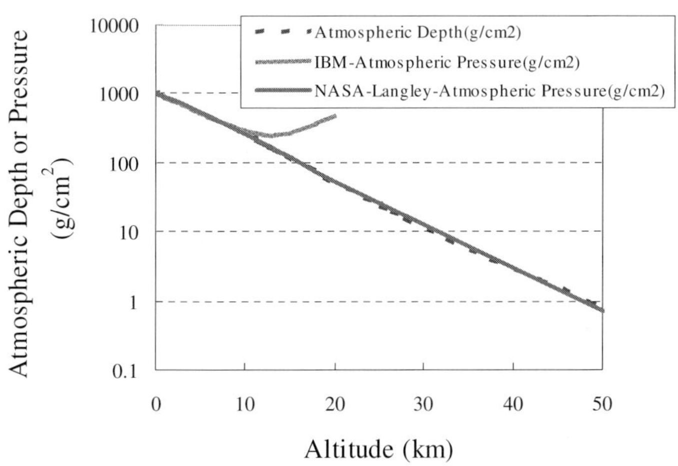

Fig. 3.8 Comparison of the three models for atmospheric pressure estimation.

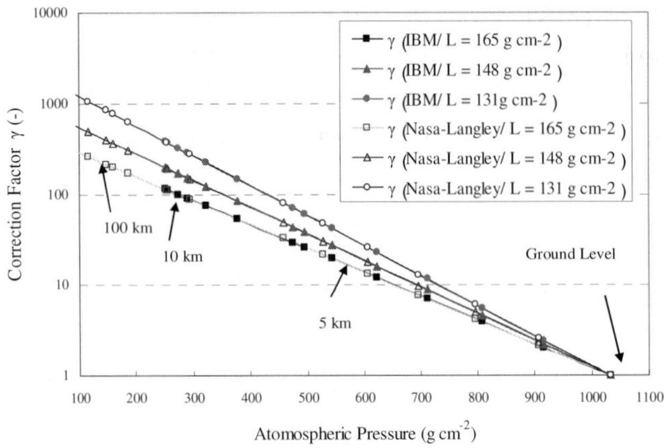

Fig. 3.9 Correction factor for atmospheric pressure with variation of the effective attenuation length *L*.

3.4.3 *Day-, night-time and monthly variation of the neutron dose at ground level*

As described in Sections 2.1, 2.4.5, and 3.4.1, cosmic-ray neutrons are largely influenced by the Sun, including the solar wind. In the day-time region the surface of the earth faces the Sun and the solar wind can compress the geomagnetic field above the region (**Fig. 3.10**) [3.18]. As a result, the protons (and therefore cosmic-ray neutrons) could have variations within a day because of the asymmetrical modification of the geomagnetic field.

Figure 3.11 shows the averaged neutron dose rate from July 2001 to December 2002 at Sendai, Japan [3.19] using a NSN2 type rem counter by Fuji Electric Co., as described in Sec. 2.2.3 (1). Data from the day-time were averaged from 6:00 – 18:00 and those from the night-time from 18:00 – 6:00. From October 4 to 26, 2001, the rem counter was set inside a concrete building with a thick upper concrete ceiling of about 1 m. A larger variation of the neutron dose rate can be seen in day-time than in night-time. The averaged neutron dose rates during day-time and night-time of each month except October 2001 are summarized in **Table 3.3**. **Figure 3.12** shows the variation of the monthly averaged neutron dose rate (A) and of the ratio of the night-time and day-time data (B)

[3.19]. In **Fig. 3.12(A)**, the data acquired inside the concrete building (October 2001) is excluded. In spite of the asymmetrical modification of the magnetic fields around the earth by the solar wind, neutron dose rate varies by no more than 3 % between day-time and night-time.

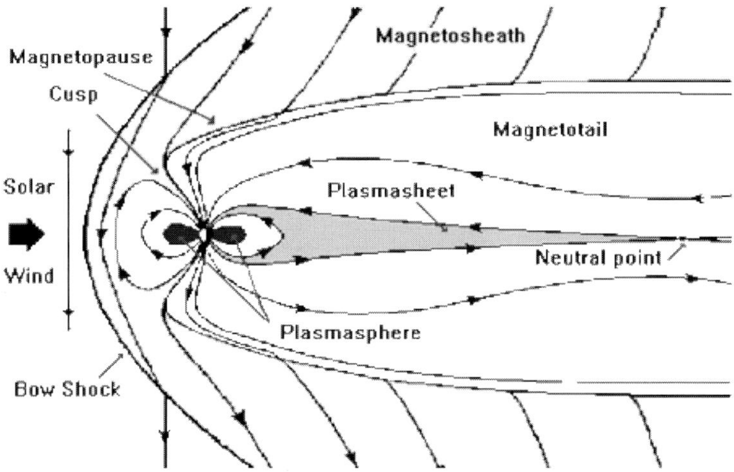

Fig. 3.10 Modification of the geomagnetic fields by the solar wind [3.18].

Table 3.3 Monthly averaged neutron dose rates obtained by a rem counter during day-time (6:00-18:00) and night-time (18:00-6:00). In Oct. 2001, the rem counter was placed inside a concrete building [3.19].

Year	2001												Average (nSv/h)
Month	1	2	3	4	5	6	7	8	9	10	11	12	
Day-time	-	-	-	-	-	3.38	3.53	3.34	3.28	(1.06)	3.42	3.40	3.39
6:00-18:00	-	-	-	-	-	2.5	2.4	2.5	2.6	(10)	2.5	5.1	
Night-time	-	-	-	-	-	3.36	3.42	3.31	3.26	(1.07)	3.49	3.45	3.38
18:00-6:00	-	-	-	-	-		2.5	2.0	2.4	(9.1)	3.0	5.7	
Year	2002												Average (nSv/h)
Month	1	2	3	4	5	6	7	8	9	10	11	12	
Day-time	3.39	3.48	3.62	3.49	3.47	3.49	3.31	3.38	3.38	3.44	3.48	3.23	3.43
6:00-18:00	3.3	3.1	3.7	1.9	2.7	3.1	2.6	3.3	2.1	3.2	2.6	3.3	
Night-time	3.36	3.46	3.67	3.51	3.49	3.45	3.31	3.35	3.36	3.45	3.46	3.21	3.42
18:00-6:00	3.5	3.2	3.3	2.6	2.7	2.6	2.6	3.0	2.4	3.0	2.8	4.0	

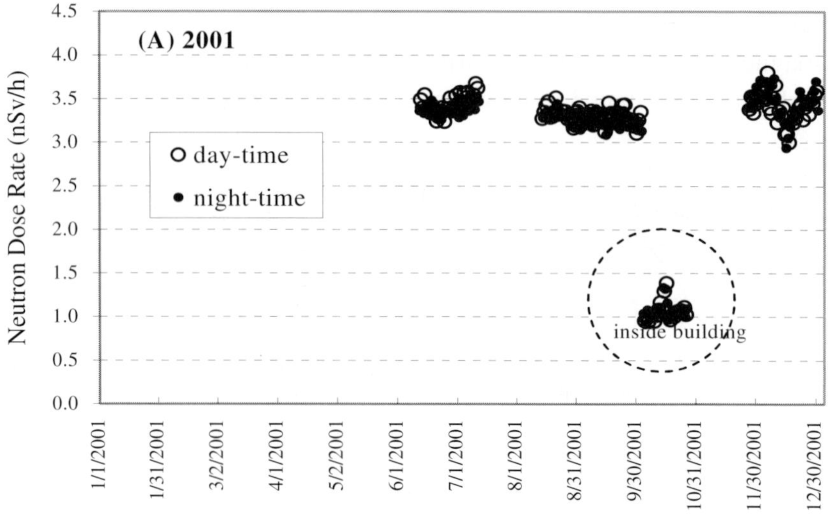

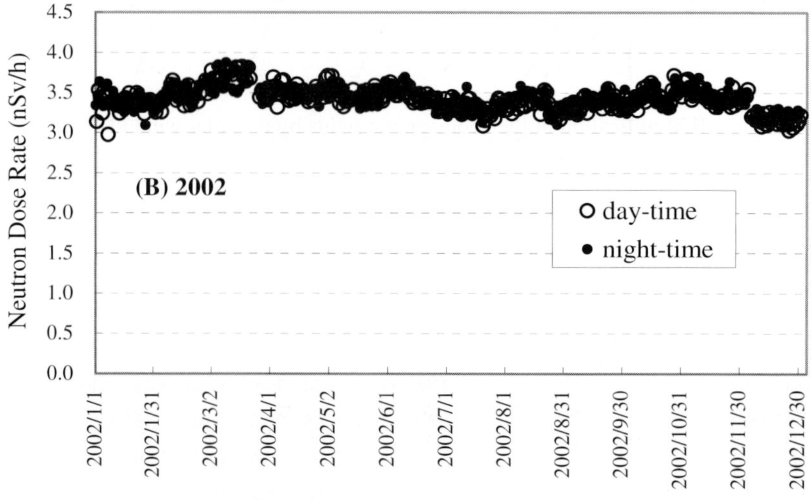

Fig. 3.11 Variation of averaged neutron dose rates) obtained by a rem counter [3.19] during day-time (6:00-18:00) and night-time (18:00-6:00).

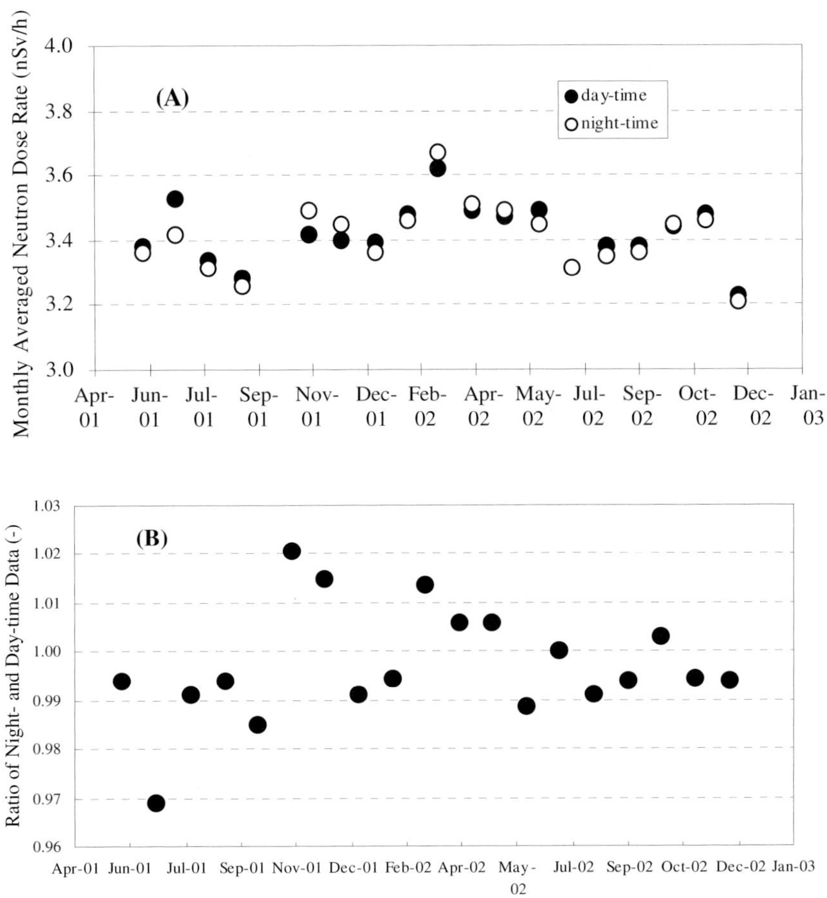

Fig. 3.12 (A) Monthly variation of averaged neutron dose rates during day-time (6:00-18:00) and night-time (18:00-6:00). (B) Monthly variation of the ratio of averaged neutron dose rates during day-time and night-time [3.19].

From the data of **Figs. 3.11**, **3.12** and **Table 3.3**, variation of neutron dose rate within a year can be identified; neutron dose rates in August and September are less than the yearly averaged value (3.45 nSv/h) by about 8%, and those in November and December are more than the average value by about 3%. Such a variation is most likely caused by meteorological conditions, like humidity. Thus, under conditions of relatively high temperature and humidity, such as the summer in Japan, neutron dose rate can be relatively low, and in the winter it can be high.

3.4.4 *Monitoring of neutron dose during Real-Time SER testing*

As described above, temporal and spatial variation on terrestrial neutron dose rate exists at all locations of Real-Time SER testing. In addition, the roof, ceiling and wall of a building used for the Real-Time SER test attenuate neutron dose. As a result, the SER value completely depends on the test period and the structure of the building where the test equipment is set. Terrestrial neutron dose rate data measured during test periods play an important role for comparing the SER data of various devices obtained under different test conditions.

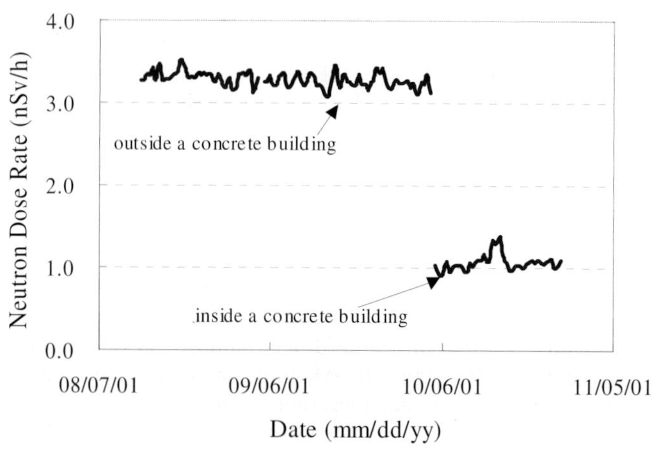

Fig. 3.13 Neutron dose rate both inside and outside a concrete building obtained using a rem counter at Tohoku University, Sendai, Japan [3.19].

Figure 3.13 shows the monitored results of neutron dose rates both outside and inside a concrete building with a thick upper concrete ceiling of about 1 m from August to October, 2001; the averaged neutron dose rate was 3.3 and 1.1 nSv/h, respectively, [3.19]. These data were obtained at Tohoku University, Sendai, Japan using the portable neutron rem counter shown in **Fig. 3.14** (NSN2 type rem counter by Fuji Electric Co. as described in Sec. 2.2.3 (1)) and in the connection diagram shown in **Fig. 3.15**. The monitored results showed a neutron attenuation of about one-third after penetration through 1 m of concrete. The neutron rem counter has high sensitivity for low-energy neutrons by using a high pressure ^{3}He proportional counter, but it has a very low sensitivity for neutrons above 15 MeV.

The NSN2 type rem counter has an output terminal with a DC voltage V_{out} (0 to +1 V) used for recording data. Seven logarithmic decades of neutron dose rate from 10 nSv/h to 9.999 mSv/h are converted to volts by a circuit implemented in the rem counter. Thus, the recorded voltage can be re-converted into neutron dose rate by employing the following equation:

$$\textbf{Neutron Dose Rate} = 10^{\left(\frac{V_{out}}{0.1428}-6\right)} \ \textbf{(mSv/h)} \tag{3.13}$$

As an interface between a rem counter and a lap top PC, the PCMCIA type data acquisition card NR-110 by Keyence Co. can be used. The counts of the rem counter can be stored automatically every 30 minutes, for instance, into a designated Micro Soft Excel™ file using a program named Flex Logger™. This rem counter system is not influenced by electronic noise (such as an instantaneous stoppage of the power supply). It also has no response for photon fields below 100 mSv/h
Main Specifications of NSN2 are as follows:

1) Detector : ^{3}He proportional counter
2) Moderator : polyethylene
3) Sensitivity : 4.5 $S^{-1}/(\mu Sv\ h^{-1}) \pm 20\%$, (ref.^{252}Cf)
4) Measurement Range : 10 nSv/h to 9.999 mSv/h (for dose rate)
5) Energy Range : 0.025 eV to 15 MeV

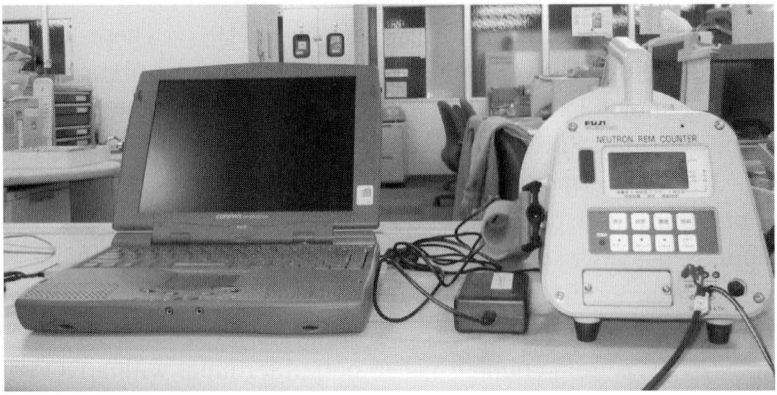

Fig. 3.14 An example of a portable neutron rem counter connected with a laptop PC.

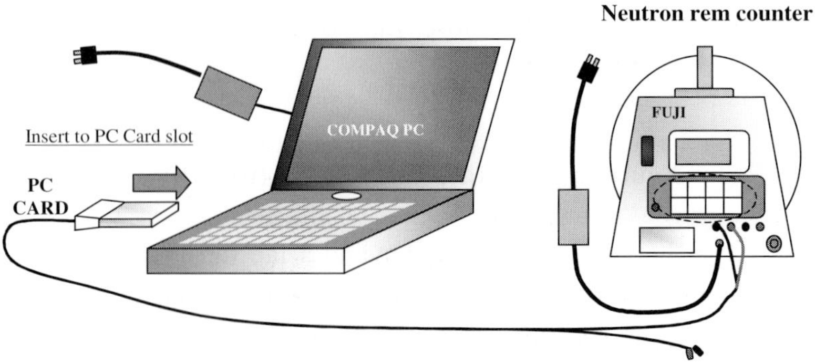

Fig. 3.15 Connection diagram between PC and neutron rem counter.

3.5 Real-Time SER Pre-test Preparations

3.5.1 *Sample selection*

Device variability for RTSER is generally small for devices produced with the same masks and fabrication steps. In general, a minimum requirement to establish a product's RTSER is to run the test with roughly equal numbers of nominal devices from at least three different non- consecutive lots. If a product is tested that has onboard error-correction codes/circuits (ECC), tests should be run under two conditions; both with ECC enabled and with ECC disabled, if it is possible. (Note: In general, ECC is effective at masking single bit cell upsets. But it is ineffective with non-cell related events, such as neutron-induced latch up or other single-event functional interrupts.) Caution should be taken in that the ECC function becomes less effective when the data pattern of device remains on for a long period of time, since the accumulation of single bit errors increases the possibility of multi-cell errors.

3.5.2 *DUT preparation and orientation*

The tested package material used for the RTSER test must be identical to the package used for production components. This is a critical issue for the alpha component of RTSER. If a finned heat sink is required, this could impact the neutron flux reaching the DUT. Since there is some angular dependence for high-energy neutrons, tested samples should be symmetrically arranged and laid out on the test board. The path of atmospheric neutrons must be taken into consideration in the Real-Time SER setup, along with the effects of building shielding. For example, a vertical stacking of SRAM DIMMs that are each oriented in a horizontal position would cause a neutron to traverse many SRAMs before reaching the final device. The possibility of generating secondary particles, as well as the attenuation of flux, could lead to some potentially-anomalous results. Therefore, the layout of the DUTs in RTSER testing should avoid this orientation. It is also important for RTSER testing to eliminate any external electromagnetic noise or power disturbances.

3.5.3 *Test program verification*

Before starting RTSER, the test program should be run to verify the proper operation of all the DUTs at the testing site, after placing the desired population of DUTs in their sockets on each DUT board. If necessary, one should make necessary adjustments to the system and DUTs to insure that the test program executes flawlessly on all DUTs in the test system. If possible, running the diagnostic test program periodically is one method to insure test quality and to retain all error information records.

3.5.4 *Effective neutron flux at the test location*

When performing RTSER testing, it is necessary to either measure the neutron flux directly or make estimates based on the amount of building shielding and the incident terrestrial neutron flux. Measurement of the actual neutron flux is rather involved because several detection schemes are required to obtain counts over a large neutron energy range. The results should be normalized to the NYC flux in the final report for purposes of standardization. A qualitative estimate of sensitivity to thermal neutrons should also be made. If the testing is to be done at high altitude, several factors should be taken into consideration when designing the experiment:

1) The cosmic ray dosimetry method used, such as a portable rem counter or an equivalent system.
2) The electrical power features and stability of the facility. The test setup often requires an independently stabilized, uninterruptible power supply to run the test. Note that the overall power consumption for a large sample size can be significant. Make sure that the facility can provide the required power with good stability.
3) The lab should be protected against electrostatic discharge from lightning (*i.e.*, presence of a Faraday cage, a lightning rod, *etc.*) or a power outage due to snowstorm when the lab is located near the top of a mountain.
4) The environmental conditions that influence the cosmic ray flux and how the flux is monitored (*i.e.*, temperature, hygrometry, average pressure, average accumulation of snow, *etc*).

3.5.5 *Test locations of Real-Time SER testing*

In order to verify the altitude dependence of neutron-induced SER, two locations both in Japan and in the USA were chosen for RTSER testing. In the first stage, verification experiments on the altitude dependence and concrete attenuation length were implemented at

3 locations (86 m, 766 m, 1988 m high) in Japan, as illustrated in **Fig. 3.16.** These three locations were all near 26 degrees of geomagnetic latitude. The first stage of RTSER testing used more than 1000 pieces of low power 180 nm SRAM devices. In the second stage, verification experiments on altitude dependence were implemented at 2 locations (0 m, 1700 m high) in the USA, as shown in **Fig. 3.17**. These locations were both near 44 degrees of geomagnetic latitude. The second stage of RTSER testing used more than 500 pieces of high-speed 130 nm SRAM devices.

(a) 766m high **(b) 1988m high**

Fig. 3.16 RT SER test locations in Japan.

(a) 0m high **(b) 1696m high**

Fig. 3.17 RT SER test locations in the USA.

3.6 The Impact of Noise on Real-Time SER and Neutron Dose Rate: An Example of Field-testing

Table 3.4 Neutron dose rate and Cutoff Rigidity corresponding with each location.

Country	Test location		Altitude (m)	Neutron Dose rate (uSv/hour)	Cutoff Rigidity (GV)	Technology node(nm)
Japan	Tokyo		86	0.001038	11.61	180nm device
	Nagano-pref.		766	0.001632	11.39	
	Nagano-pref.		1988	0.003	11.35	
USA	Location (A)	(except 2/1-2/18)	1696	0.00458	3.13	130nm device
		(2/1-2/18 only)		-		
	Location(B)		0	0.00147	3.58	

When the Renesas Technology Corp. studied the altitude dependency on RTSER and the relation between RTSER and the neutron dose rate in the USA [3.20], we found an unusual phenomenon during the testing. It is believed that the abnormal data may possibly be due to the geomagnetic noise generated near the topographical location at the time of testing.

Five locations were chosen for RTSER testing, with three locations in Japan and two locations in the USA. **Table 3.4** shows neutron dose rate and cutoff rigidity information at each of test locations. The five locations were chosen for their accessibility and convenience in moving equipment to and from the testing sites, and their ample space for RTSER testing. The first priority in the selection of devices for testing was deep sub-micron SRAM devices, for reasons of decreasing critical charge [3.21, 3.22].

At each location, various SER testing had been conducted in order to study the correlation between the results and neutron-flux attenuation behind various thicknesses of concrete, as well as the dependence of the results on atmospheric depth and the geomagnetic rigidity cut-off. In this section, the dependence of SER testing results on flux attenuation through concrete (Sec. 3.6.1) and the dependence on altitude (Sec. 3.6.2) are described.

3.6.1 *Concrete attenuation length*

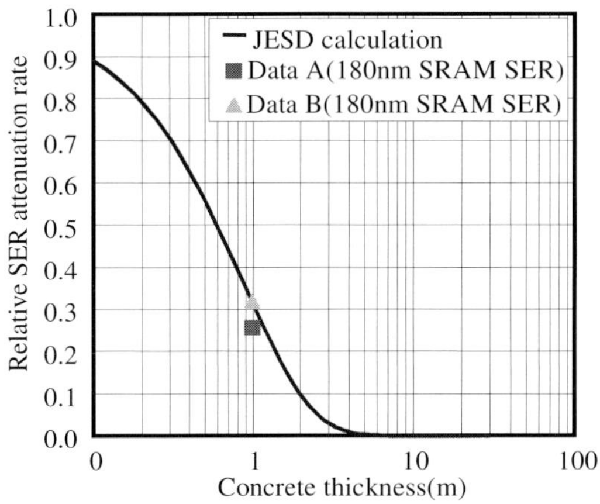

Fig. 3.18 Concrete attenuation rate and SER.

It is known that the concrete attenuation length, which is introduced in JESD89A document [3.2] published by EIA/JEDEC (Electronic Industries Alliance/Joint Electron Device Engineering Council), is recognized as one of the important attenuation factors for SER. Most applications are indoors, and building materials attenuate the incident flux that strikes the electronics.

In order to verify the effect of concrete attenuation length between SER results obtained by the JESD89A calculation method and by actual device testing, measurements were conducted both outside and inside using two kinds of 180 nm CMOS SRAM devices (Data A and B in **Fig. 3.18**), at the Tokyo testing site. The soft error's FIT rate was measured by using more than 1000 pieces for up to 1000 hours of operation in test locations both outside (equal to 0m concrete thickness) and inside.

As shown in **Fig. 3.18**, the two SER attenuation rates (Data A and B) are 0.26 and 0.32, respectively. The data are well fit with the JESD89A calculation. In general, for concrete of thickness t (m), the attenuation length is calculated by

$$\Psi = \Psi_0 \times \exp(-1.17 \times t) \tag{3.14}$$

where Ψ and Ψ_0 are the attenuated and initial fluxes, respectively [3.2].

For the purpose of calculating the attenuation effect, if there are thick concrete walls as well as floors and roofs, and if an accurate differential neutron flux is desired, a high-energy radiation transport code, such as MCNPX [3.23], FLUKA [3.24], or GEANT [3.25], should be used to model radiation transport through the building. The outdoor reference neutron spectrum, as described in JESD89A (see Appendix 3, pp. 64–67), may be used as the incident radiation for such modeling. As an alternative to performing such calculations, if a detection system that is only sensitive to high-energy nucleons is available, the relative flux of high-energy nucleons inside and outside the building could be measured. Large amounts of steel (such as aboard ships) or other materials with high atomic number (e.g., lead shields) can significantly distort the neutron spectrum. Each incident high-energy neutron can liberate several neutrons with energies about 1 MeV, sometimes creating an increase of a factor of 2 or more in that region of the spectrum. However, steel shielding above the point of interest attenuates the portion of the spectrum with E > 10 MeV. Steel (or iron) also absorbs thermal neutrons, and a room or enclosure with steel walls will have a significantly reduced thermal neutron flux.

3.6.2 *Verification of the altitude dependence at field-testing*

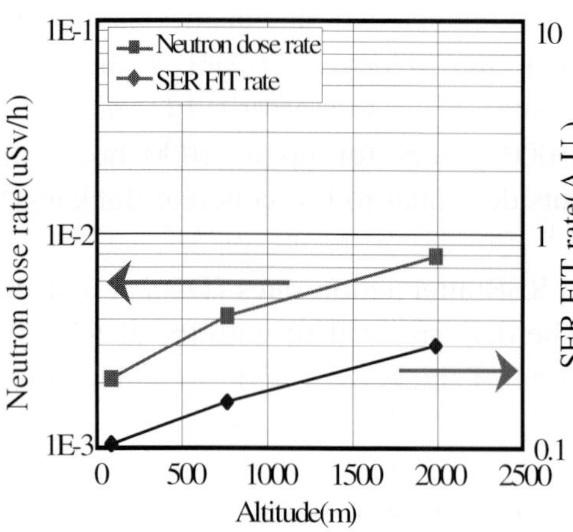

Fig. 3.19 Altitude dependence for SER in Japan.

To determine device's dependence on neutron dose rate and RTSER on deep sub-micron SRAM at different altitudes, SER testing was conducted at 3 locations in Japan at altitudes of 86 m, 766 m and 1988 m. Also, we conducted RTSER at 2 locations in the USA, at altitudes of 0 m and 1696 m, as outlined in **Table 3.4**.

(1) For RTSER testing in Japan, more than 1000 samples of 180 nm low-power SRAM devices were used under battery back-up mode.
A good correlation was found on the altitude dependence between SER and neutron dose rates, as shown in **Fig. 3.19**. The left vertical axis shows neutron dose rate, and the right-hand axis shows the SER FIT rate (A.U). The results show that the SER susceptibility for deep sub-micron SRAM devices (180 nm or less) shows a strong dependence on the neutron dose rate that increases along with the altitude. This dependence is true as long as the effects of alpha rays and thermal neutrons are negligibly small for deep sub-micron SRAM devices.

(2) For RTSER testing in the USA, more than 500 devices of 130 nm SRAM devices were tested under dynamic operation mode for up to 3000 hrs at 2 test locations. The neutron dose rates at location (A) and location (B) were measured every hour using a rem counter with a connecting PC for data storage. The measured altitude dependence of SER in the USA is shown in **Fig. 3.20**. The trend of the data here is consistent with trend of neutron dose rate in Japan. The data

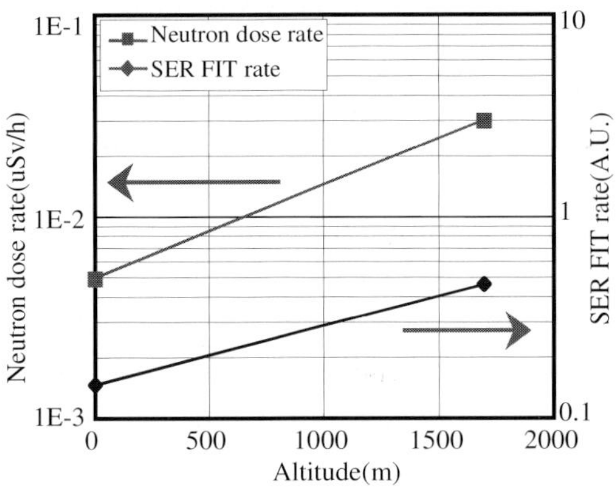

Fig. 3.20 Altitude dependence of SER in the USA.

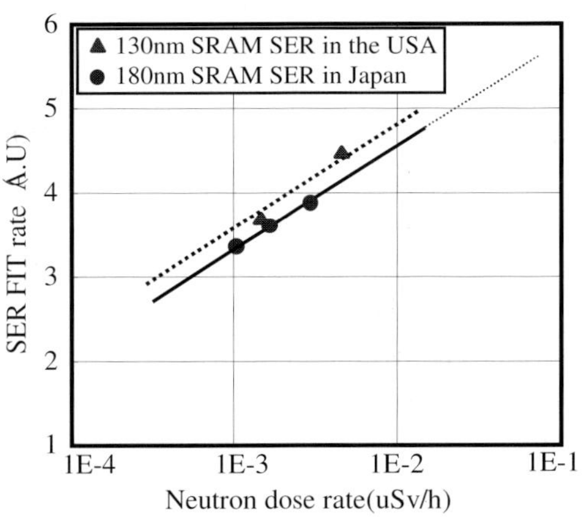

Fig. 3.21 Neutron dose rate vs SER FIT rate [3.26].

confirm that the neutron-induced soft-error dependence on altitude is obviously one critical issue for soft error.

3.6.3 *Correlation between neutron dose rate and neutron-induced soft error in the field [3.26]*

In order to further understand the correlation between neutron dose rate and neutron-induced soft error, two kind of SRAM devices were implemented for RTSER field testing at 3 locations in Japan and 2 locations in the USA, as mentioned in Sec. 3.6.2.

After plotting the experimental SER data for both the 130-nm SRAM device and 180-nm SRAM device in **Fig. 3.21**, it was verified that the altitude dependence of 180 nm SER for neutron doses in the range of 10^{-3} μSv/h to 10^{-2} μSv/h was the same as that of 130 nm SRAM SER. The difference in increasing SER error rates for 130-nm devices and 180-nm devices might be due to the susceptibility of the devices or the influence of geomagnetic difference between Japan and the USA, as described in Sec. 2.2.4.

3.6.4 *Neutron dose equivalent rate in the environment*

In this section, we discuss how to distinguish the effects between the normal neutron dose rate and a peculiar neutron dose rate due to unknown source, as shown in **Fig. 3.22** [3.26]. During RTSER testing from the end of November to the middle of the following year, a peculiar phenomenon was observed. The phenomenon was observed between the beginning of February and the middle of February. The peculiar phenomenon may have been due to the disturbance of the mesosphere, a disturbance in meteorology, or a strong electromagnetic wave, created by an unknown source passing over the testing site (Location A) in the USA.

The anomalous neutron dose rate is comprised of at least three peaks. The first peak occurred from the beginning of the month through the first 3 days. The second peak, which is much higher than the first peak, started from the third day, and lasted through ninth day. The last peak, which is higher than the second peak, occurred from ninth day through the fifteenth day. The exact cause of the three peaks is unknown. The difference in the neutron dose rate between the abnormal period and the normal period is clearly seen in **Fig. 3.22**. The soft error rate, bit multiplicity, and differential SER were checked as a function of time to determine if any differences could be seen between the periods of normal and abnormal neutron dose rates.

Before proceeding with the analysis, it is necessary to understand the bit configuration for the 130 nm SRAM memory cell array along with the indicated bit-multiplicity, according to the definition of SBU/MCU [3.27–3.29], as shown in **Fig. 3.23**.

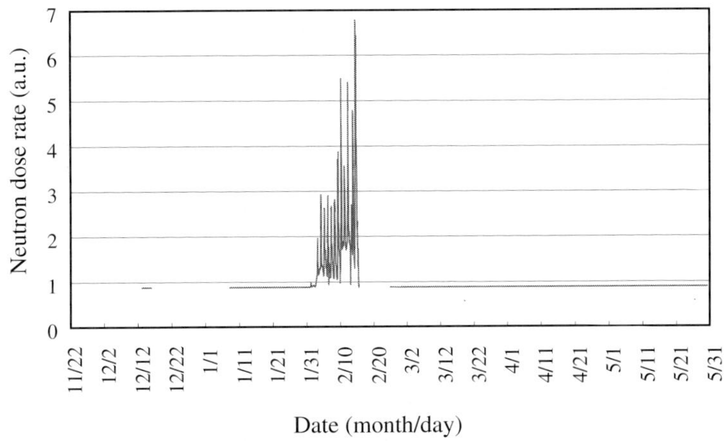

Fig. 3.22 An anomalous neutron dose rate over a limited period [3.26].

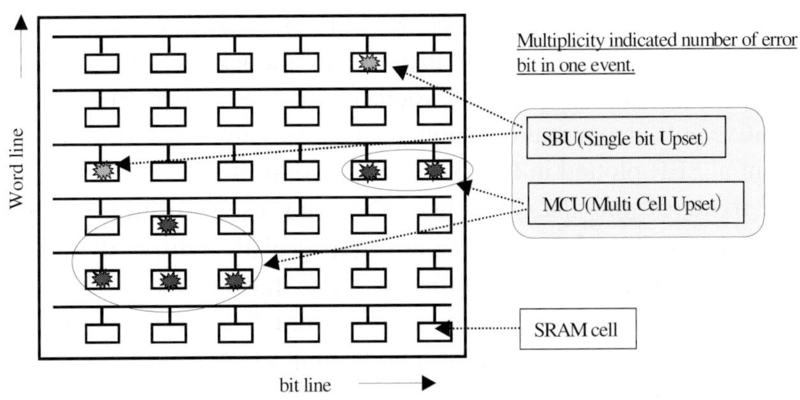

Fig. 3.23 Memory cell layout and definition of multiplicity [3.26].

The horizontal direction shows the bit line and the vertical direction shows the word line along the source/drain direction of the NMOS transistor in the SRAM cell. As for the definitions of SBU and MCU, refer to chapter 1.

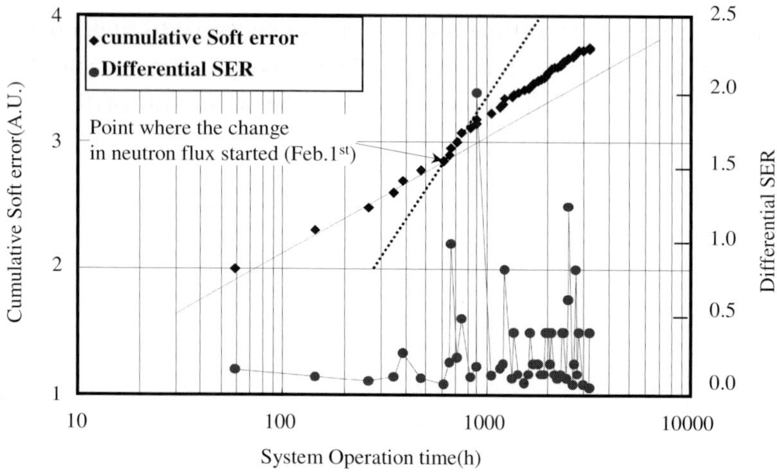

Fig. 3.24 Cumulative soft error and differential SER in the USA [3.26].

The neutron monitoring data analysis should also be done to investigate the correlation between the neutron dose rate change illustrated in **Fig. 3.22** and the atmospheric conditions around the test location.

After a statistical analysis of RTSER data and of the neutron monitoring data, it was found that the abnormal phenomena of the neutron dose rate were well synchronized with increases in both cumulative soft errors and differential SER, as shown in **Fig. 3.24**. The differential SER plotted in **Fig. 3.25** is defined in Eq. (3.15).

The trend of the data in **Fig. 3.24** suggests a strong correlation between the neutron dose rate and differential SER. Further evidence of the effect of the abnormal neutron dose rate can be seen in Fig. 3.25, where the cumulative soft error rate is plotted as a function of bit multiplicity for two conditions; (1) normal neutron dose rate, and (2) during the period of abnormal dose rate.

$$\text{Differential SER} = \frac{\delta f}{\delta t} = \frac{f_1 - f_0}{t_1 - t_0} \tag{3.15}$$

where, t_1 and t_0 are time and f_1 and f_0 show the number of soft errors corresponding with t_1 and t_0, respectively. As is shown in **Fig. 3.24**, the increases in both the cumulative soft errors and differential SER were well synchronized with the abruptly increasing neutron dose rate. As is shown in **Fig. 3.25**, there is a factor of 2 differences for multiplicity between soft error events during the normal period and the abnormal period. However, this factor of two did not synchronize with the change in the anomalous neutron dose rate recorded by the rem counter. It has been suggested that the cause may be possibly due to a neutron energy spectrum with energies below 10 MeV. These low-energy neutrons have a low probability to drive the memory bit data to flip over.

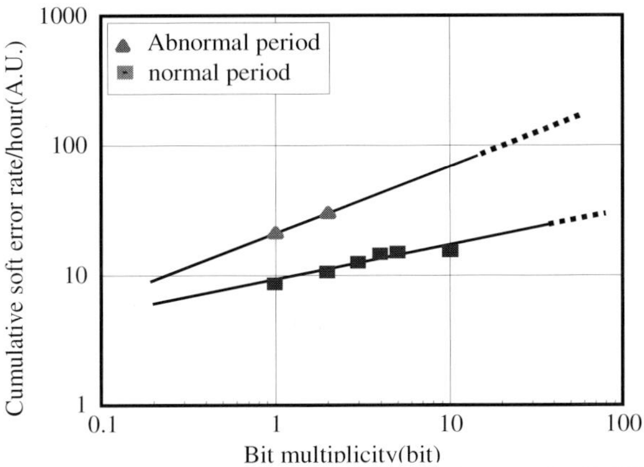

Fig. 3.25 Bit multiplicity vs SER at the USA test site [3.26].

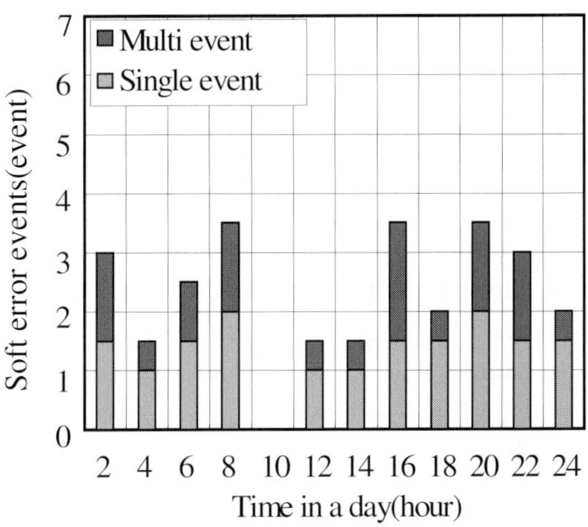

Fig. 3.26 MCU and SBU at the USA test site.

Another approach for data analysis is to rearrange the data about the frequency of occurrence every two hours. Such data for a day under normal neutron dose rate is illustrated in **Fig. 3.26**. The data indicates no time dependence between the day's neutron dose rate and the night's dose rate, even though the literature [3.30] (see **Fig. 3.27**) indicates the plasma density at night is less than day's in the Ionosphere. (In general, it is known that neutron density in the atmosphere increases with decreasing plasma density in the Ionosphere because the plasma itself shields against heavy ions from galactic cosmic rays.) [3.18]. Ionization appears at a number of atmospheric levels, producing layers or regions which may be identified by their interaction with radio waves. These layers are known as the D, E, and F layers, and their locations are shown in **Fig. 3.27** for both night and day conditions at mid-latitudes. The first ionospheric layer found was the so-called E layer, at about 110 km in altitude. It is interesting to note that this shielding effect works in other ways. For example, the auroral kilometric radiation created by trapped particles high above the ionosphere does not reach the ground because of the ionospheric E layer. Above the E layer, a F layer consisting of two parts can be found: F1, at about 170 km, and F2, at about 250 km. The F layer also reflects radio-waves. However, the lowermost region of the ionosphere below 80 km altitude, the D layer, principally absorbs radio waves.

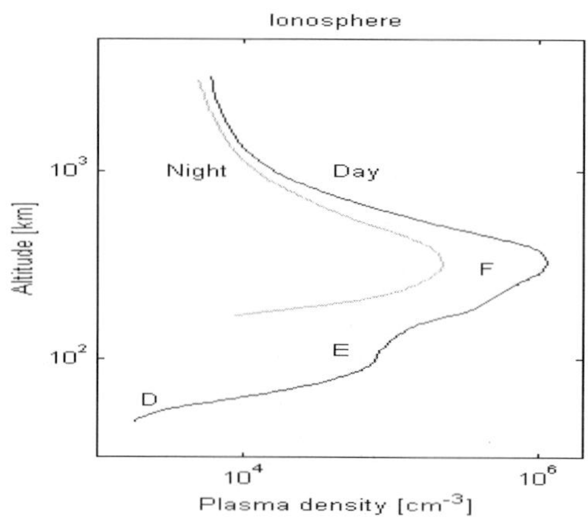

Fig. 3.27 Variation of the plasma density in the Ionosphere for day and night [3.20].

Within the auroral oval, the night-time E layer plasma densities can be much higher than indicated in **Fig. 3.27**. Densities are also variable because of the spatial and temporal structure in the trapping of ionizing particles. The E layer plasma density profiles can also be drastically altered due to the occasional formation of the so-called sporadic E layers. At F layer altitudes, one encounters features such as polar cap ionization patches and different types of troughs.

Note that there is also a clear solar cycle effect that can be seen: the average densities are higher during solar maximum years than during the minimum years.

3.6.5 *Comparison of MCU ratio between RTSER and neutron-induced SER [3.26]*

After RTSER testing had been completed, the data were compared with similar ASER (Accelerated soft error rate) neutron-induced data taken at LANSCE (Los Alamos Neutron Scattering Center) with a white neutron beam with energies up to 800MeV, and at TSL (The Svedberg Laboratory) with 180MeV and 100MeV quasi-monoenergetic neutrons. The relation between the bit multiplicity and the cumulative probability function, obtained from the data taken with the white beam test, is similar with the RTSER results, as can be seen in **Fig. 3.28**.

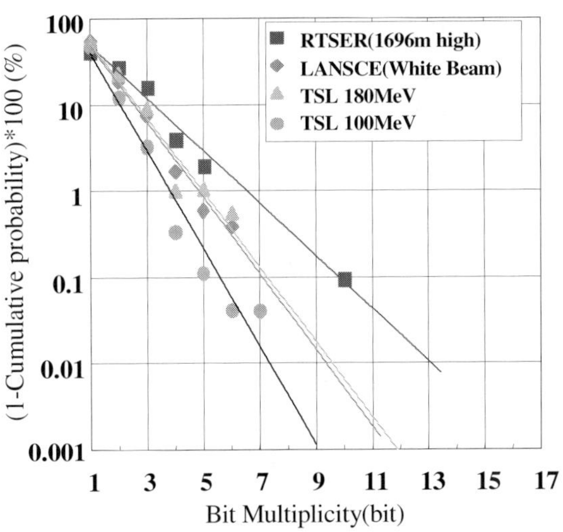

Fig. 3.28 MCU distribution functions both of RTSER (at 1696 m, USA) and ASERs obtained at LANSCE and TSL [3.26].

After examining the data for MCU on altitude dependence, the analysis indicates that there is no difference in the standard deviation vs bit multiplicity for SER testing at 0 m and 1696 m, as illustrated in **Fig. 3.29**. This fact is consistent with the fact that the neutron energy spectrum does not change much with altitude, as described in Chapter 2.

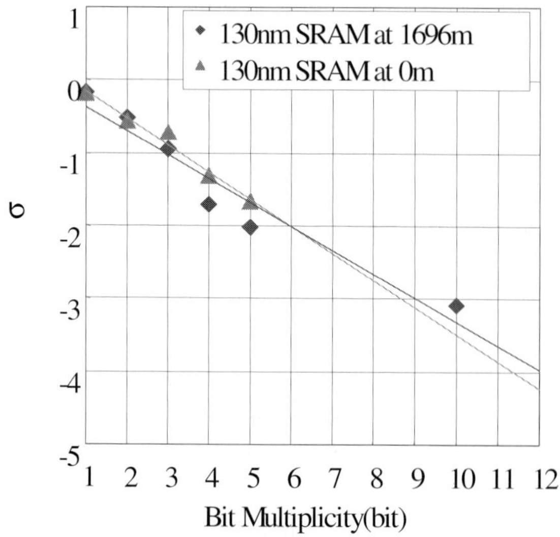

Fig. 3.29 Comparison data for MCU ratio of RTSER done by differential altitude in USA such as 0 m and 1696 m [3.26].

In order to determine a suitable neutron spectrum to use for testing, several data sets from ASERs and RTSERs using both 130nm and 180nm SRAMs were compared with each other. The results are shown in **Fig. 3.30** for the MCU/SEU ratios from RTSER measurements (over 3000 hours, using 500-1500 SRAM devices) and from ASER measurements. It demonstrates that MCU/SEU ratios obtained from the RTSER measurements cross over the data of ASER measurements at a quasi-monoenergetic peak neutron energy between 20 and 40MeV. These data show that the MCUs obtained from the ASER test using high energy neutrons (more than 50MeV) tend to lead to an overestimation of the MCU/SEU ratio compared to the RTSER data. On the other hand, it indicates that the MCU/SEU ratio at neutron energies between 20 and 40 MeV is consistent with the MCU/SEU ratio obtained from RTSER's, and then the most suitable neutron energy to use in the ASER testing is the energy which overlaps the MCU/SEU ratio obtained from RTSER.

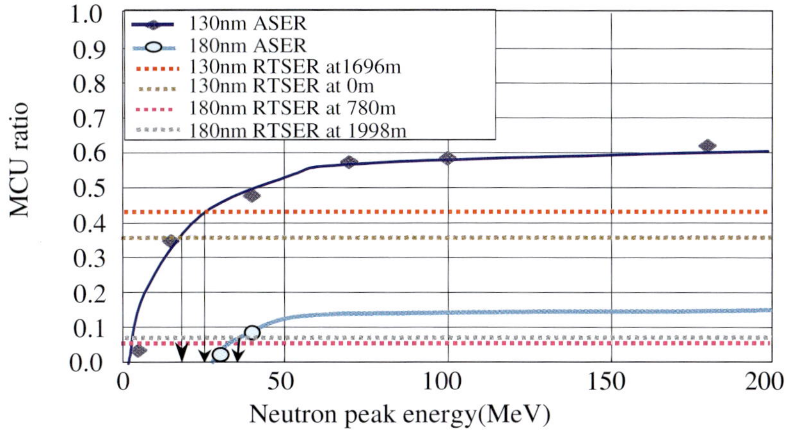

Fig. 3.30 Verification of justifiable MCU ratio between RTSER and the ASER's energy [3.26].

3.6.6 *Analysis of MCU and anomalous noise results from SER testing at the USA test sites [3.26]*

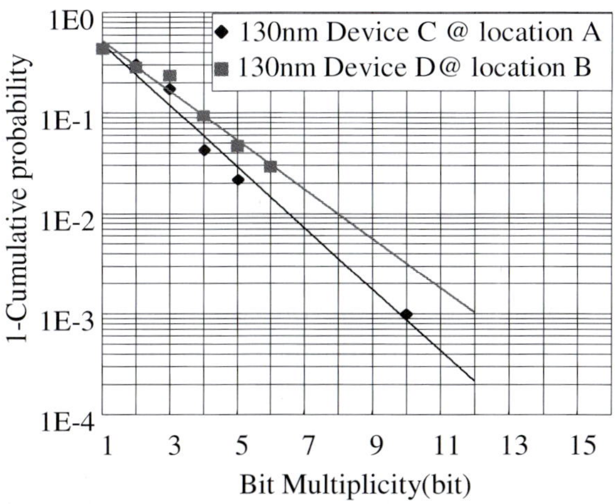

Fig. 3.31 Error-bit multiplicity during RTSER testing at locations A and B as shown in Table 3.4.

The reliability of data from RTSER testing in regards to its duplicability was determined by using more than 700 pieces of 130 nm SRAM (Device C) at location A (altitude 1698 m) and another 130 nm SRAM (Device D) at location B (altitude app. sea level) for more than 2500 hours of testing (**Fig. 3.31**). Eqs. (3.16) and (3.17) were obtained by simulations of the data in **Fig. 3.31**. It was estimated that the soft error rate had an accuracy of 95.3% for the 1-bit probability, as obtained from the comparison between Eq. (3.16) and Eq. (3.17), and that the SEU/MCU ratio for the probability of up to 5 bits had an accuracy of 73.7% (= (Plot B-Plot A)/Plot B = 1-Plot A/Plot B)) from Eq. (3.18).

$$\text{Plot A (130nm Device C at location A)} \rightarrow 0.9863 \times exp(-0.7024 \times \chi) \tag{3.16}$$

$$\text{Plot B (130nm Device D at location B)} \rightarrow 0.9023 \times exp(-0.5643 \times \chi) \tag{3.17}$$

$$\text{error } (\%) = \left[1 - \frac{0.9863 \times \sum_{\chi=1}^{5} exp(-0.5643 \times \chi)}{0.9023 \times \sum_{\chi=1}^{5} exp(-0.1563 \times \chi)} \right] \times 100 \tag{3.18}$$

where χ is number of a bit multiplicity and the error indicates a gap between Plot A and Plot B.

3.6.7 *Relation between the influence of solar wind and the change in neutron dose rate*

According to the Neutron Monitor Center in Moscow [3.31], the solar wind was in a normal state with no significant cosmic ray variation throughout the year of testing, at least in the macroscopic point of view, as shown in **Fig. 3.32** [3.31]. A closer look at the data was taken for the specific dates when the anomalous dose rates were observed, however; no large changes were observed (See **Figs. 3.33**, **3.34**) [3.32, 3.33]. Therefore, it may be that the anomalous neutron dose rate change may have been due to some local phenomena such as an unknown source of anomalous noise occurring in a very limited area around the monitoring location.

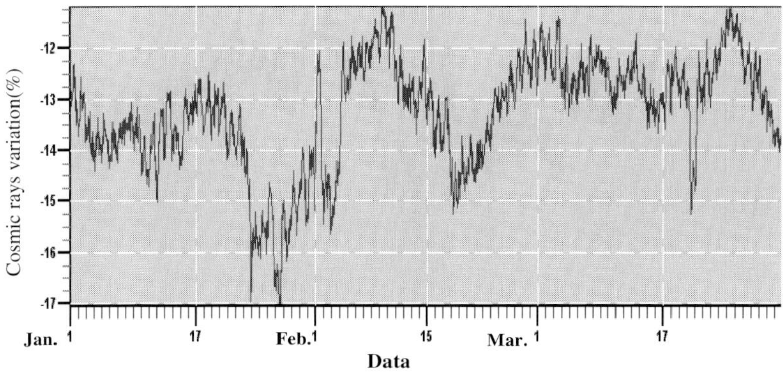

Fig. 3.32 Moscow cosmic ray station neutron monitor data [3.31].

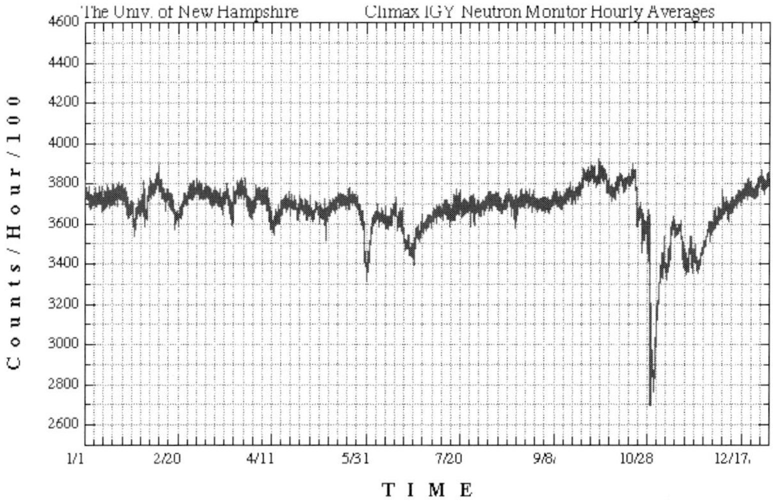

Fig. 3.33 Climax IGY Neutron monitor data through a whole year [3.32].

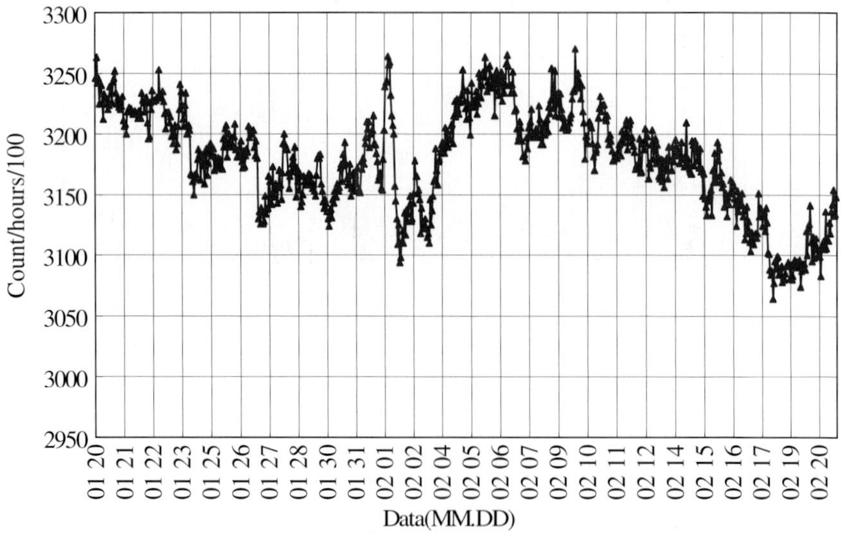

Fig. 3.34 Bartol Research Swarthmore Newark neutron monitoring data [3.33].

3.6.8 *Verification of proper operation of the rem counter after the SER test*

To determine whether or not the rem counter was properly operating during testing at location A in the USA (see Table 3.4), we checked its performance and reliability both after the system had been moved to location B at sea level and in Japan after the system had been returned from the USA. It was operated both at location B at sea level in the USA and at 86m in Japan under the same test conditions. The results of those tests are shown in **Fig. 3.35** and **Fig. 3.36**, indicating that the system was functioning properly.

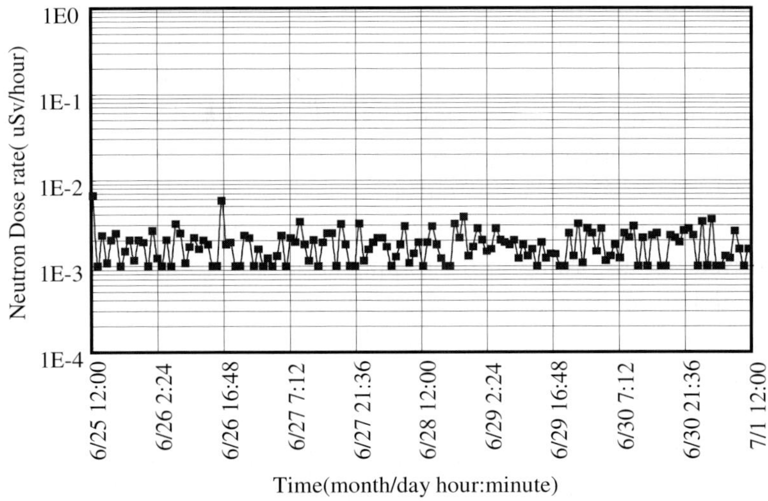

Fig. 3.35 Rem counter system operation check at location B in the USA (after the system was brought back from location A).

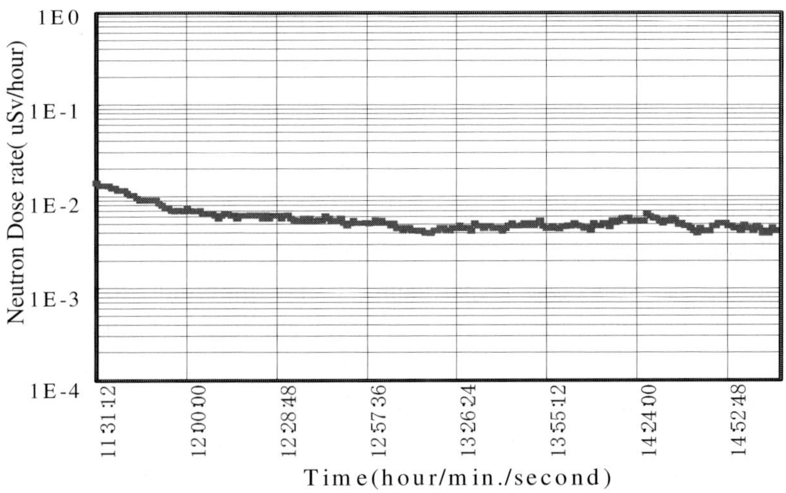

Fig. 3.36 Rem counter system operation check in Japan (after the system was brought back from the USA).

3.7 Summary

In the preceding sections, the meaning, method and results from an example of Real-Time SER testing were described. The results are summarized as follows:

(1) The Real-Time SER evaluation of semiconductor devices should be carried out according to their application. For example, a low power SRAM used in mobile applications is, in most cases, in retention mode; therefore, the Real-Time SER should be measured under static conditions.
(2) Since the acquired value of Real-Time SER strongly depends on where and when the field test was performed, an assessment of the environmental conditions during the field test must be taken. Such an assessment is absolutely required for comparison between Real-Time SERs or FIT rates that are measured in different places and during different periods.
(3) The neutron dose rate around the test location should be monitored by using a rem counter or associated system for the purpose of investigating the influences on neutron induce Real-time SER. If "noises" such as a solar flare or an accidental disturbance are observed during a specific period, the error data obtained in that period should be eliminated in order to calculate the typical Real-Time SERs of the tested devices.
(4) When an external disturbance such as an unknown noise occurs during a SER test experiment, it is important to detect and distinguish the external noise element amongst the entire real SER data set. If possible, further analysis should be performed in order to understand the nature of the external disturbance.
(5) It is found out that the MCU/SEU ratios obtained from the RTSER measurements cross over the data of ASER measurements for a quasi-monoenergetic peak neutron energy between 20 and 40 MeV. This implies that the most suitable neutron energy to use in ASER testing is in the range of 20 to 40 MeV, which overlaps the MCU ratios obtained from RTSER.

(6) It is highly recommended to obtain RTSER data of a particular silicon technology node at intervals in order to confirm whether or not the ASER results can give the representative SER value.

The "standard" Real-Time SER test platform in Europe, the "Altitude SEE Test European Platform" (ASTEP), located in the French Alps at 2552 m, has been in operation since March 2006 [3.34]. The platform will produce data that will play an important role in the SER database of devices and in the validation of simulation models. For reference, the recent SER trends for DRAM and SRAM sub-micron devices are summarized in Ref. [3.35].

Chapter 4

Neutron Irradiation Test Facilities

Mamoru Baba and Yasuo Yahagi

This chapter reviews the properties and utilization of neutron facilities available for neutron irradiation tests of semiconductor microelectronics devices. The neutron facilities listed in JESD89-Rev (http://www.seutest.co, revised version of JESD89 Standard.JESD89A) that are available for semiconductor irradiation tests are described.

4.1 Overview of Neutron Sources used in Neutron Irradiation Test Facilities

The neutron fields which are used for neutron irradiation tests are divided into three types according to the method of neutron production and the resultant neutron spectrum [4.1, 4.2];

(1) accelerator-based monoenergetic and quasi-monoenergetic neutron fields,
(2) accelerator-based spallation neutron fields having continuous neutron spectra, and
(3) reactor-based neutron sources with a continuous spectrum of neutrons.

The facilities listed in JESD89-Rev that are available for SEU studies are summarized in **Table 4.1**.

Reactor-based facilities are used mainly for irradiation with low energy neutrons, including thermal neutrons obtained from nuclear reactors. They were effective for the testing of devices that included boron compounds. Presently, however, boron containing devices are relatively few and irradiation is primarily done now using accelerator-based high energy neutron fields, via (1) and (2).

In this section, therefore, only accelerator-based neutron sources are described.

Table 4.1 Neutron source facilities for semiconductor-irradiation test listed in Ref. [4.2].

Type	Facility	Neutron energy (MeV)	Source reaction
Monoenergetic	Boeing (USA)	14	d-T
	US Naval Academy (USA)	2.5 & 14	d-T, d-D
	Indiana University (USA)	Thermal-5	^{9}Be(p,n)
	ASP	3-14	d-T
	National Physics Laboratory (UK)	0.08-5, 15.5-18.0	Sc,^{7}Li,T(p,n), D,T(d,n)
	FNL, Tohoku Univ. (Japan)	0.08-7.5, 13.5-18.0	Sc,^{7}Li,T(p,n), D,T(d,n)
Quasi-monoenergetic	Crocker Nucl; lab, UC Davis (USA)	20-65	^{7}Li(p,n), ^{9}Be(p,n)
	CYCLONE, UC Louvain (Belgium)	20-70	^{7}Li(p,n)
	TSL, Uppsala Univ. (Sweden)	22-173	^{7}Li(p,n)
	TIARA, JAEA (Japan)	20-75	^{7}Li(p,n)
	RCNP, Osaka Univ. (Japan)	<395	^{7}Li(p,n)
	CYRIC, Tohoku Univ. (Japan)	20-75	^{7}Li(p,n)
Spallation	LANSCE (USA)	Thermal-800	W(p,n)
	TRIUMF (Canada)	Thermal-400	W(p,n)
	RCNP Osaka Univ. (Japan)	Thermal-392	Pb(p,n)
Reactor	National Institute for Standards (USA)	Fission neutrons	Fission
	Reactor Institute Delft (Netherland)	Fission neutrons	Fission

For SEU studies, neutrons with a monoenergetic or quasi-monoenergetic energy spectrum are used, along with spallation neutron sources that provide a continuous spectrum of neutrons. Monoenergetic neutron fields are used for the study of energy dependence of SEU cross sections, while spallation neutron fields are useful for experiments to measure a single event rate (SER) in a simulated natural terrestrial environment.

The recent development of high-density, nano-scale technology has led to a decrease of the critical charge of devices (charge which is required to change the state of devices). As a consequence, the SEU threshold is getting lower and the sensitivity to SEU is becoming much greater [4.1]. For this reason, the test of modern technology devices requires a monoenergetic neutron beam down to a few MeV, as well as medium energy neutrons in the region of ten's to hundred's of MeV. In JESD89 Standard from JEDEC [4.2], several monoenergetic neutron facilities are listed.

Up to around 20 MeV, purely monoenergetic neutrons are obtained by combining various light ion induced reactions, except for 8–13.5 MeV, where background neutrons due to breakup reactions accompany the primary neutrons. Above ~20 MeV, purely monoenergetic neutrons are not available because of the occurrence of multi-body breakup reactions [4.3], and only quasi-mononergetic neutron sources are available. In this energy region, the ^{7}Li(p,n) reaction provides the most prominent monoenergetic component with highest intensity [4.3]. It has been used as a unique "monoenergetic" neutron source for energies up to around 500 MeV. The ^{9}Be(p,n) reaction is also used as a quasi-monoenergetic neutron source in the region of ten's to hundred's of MeV, although the peak intensity is a bit inferior to the ^{7}Li(p,n) source.

The structure and nature of the sources above 20 MeV are entirely different from those for lower energies. As such a description is given separately for sources below 20 MeV and above 20 MeV.

4.2 Monoenergetic Neutron Source below 20 MeV

In the region of energies between a few MeV and 20 MeV, monoenergetic neutrons are obtained by combining ^{7}Li(p,n), ^{3}H(p,n), ^{2}H(d,n), and ^{3}H(d,n) reactions [4.3–4.5]. **Figure 4.1** illustrates the relation between neutron energy and the incident beam energies for the above mentioned four neutron production reactions. Between ~8 MeV and ~13.5 MeV, the ^{2}H(d,n) reaction is the most intense neutron source but it is not monoenergetic due to breakup neutrons via the ^{2}H(d,np) reaction. Other sources such as the ^{9}Be(α,n), ^{14}N(d,n), ^{15}N(d,n) reactions are also used to obtain quasi-monoenergetic neutrons but their intensities are significantly lower than the ^{7}Li, ^{9}Be(p,n) reactions for a given energy spread [4.3].

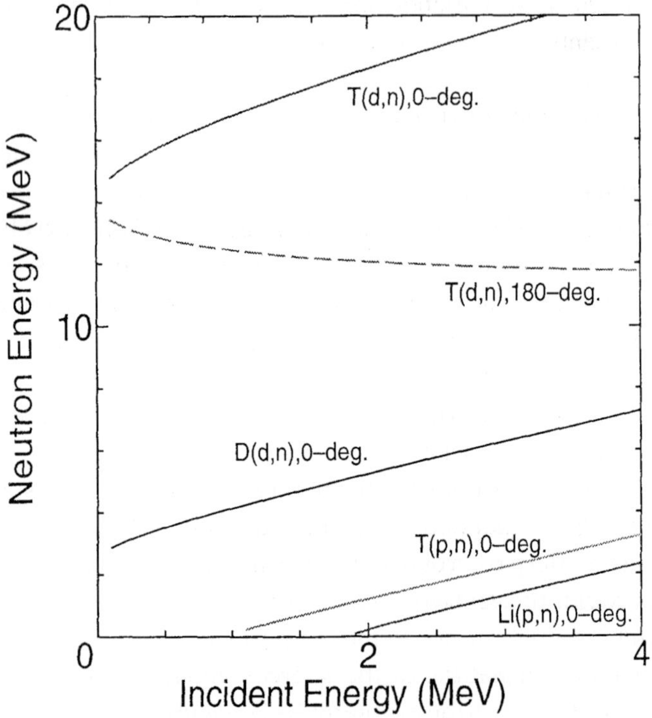

Fig. 4.1 Neutron energy vs incident energy for production of neutrons below 20 MeV [4.4, 4.5].

These neutron sources have been installed in various laboratories around the world and play an important role in the course of development of nuclear energy from fission and fusion reactors [4.3]. Some of them are also used effectively in SEU test irradiations. The general properties of neutron sources are summarized in Ref. [4.3], and the neutron yields and energies are tabulated as a function of incident energy and outgoing angle in Refs. [4.4, 4.5].

Here, a description is given for 14 MeV neutron sources which have been installed extensively around the world and have been used for irradiation tests of semiconductors. A description is also given variable energy monoenergetic sources at the Fast Neutron Laboratory (FNL), Tohoku University, and the National Physics Laboratory (NPL), as typical examples of a monoenergetic neutron facility that is used for SEU experiments.

4.2.1 *14 MeV neutron source*

In the JESD89 Standard report [4.2], the following facilities were cited: Boeing Radiation Effects Lab., Texas A&M, Eastern Kodak Research Center, Dow Chemical, US Naval Research Academy in the USA, Shell Research & Tech, Delft Univ., and French Central Agency of Atomic Energy, Valduc. The last two facilities are capable of producing 2, 5 and 6 MeV neutrons.

Neutrons around 14 MeV can be produced via the T(d,n) reaction using an inexpensive Cockcroft Walton accelerator with a range of a few hundred's of kV and conventional tritium-titanium metal targets. The neutron intensity generated by these reactions is fairly high owing to rather large reaction cross-section. For that reason, such neutron generators were constructed extensively and used effectively for SEU experiments.

As a typical example, the d-T neutron source at the Agency of Atomic Energy, Valduc can provide a neutron flux around 10^{10} cm^{-2} s^{-1} for use in SEU experiments [4.6].

4.2.2 *Variable energy sources; Fast Neutron Laboratory (FNL), Tohoku University [4.7]*

(1) Neutron production

At the FNL, neutrons between 8 keV and 18 MeV are utilized for various applications, including SEU experiments. Neutrons are produced using proton and deuteron beams accelerated by a 4.5 MV electrostatic accelerator (Dynamitron) that is equipped with a nano-second pulsing system.

The layout of the FNL is illustrated in **Fig. 4.2**. The neutron field is set up at the 30-deg port located in the central part of an experimental room that is about 20 m by 25 m wide. The location is advantageous for present applications due to the low room-scattered background and the

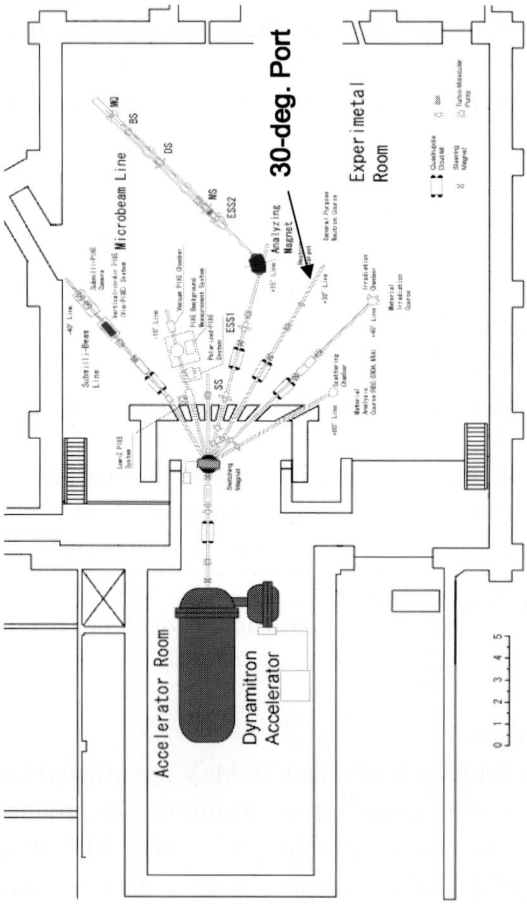

Fig. 4.2 Layout of the Fast Neutron Laboratory (FNL) [4.7].

wide space for TOF measurements of neutron spectra. The neutron production target is ~1.5 m above the ground. The room-scattered background is not so serious, as described later. Devices Under Test (DUTs) are placed a few to ten's of cm from the target along the 0-deg line with respect to the incident beam. The distance was chosen according to the sensitivity and volume (size) of the DUTs. The utilization of 0-deg. neutrons minimizes the effect of neutron scattering in the target [4.7, 4.8].

The neutron sources used at the FNL are summarized in **Table 4.2** together with the associated neutron intensities.

Table 4.2 Summary of neutron source at FNL, Tohoku University [4.7].

Neutron energy (MeV)	Source reaction	Target/ backing	Energy spread (keV)	Neutron fluence*	Beam current (μA)
0.55 MeV	$^{7}Li(p,n)^{7}Be$	LiF/Pt	~50	3.2×10^{4}	10
1.0 MeV	$^{3}H(p,n)^{3}He$	T-Ti/Cu	~100	1.2×10^{4}	10
2.0 MeV	$^{3}H(p,n)^{3}He$	T-Ti/Cu	~ 80	2.2×10^{4}	10
5.0 MeV	$^{2}H(d,n)^{3}He$	D_2 gas/Pt	~250	2.2×10^{5}	4
15 MeV	$^{3}H(d,n)^{4}He$	T-Ti/Cu	~500	8.0×10^{4}	10

*n cm^{-2} μC^{-1} at 10 cm from the target center.

As a lithium target, LiF is preferred to lithium metal because of its easier handling and stability under proton bombardment, although its neutron yield per unit energy interval is inferior. The tritium-loaded titanium (T-Ti) targets for the T(p, n) and T(d, n) sources are commercially available, with nominal thickness of –0.75 or –2.5 mg cm^{-2}. The deuterium target is a gas cell with a window foil of Mo (–5 μm thick) or Havar (–2.2 μm thick), and a Pt beam stop (0.3 mm thick). The gas pressure is 1–1.2 atm. A gas target provides neutrons with a much cleaner spectrum and a higher neutron yield compared with a metal target, while the beam current is limited to around ~4 μA. The target thicknesses are chosen as a compromise between neutron yields and the energy spread. The energy spread deduced from TOF measurements is generally less than 10%, as presented in **Table 1**. The relatively large energy spread of 15 MeV neutrons is because they are used at 0-deg with respect to the incident beam (D_3^+) of 1.5 MeV (0.5 MeV per D_1^+). The energy spread can be reduced to less than ~100 keV by using neutrons emitted around 90-deg, if necessary.

The neutron production targets were mounted in thin-walled (<0.5 mm thick) Cu chambers in order to reduce the scattering of primary neutrons by the chamber. The targets were also cooled by an air jet to avoid neutron moderation. The backing plates and beam stop were soldered to the chamber, thereby eliminating the need of a massive flange in the vicinity of the target. For the same reason, the T-Ti target was encased in a dual-walled chamber. In addition, the beam line is pumped by turbo-molecular pumps and sputter-ion pumps to avoid hydrocarbon contaminants on the target. The ion beam from the Dynamitron was shaped to a ~7-mm-diameter using an aperture made of Ta which has a high reaction threshold. Except for the D_2 gas cell, the targets can be bombarded by a beam current up to 10 μA without a noticeable reduction of the neutron yields. With these source strengths, an irradiation test of semiconductor can be done within a few to several hours, although a higher neutron intensity is always desirable.

(2) Neutron intensity and spectrum

The neutron energy and the spectrum was measured with a Time-of-Flight (TOF) method using a pulsed beam. The TOF measurement was done by employing an NE213 scintillation detector, 5.08 cm-diameter and 5.08 cm-thick, coupled to a fast photomultiplier. The detector bias was set to lower than 0.3 MeV proton energy in order to measure the amount of low energy contaminant components. The flight paths used were 2 to 12 m, depending on the neutron energy [4.7].

Typical examples of energy spectra are shown in **Fig. 4.3**. (The peak widths in the figure are larger than those in **Table I** because of effect due to the detector's intrinsic timing resolution.) The spectrum is clean with little contribution from background neutrons. Contaminants are most serious in the 15 MeV source because of the production of low-energy parasitic neutrons via the D(d, n) and C, O(d, n) reactions. The amount

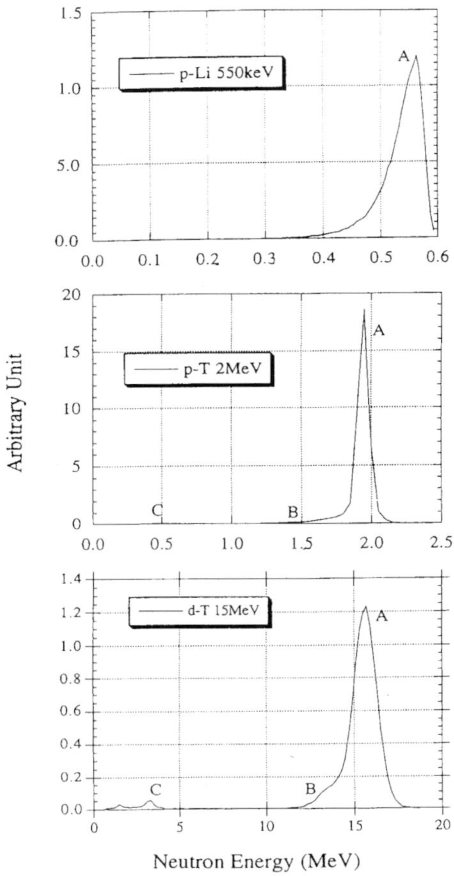

Fig. 4.3 Neutron energy spectrum at FNL, Tohoku University [4.7]: A is the primary neutron spectrum, B is neutrons scattered in the target, and C is the neutrons produced by parasitic reactions.

of contaminants is usually ~5%, at most, relative to the amount of primary neutrons. For lower energy sources, especially (p,n) sources, the contaminant component is much less. There is additional background due to neutron scattering through the target. The intensity and spectrum of these neutrons were estimated using a Monte Carlo method and proved to be less than 2% (highest at 15 MeV) [4.7]. Therefore, in SEU experiments, the effect of these background neutrons can be ignored because of their relatively small amount and their low energies. In fact, their energies will be lower than the SEU threshold for most devices. If necessary, the measured contaminant components can be taken into account.

The reproducibility of the neutron energy and spectrum proved to be very good, and as a result the neutron energy could be set without a TOF measurement.

(3) Neutron fluence determination [4.7]

The neutron fluence is determined with a ^{235}U fission chamber (FC) and a proton-recoil-counter or -telescope, generally good to within 5%. **Figure 4.4(a)** and **4.4(b)** illustrate a schematic view and a typical pulse-height distribution of the FC [4.7], respectively. The FC is a parallel plate ionization chamber of the back-to-back type, and is operated in a gas flow mode. Its detection efficiency is close to 100% and is known very accurately. Furthermore, the FC has a great advantage in that it can be applied over a wide range of neutron energies without changing the detector conditions and electronics setting. This is because the pulse-height distribution is almost independent of the neutron energy.

The chamber has a little sensitivity to room-scattered neutrons, but generally there is a contribution of only a few % relative to the primary beam when it is used 10 cm from the target. This value was found to be a systematic function of neutron energy, and it can be estimated from the systematics without doing an additional shadow bar measurement. The back-to-back structure allows for the use of two fission foils with different foil thickness to confirm the consistency of the measurement.

Therefore, the neutron fluence for each measurement is simply determined by the following equation:

$$\phi = \frac{C_{FC}}{\sum_i N_i \sigma_i} \frac{f_{corr}}{C_{mon}}, \tag{1}$$

where C_{FC} is the fission fragment counts, N_i and σ_i are the number of atoms and fission cross sections of nuclide i, respectively, f_{corr} is the correction factor, and C_{mon} is the monitor counts. The factor N_i is determined by α-counting of the fission foil with a well-defined

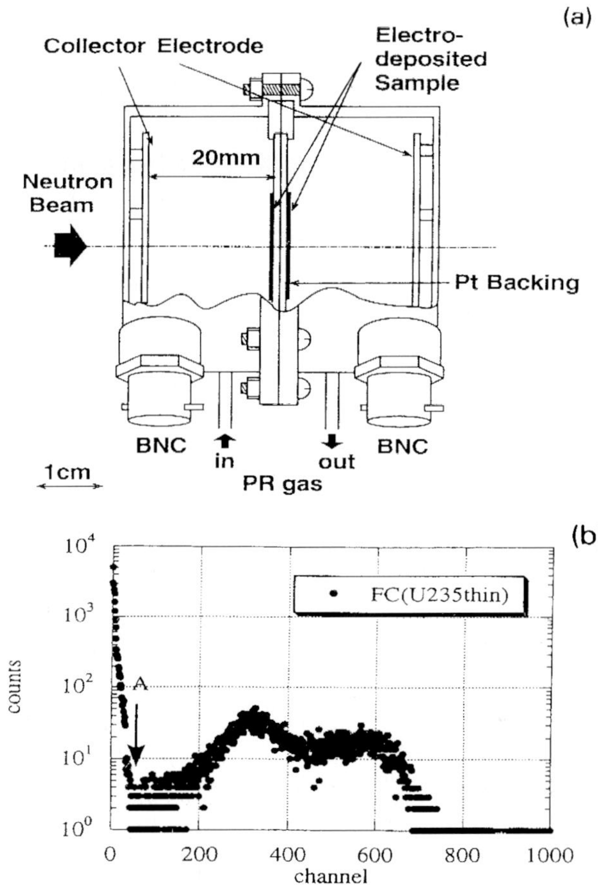

Fig. 4.4 (a) Fission chamber for neutron flux measurement at FNL, and (b) pulse height spectrum of the fission chamber [4.7].

geometry (low-geometry counter) [4.7, 4.8]. The factor f_{corr} is 0.97–1.00, depending on the geometry and the energy.

(4) SEU Irradiation [4.9]

The present neutron field at the FNL has been used for SEU irradiation experiments. A typical experimental geometry is shown in **Fig. 4.5**. The DUT is placed at several ten's of cm from the target to ensure uniform irradiation over the device. As such, the neutron flux on the

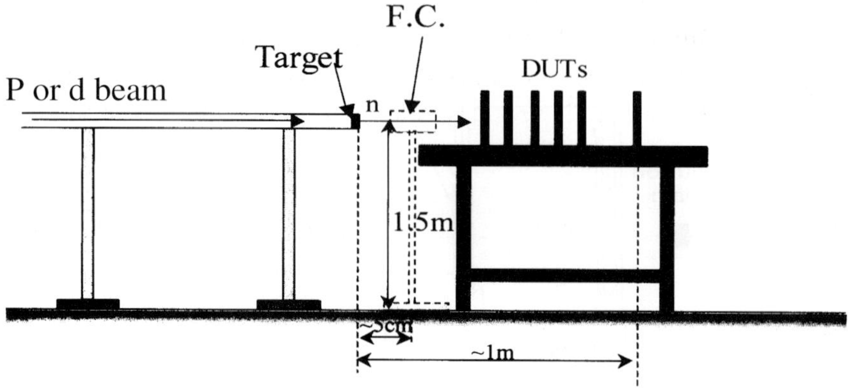

Fig. 4.5 Experimental setup for irradiation experiment at FNL; FC is a fission chamber for flux measurement and DUTs are "devices under test" [4.9].

device surface is much lower than the flux reported in **Table 1**. Nevertheless, the test can be done within a few to several hrs of continuous irradiation, with SEU rate observed online.

The neutron fluence on the device is monitored with a relative fluence monitor that was calibrated relative to the fission chamber (FC) prior to the irradiation for each energy. The relative fluence monitor used here is either a hydrogen proportional counter or an organic scintillator placed around 45-deg and a few meter from the target.

The combination of this field with that at the Tohoku University CYRIC cyclotron can provide a unique neutron irradiation filed from less than 1 MeV up to 80 MeV.

4.2.3 *Variable energy source at the National Physics Laboratory (NPL) [4.10]*

Well-characterized monoenergetic neutron sources similar to FNL are routinely used in the National Physics Laboratory (NPL) in the UK, using neutron production reactions of $^{7}Li(p,n)^{7}Be$, $^{3}H(p,n)^{3}He$ and $^{2}H(d,n)\,^{3}He$. The ion beams for neutron production are obtained with a 3 MV Van de Graff accelerator. The experimental room adopts a low scatter gridded-ceiling structure to reduce room scattered neutrons.

The neutron fluence is determined using a calibrated long counter. The neutron flux (cm^{-2} s^{-1}) and corresponding ambient dose equivalent (mSv h^{-1}) are summarized in **Table 4.3**. The neutron energies listed are the "standard energies" recommended by the ISO (International Organization for Standardization).

Table 4.3 Property of the neutron sources at the National Physics Laboratory (NPL) [4.10].

Neutron Energy (MeV)	Reaction	Approximate maximum rates at 1m	
		Neutron Flux (cm^{-2} s^{-1})	Ambient dose equivalent (mSv h^{-1})
0.144	^{7}Li(p,n)^{7}Be	1.0×10^3	450
0.250	^{7}Li(p,n)^{7}Be	6.0×10^2	440
0.565	^{7}Li(p,n)^{7}Be	1.6×10^3	2000
1.2	^{3}H(p,n)^{3}He	2.0×10^2	300
2.5	^{3}H(p,n)^{3}He	6.0×10^2	900
5.0	^{2}H(d,n)^{3}He	6.0×10^2	880

4.3 Quasi-monoenergetic Sources above 20 MeV

4.3.1 *^{7}Li(p,n) and ^{9}Be(p,n) neutron sources*

In the energy region above 20 MeV, only a quasi-monoenergetic source is available because of the occurrence of multi-body breakup processes due to the higher incident energy. In this energy region, the ^{7}Li(p,n) reaction is the most intense quasi-monoenergetic neutron source, and has been used at various laboratories. Among those laboratories are the Crocker Nuclear Laboratory, University of California Davis (UC Dais), (En ≤ 65 MeV) [4.11], Institute de Physique Nucleaire, Louvain-la-Neuve (≤ 75 MeV) [4.12], TRIUMF (Tri-University Meson Factory), Vancouver (200–500 MeV) [4.13], The Svedverg Laboratory (TSL), Uppsala University (50–150 MeV) [4.14, 4.15], and Cyclotron Laboratory, Indiana University, Bloomington (30–200 MeV) [4.16]. All of these facilities are cited in the JESD89 Standard [4.1] as quasi-monoenergetic neutron facilities useful for SEU experiments.

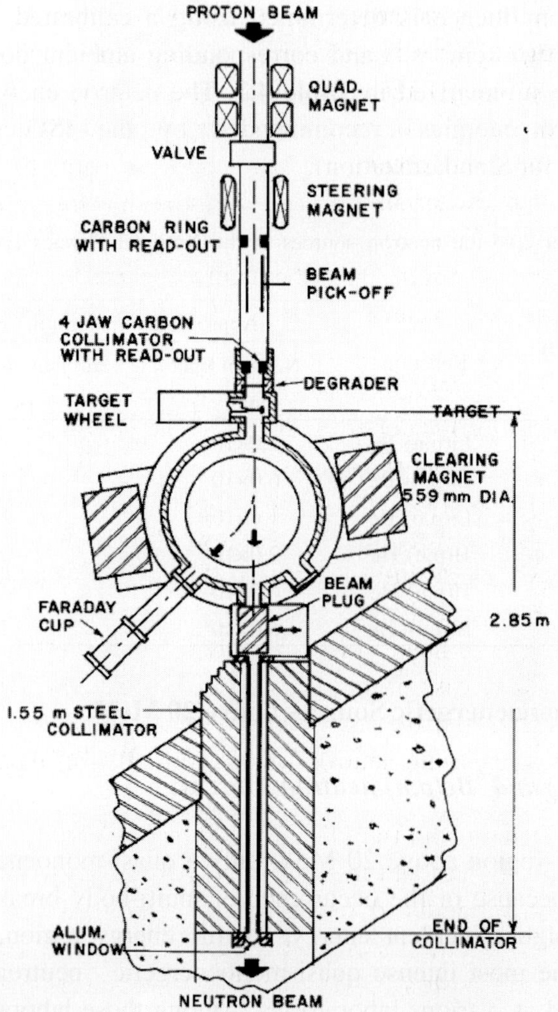

Fig. 4.6 Lay out of the ^{7}Li(p,n) source of UC Davis [4.11].

As a typical example of the ^{7}Li(p,n) source, **Fig. 4.6** illustrates the layout of the ^{7}Li(p,n) source at UC Davis [4.11]. This laboratory conducted pioneering work using the source for nuclear physics and nuclear data research. Neutrons are produced using a metal lithium target with an appropriate thickness to obtain a monoenergetic component with energy widths suitable for the experiment. Under that condition, most of the incident protons pass through the lithium target without an reaction in the lithium target. These protons are deflected out of the neutron beam with a bending magnet into a shielded beam dump to avoid the large amount of background produced when they bombard the end of the beam tube. Furthermore, to obtain a clean neutron beam, there should be adequate shielding materials between the experimental area and both the lithium target and the beam dump. For that reason, generally, the experimental area is distant from the neutron source by anywhere from a few meters to several meters, and as such it leads to limited neutron flux in the experimental area.

For example, as shown in **Table 4.4** (which summarizes parameters of lithium sources at various laboratories), the maximum neutron flux of the UC Davis source is about 6×10^5 n cm^{-2} s^{-1} for a 1 MeV-thick target, even with a rather high beam current of 10 μA. This is because of long distance (~3.5 m) between the ^{7}Li target and the experiment point. The sources at UCL Louvain-le-Neuve [4.12], TSL, Uppsala University [4.14] and TIARA [4.17] were constructed in a similar manner in order to facilitate studies of neutron-induced reactions that require low neutron backgrounds. In the case of the original TSL's source [4.14], the distance between the target and the experimental area was as long as 11 meter [4.14], and as a result the neutron fluxes available for experiments were limited.

After publication of the JESD89 Standard from JEDEC [4.1], upgrades of old sources and installation of new sources were undertaken to promote neutron-induced experiment, including SEU tests, in TSL (25–180 MeV) [4.15], TIARA, JAEA (30–85 MeV) [4.17], RCNP Osaka Univ., (≦392 MeV) [4.18], and Tohoku University (20–85 MeV) [4.19].

For the new source at TSL, the distance between the neutron production target and experimental point was shortened to approximately 3 m in order to increase the neutron flux by more than one order of magnitude. The new source at Tohoku University [4.20] was constructed to achieve high neutron flux as the first priority. The shortest distance there between the target and the irradiation point can be as close as 1 m, which is achieved by adopting a modular shielding structure Owing to that design, the neutron flux is as high as 1×10^6 n cm^{-2} s^{-1} for a 1 MeV-thick ^{7}Li target and 3 μA beam current. Below 50 MeV, a higher beam current is available by use of negative ion beam. More details of these sources are described in Sec. 4.3.3.

As a quasi-monoenergetic neutron source, the ^{9}Be(p,n) reaction is also useful because of a prominent peak component. However, the peak neutron yield is smaller than the ^{7}Li(p,n) reaction for a given beam current and energy spread [4.3].

Table 4.4 Comparison of ^{7}Li(p,n) neutron sources.

	Energy	ΔE	Distance	Flux (cm^{-2}·s^{-1})	Current	Reference
UC Davis	40~60 MeV	1 MeV	3 m	6×10^5	10 μA	4.11
UCL	20-65 MeV	2 MeV	3.3 m	10^6	10 μA	4.12
TRIUMF	70-200 MeV	0.7MeV	~ 1 m	10^5	0.3 μA	4.13
TSL*	25-180 MeV	1 MeV	3 m	3×10^5	10 μA	4.14, 4.15
TIARA	30-85 MeV	2 MeV	5.2 m	1.2×10^5	3 μA	4.17
RCNP	392 MeV	1 MeV	~ 1 m	3×10^5	1.0 μA	4.18
CYRIC*	50-85 20~50 MeV	1 MeV	1.0 m	1.0×10^6 $\sim10^7$	3μA(H$^+$) 10 μA(H$^-$)	4.19,4.20

*For the case of a new source.

4.3.2 *Neutron spectrum and intensity of the ^{7}Li(p,n) source*

Figure 4.7 shows neutron spectra measured by Nakao *et al.* of the ^{7}Li(p,n) reaction for various proton energies, using the TOF method at RIKEN [4.21]. As shown in **Fig. 4.7**, the neutrons from the ^{7}Li(p,n) reaction consist of a monoenergetic peak due to the ^{7}Li(p,n$_{0,1}$) process feeding into the ground and the first excited states of ^{7}Be, and a continuum attributable to breakup processes like the ^{7}Li(p,n^{3}He)α reaction [4.22]. The fraction of continuum neutrons is around 50 % or more for incident energies below 100 MeV [4.21, 4.22]. The goal of

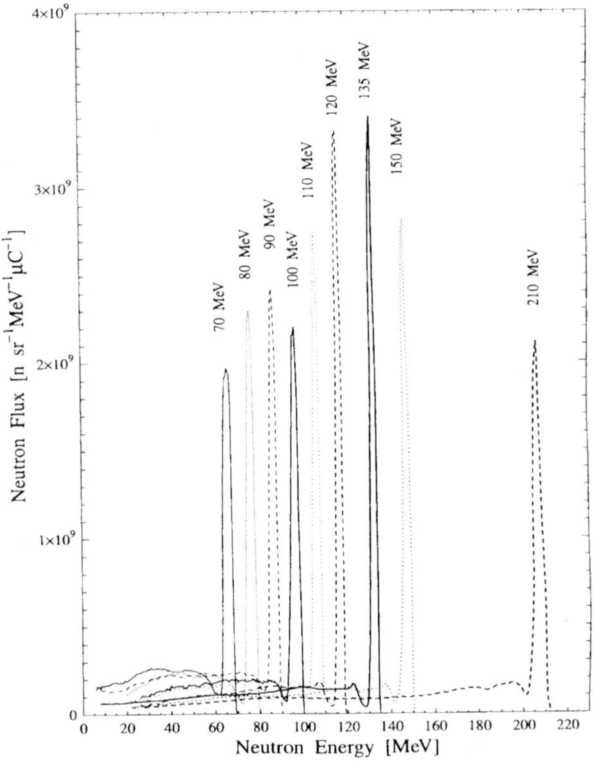

Fig. 4.7 Neutron spectrum from the ^{7}Li(p,n) source for various proton energies [4.21].

most experiments is to measure the SEU rate for the peak neutrons. Therefore, SEUs induced by continuum neutrons are considered as backgrounds and need to be eliminated. The flux of continuum neutrons per energy is not so high when compared with peak neutrons, but the total background should be evaluated accurately because the energy integrated fluxs are comparable. To make a correction for background contributions, the shape of the continuum neutrons and the SEU cross section or SEU rate should be known accurately. The SEU cross sections are deduced by unfolding the SEU rate at various peak energies. Details are described in next section.

The spectrum shape of continuum neutrons were studied at TIARA, JAEA and RIKEN with TOF and proton-recoil telescope methods [4.22].

(a)

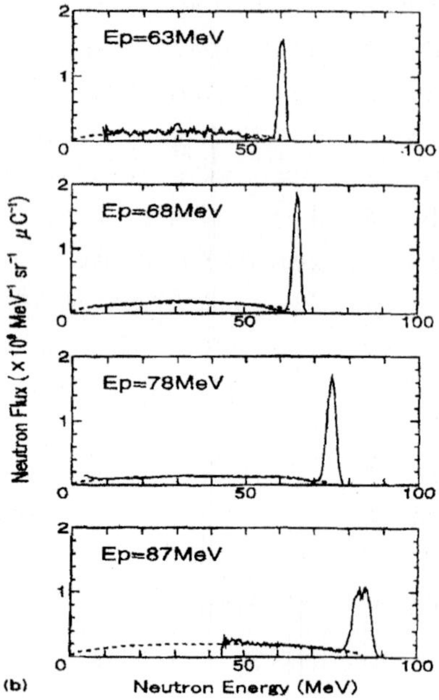

(b)

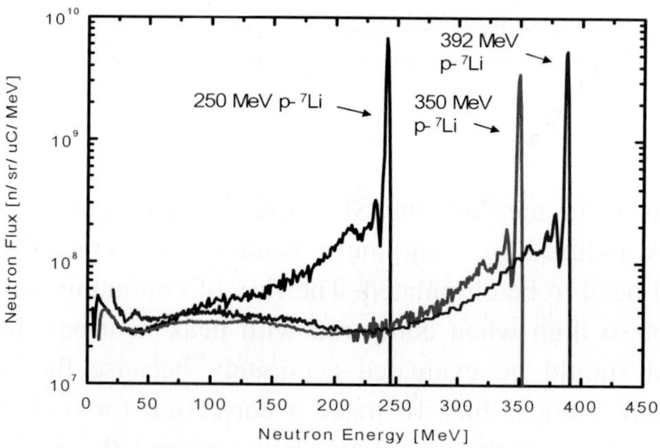

Fig. 4.8 (a) Spectra of continuum neutrons of ^{7}Li(p,n) reaction below 80 MeV, in comparison with the three-body phase-space distribution [4.22]. (b) Spectra of continuum neutrons of the ^{7}Li(p,n) reaction in 250-392 MeV region [4.24].

The shape was found to be represented by a phase-space distribution [4.23] corresponding to the above-mentioned ^{7}Li(p,n^{3}He)α process. The comparison between the experiment and calculation is shown in **Fig. 4.8(a)**. The spectra for proton energies in the ten's of MeV region are described fairly well by the simple model. However, as can be seen in **Fig. 4.7** and **Fig. 4.8(b)** [4.24], the spectrum becomes harder and deviates from the phase-space distribution above hundred's of MeV, probably due to the contribution of quasi-free scattering. Therefore, in this energy region, it is necessary to consider additional components. Prokofiev *et al.* provided a semi-empirical formula to describe the continuum spectrum at higher proton energies [4.24].

The differential cross-section of the peak neutron production processes, ^{7}Li(p,n$_{0,1}$), also has been studied extensively [4.22, 4.26]. In **Fig. 4.9**, the angular distribution of the ^{7}Li(p,n$_{0,1}$) reaction in the center-of-mass system is shown as a function of momentum transfer q, which is defined as follows [4.26]:

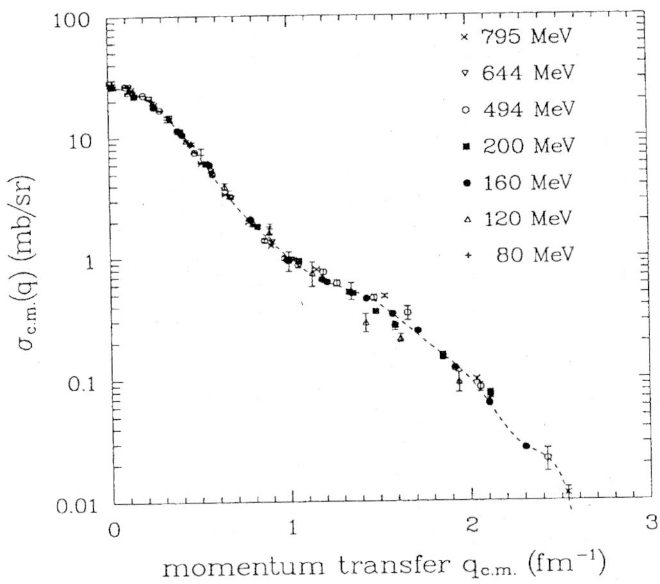

Fig. 4.9 Differential cross section of the Li(p,n$_{0,1}$) reaction, as a function of momentum transfer. (See the text for detail) [4.26].

$$q^2 = p_i^2 + p_f^2 - 2p_i p_f \cos\theta,$$

where p_i and p_f are the momentum vectors of incident and outgoing neutrons, respectively, and θ is the scattering angle. By using the momentum transfer as the angle-dependent parameter, the differential cross-sections can be reproduced from 80 MeV to 795 MeV, as shown in **Fig. 4.9**. In addition, the cross section at 0-deg was found to be nearly constant (around 38 mb/sr) in the center-of-mass (CM) system above 30 MeV, as shown by Taddeuchi *et al.* **(Fig. 4.9)** [4.26, 4.27]. Therefore, the yield of peak neutrons can be estimated by calculation, if the target thickness is known.

4.3.3 *Utilization of ^{7}Li(p,n) sources for SEU experiments*

In this section, five examples are described for SEU irradiation experiments using the ^{7}Li(p,n) neutron source. In these experiments, a high neutron intensity is essential, and background rates should be kept low. To correct for background SEUs induced by continuum neutrons, a high degree of statistical accuracy is important. Therefore, a higher neutron flux is essential for efficient experiments. Source parameters for various facilities are shown in **Table 4.4**.

(1) TSL (The Svedberg Laboratory) of Uppsala University, Sweden [4.15, 4.29]

Figure 4.10 and **Fig. 4.11** show the experimental hall and a typical experimental setup with the new neutron source at the TSL laboratory of Uppsala University [4.15]. Protons accelerated by the Gustav-Werner cyclotron irradiate an enriched ^{7}Li (99.98%) target set in the beam transport system in Blue Hall. In this accelerator, protons above 100 MeV are accelerated in a frequency-modulation (FM) mode. As a result, the beam current and neutron flux are lower by about one order of magnitude than those below 100 MeV. Typically, the proton beam intensity is 5 μA and 100 nA for 95 MeV and 174 MeV, respectively, and the target thicknesses are around 8 mm and 15 mm.

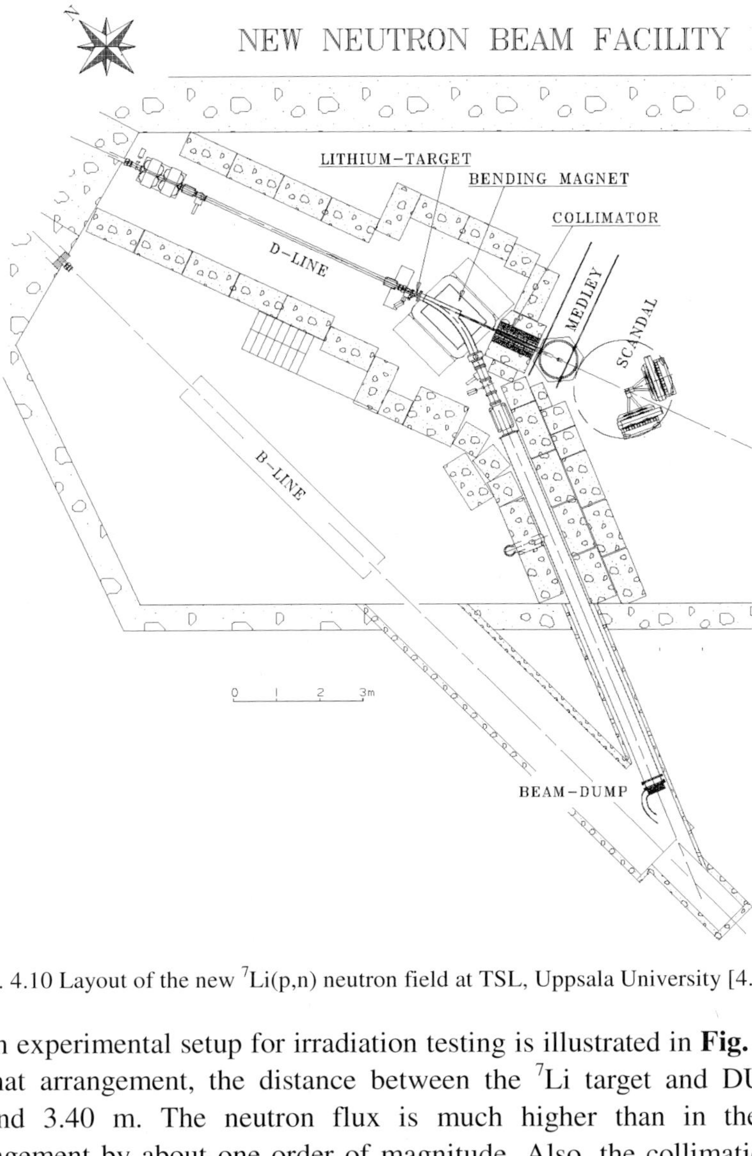

Fig. 4.10 Layout of the new $^{7}Li(p,n)$ neutron field at TSL, Uppsala University [4.15].

An experimental setup for irradiation testing is illustrated in **Fig. 4.11**. In that arrangement, the distance between the ^{7}Li target and DUT is around 3.40 m. The neutron flux is much higher than in the old arrangement by about one order of magnitude. Also, the collimation of the neutron beam (with an iron collimator) is variable in diameter up to 30 cm in steps of 5 cm. Owing to the shorter distance from the target to sample chips, the neutron flux on the chips increased to 10^{6} n cm^{-2} s^{-1} for neutron energies below 100 MeV.

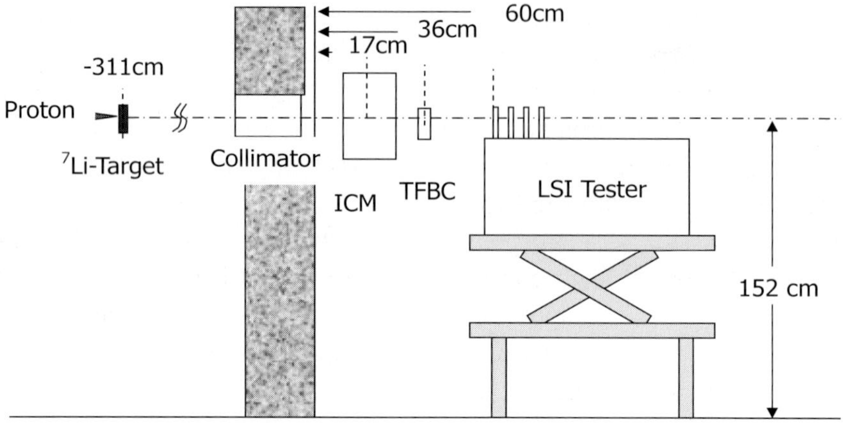

Fig. 4.11 Experimental setup for neutron irradiation experiments at TSL [4.9, 4.15].

The neutron beam is collimated by a 2 m thick, concrete wall that separates proton beam line and Blue Hall. Protons that penetrated the target are swept out from the beam to a beam dump by a sweeping magnet in order to avoid backgrounds induced by protons. The neutron fluence is monitored using a TFBC (Thin Film Breakdown Counter) [4.28] placed ~3.5 m from the ^{7}Li target. The neutron spectrum is estimated by a calculation using the method of Prokofiev *et al.* [4.25] on the basis of proton energy and ^{7}Li target thickness.

Under these conditions, SEU testing requires a few to several hours of irradiation.

(2) RCNP (Research Center for Nuclear Physics) of Osaka University, Japan [4.18, 4.24]

The schematic view of the ^{7}Li(p,n) source at RCNP of Osaka University is illustrated in **Fig. 4.12**. At RCNP, a proton beam up to around 392 MeV is available by use of an AVF cyclotron and a ring cyclotron. A proton beam is accelerated up to 65 MeV by an AVF cyclotron (K = 140 MeV) and then injected into a ring cyclotron (K = 400 MeV).

The beam current on the target is about 1.0 μA at most but the neutron flux is rather high (~3 × 10^5 cm^2 s^{-1}) because thicker neutron targets can be used owing to the higher proton energy. The proton beam is also delivered to a spallation neutron beam line (Sec. 4.4.4).

A neutron beam with peak energy below 90 MeV is obtained using the AVF cyclotron. Neutron beams higher than 90 MeV using both the AVF and ring cyclotrons. In the case of a peak energy below 90 MeV the neutron beam is collimated to 10-cm-diameter by a concrete wall (1 m thick) between the target room and the experimental room. Semiconductor chips on circuit boards are placed between 440 cm and 650 cm from the ^{7}Li-target. On the other hand, for a peak energy above 90 MeV, a bending magnet placed ~1 m from the ^{7}Li-target between the target room and the TOF room serves as a collimator. The beam size is 10 cm × 10 cm at the inlet of the TOF room.

Samples are placed in the TOF room, 3.5 m to 5.5 m apart from the ^{7}Li target (**Fig. 4.12**). Typical proton current is 650 to 950 nA. In each case, the neutron spectrum was measured by the TOF method before setting samples in the experimental room, and the neutron flux was monitored by the proton beam current at beam dump [4.18, 4.24]. In **Table 4.4**, the main properties of the neutron source are summarized on the basis of the recent measurement by Taniguchi *et al.* [4.24].

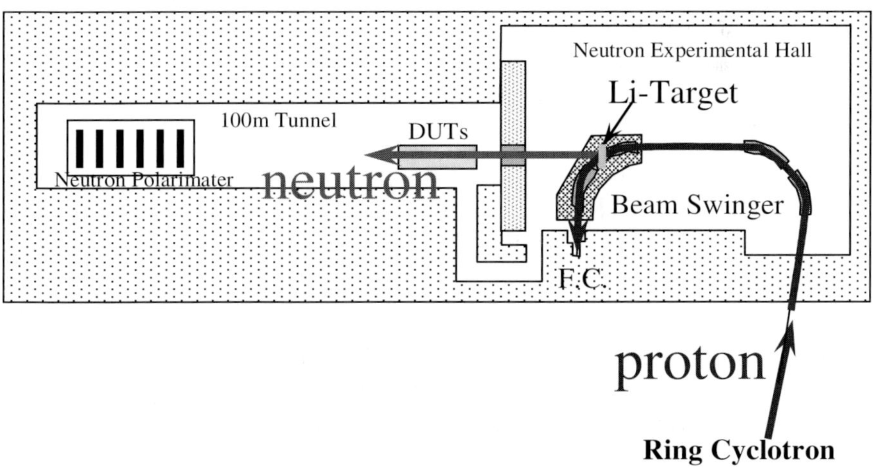

Fig. 4.12 The ^{7}Li(p,n) neutron field at RCNP, Osaka University [4.9].

Table 4.5 Characteristics of the ^{7}Li(p,n) neutron field at RCNP, Osaka University [4.24].

Proton energy [MeV]	250	350	392
Peak neutron energy [MeV]	248	348	390
Energy range of peak neutron [MeV]	236-250	344-350	383-392
Peak neutron intensity [n sr^{-1} μC^{-1}]	1.94×10^{10}	1.07×10^{10}	1.50×10^{10}
FWHM of peak neutron [MeV]	2.5	2.5	2.5
Energy range of tail neutron [MeV]	4-236	6-344	21-383
Tail neutron intensity [n sr^{-1} μC^{-1}]	1.70×10^{10}	1.35×10^{10}	1.76×10^{10}

(3) TIARA, JAEA [4.17, 4.22]

In TIARA, JAEA, a well characterized ^{7}Li(p,n) neutron source is available. As shown in **Fig. 4.13**, neutrons are produced by use of the proton beam provided by an AVF cyclotron (K = 110 MeV) and an enriched ^{7}Li target, usually 2 MeV thick, mounted on a water-cooled ladder. The neutron beam line is collimated and shielded from both the target region and beam dump with an iron collimator and a concrete wall around 3 meters long. Consequently, the experimental area is about 5.5 m from the ^{7}Li target, and the neutron flux in the experimental area is on the order of 10^{-4} cm^{-2}·s^{-1} as estimated from the data in **Table 4.4**. The energy dependent neutron flux is summarized in **Table 4.6**. Despite of the low intensity, the source has been used efficiently for SEU studies observing current wave forms during transients due to SEU [4.30].

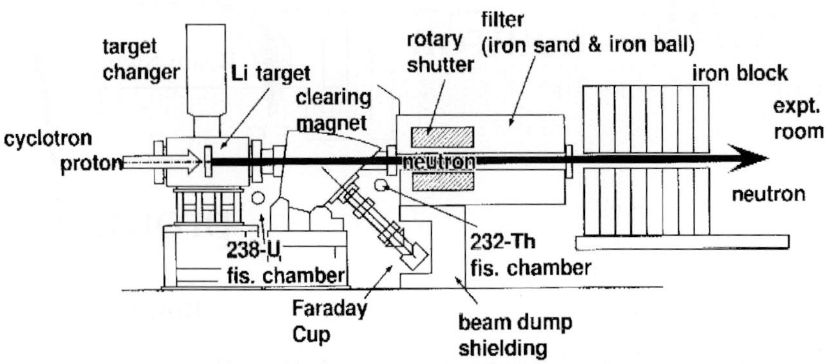

Fig. 4.13 The ^{7}Li(p,n) neutron beam facility at TIARA, JAEA [4.17, 4.22].

Table 4.6 Neutron flux of the TIARA $^{7}Li(p,n)$ neutron source [4.22].

Proton energy (MeV)	^{7}Li target thickness (mm)	Neutron energy [a)] (MeV)	Neutron flux[b)] 10^9 n (Sr μC)$^{-1}$
43	3.60	40.5	3.46 (±3.3%)
48	3.80	45.4	2.70 (±3.2%)
53	4.30	50.9	3.82 (±4.6%)
58	4.70	55.3	4.45 (±3.5%)
63	5.00	60.6	4.17 (±3.3%)
68	5.20	65.2	4.82 (±3.9%)
78	6.00	75.0	5.34 (±4.5%)
87	6.75	84.6	6.35 (±6.2%)

a) The energy of peak neutrons, b) Neutron flux in the experimental area.

(4) iThemba labs, South Africa [4.31, 4.32]

In iThema labs, South Africa, a unique neutron source is in operation by use of a 200 MeV cyclotron. The $^{9}Be(p,n)$ reaction is used as well as the $^{7}Li(p,n)$ reaction for production of quasi-monoenergetic neutrons. In this laboratory, a neutron beam delivered to non-zero degrees of, 4, 8, 12, 16, 20 deg., is available to enable a "tail correction", which is described in Sec. 4.3.4. This technique is efficiently utilized in studies on dosimetry and quality factors of neutron beams. These techniques will also be useful for the SEU studies as in Sec. 5.1.2, although thus far, the laboratory is not used to SEU studies.

(5) CYRIC, Tohoku University [4.20, 4.33]

In CYRIC, Tohoku University, a new $^{7}Li(p,n)$ source designed and constructed as shown in **Fig. 4.14**. In this design, a modular structure of shielding was adopted to enable the source-detector distance to be as short as ~75 cm. Owing to the design, as summarized in **Table 4.4**, the neutron flux available for experiments is the highest in the world, at present. For neutron-induced experiments, the area between the target and the experimental point, usually 1 m long, is shielded with conical

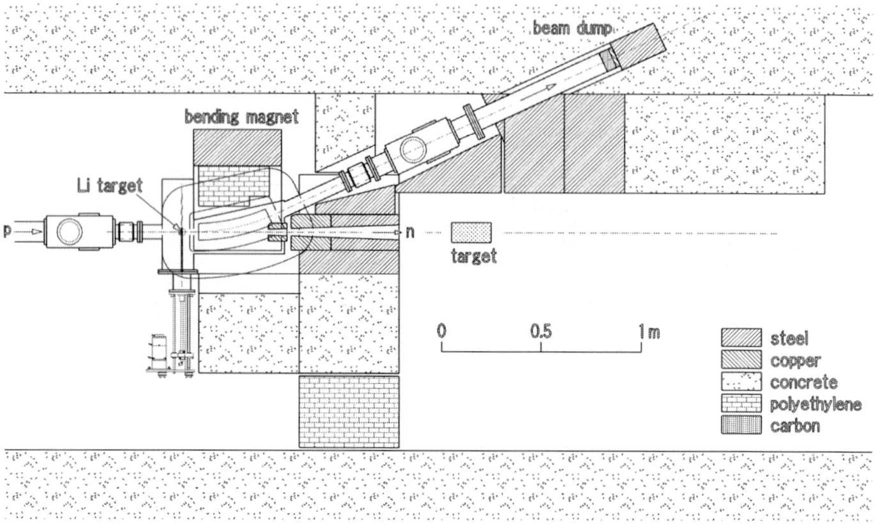

Fig. 4.14 Layout of ^{7}Li(p,n) neutron source at CYRIC, Tohoku University [4.20, 4.33].

collimators of 75 cm-thick Cu and Fe of. For effective shielding, the yoke of the dipole magnet upstream of the beam dump is used as a shielding material.

The neutron yield and spectrum is measured by a TOF method using an organic scintillator. As shown in **Fig. 4.15**, the shield proved to be effective, as shown by a good signal-to-background ratio. The signal-to-background ratio is also high enough for the for SEU experiments with semiconductor devices. The high neutron flux is very useful, as shown by the fact that the SEU could be observed within a few hour's irradiation even for DRAMs that are insensitive to SEU [4.9]. However, for such high flux, the area with uniform flux is limited within a few ten's of cm.

The spectrum is extrapolated to low energies using the three-body phase space distribution mentioned in Sec. 4.3.2. Now preparation is being made for a new arrangement that will have an additional beam channel around 20-deg, to enable an "experimental tail correction" mentioned in the next section.

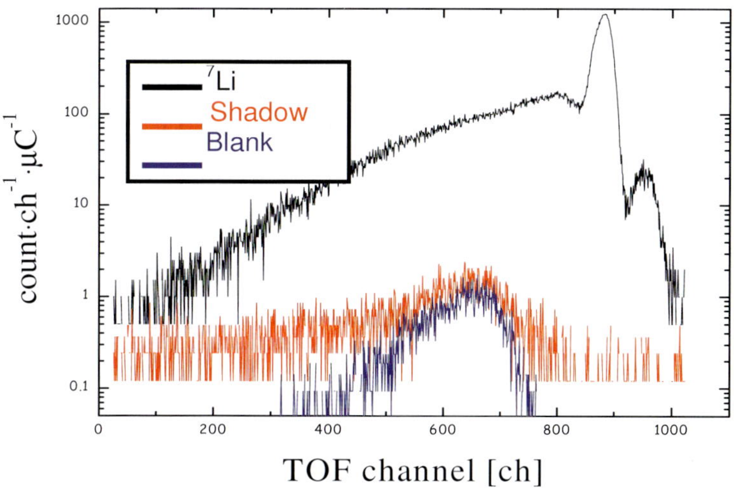

Fig. 4.15 TOF spectrum in new ^{7}Li(p,n) neutron source at CYRIC, Tohoku University [4.20, 4.33].

4.3.4 *Experimental tail correction method [4.31, 4.32]*

As noted above, the backgrounds of SEUs produced by continuum neutrons from the ^{7}Li(p,n) source should be determined and corrected for in some manner.

One method for the correction is to step the experiment from low energies just above the threshold and then make corrections using the experimental data from lower energies. The SEU cross-section can be found using an unfolding technique (see Chap. 5 for more details). But this is a very time-consuming procedure. Therefore, it is highly desirable if the correction can be done experimentally at each energy point.

For a ^{7}Li(p,n) source, the iThemba labs and PTB group reported that for 200 MeV protons, the spectral shape of the continuum background neutrons at zero-degree is close to the shape of the spectrum at 16-deg. At 16-deg, the monoenergetic neutron peak has almost completely vanished. (**Fig. 4.16**) [4.32, 4.33]. Therefore, the background due to continuum neutrons can be eliminated by subtracting the experimental data at 16-deg from the data at zero-deg. This method is also applicable to the ^{9}Be(p,n) source, as shown in **Fig. 4.17**.

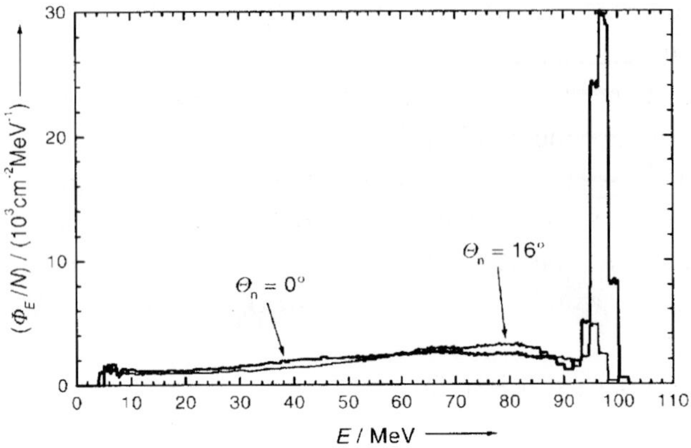

Fig. 4.16 Comparison of the ^{7}Li(p,n) spectrum at zero-deg with that at 16-deg [4.31].

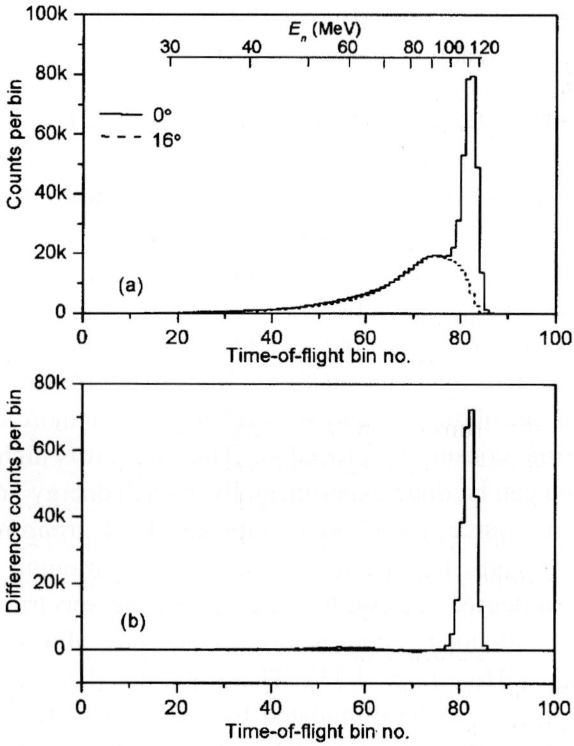

Fig. 4.17 Comparison of the ^{9}Be(p,n) spectrum at zero-deg with that at 16-deg [4.32].

This method has been successfully applied to cross-section measurements of activation reactions [4.31, 4.32]. Any results that include this correction method will also include some systematic uncertainty because the continuum spectra are not completely the same at the two angles. Nevertheless, the uncertainty is not so large because the SEU cross section has a smooth energy dependence. As a result, this method can also be applied to SEU studies using the quasi-monoenergetic neutron sources of ^{7}Li(p,n) and ^{9}Be(p,n).

4.4 Spallation Neutron Facilities

4.4.1 *Overview of spallation sources [4.3, 4.34]*

When high energy (> several tens of MeV) particles hit heavy elements, a number of neutrons are produced through a spallation reaction. In these reactions, many particles and fragments (composite particles) are produced as a result of the de-excitation of the many highly excited nuclei that are produced. In a spallation reaction, the number of neutrons per incident particle can be more than ten and the energy required to produce one neutron is much lower than in ion-induced reactions with lower incident energies and photo reactions. For this reason, MW class spallation sources are under construction in the USA (SNS: Spallation Neutron Source) [4.35] and in Japan (J-PARC: Japan Particle Accelerator Research Complex) [4.36], and more are planned in several other countries.

The neutron spectrum for spallation reaction is, as shown in **Fig. 4.18**, continuous because of many reaction processes involved. The spectrum is peaked around 1 MeV because most of neutrons are produced via the evaporation process from the decay of excited nuclei. The neutron spectra strongly depend on the neutron energy and on the mass of targets. High energy neutrons are strongly peaked in the forward directions, whereas low energy neutrons are almost isotropic. The average energy of outgoing neutrons becomes lower (*i.e.*, a soft spectrum) with the increasing target mass.

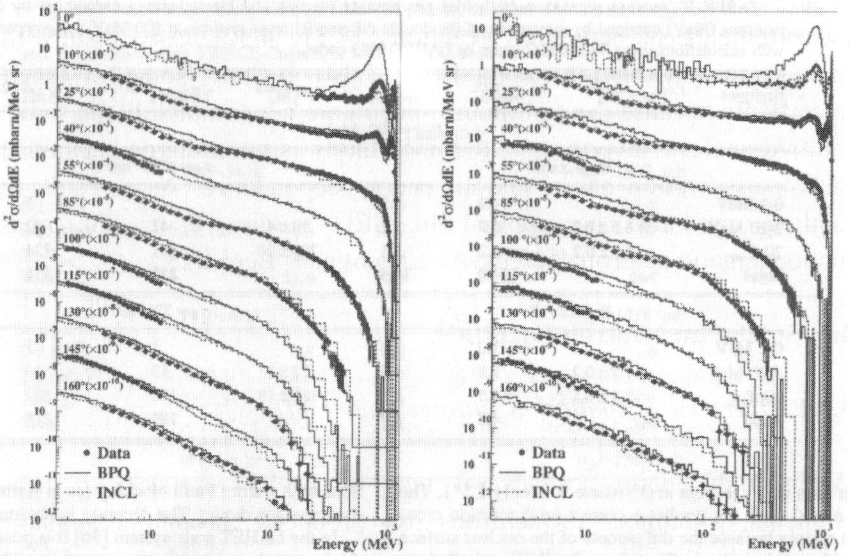

Fig. 4.18 (a) Neutron spectra by the spallation reactions of 1200 MeV protons on Pb (left) and Fe (right) [4.34].

An important feature of the spallation reaction is that the neutron spectrum produced from heavy elements in the forward direction is close to that of terrestrial neutrons at sea level (see Sec. 5.2). This is of great advantage for SEU studies because any test in a spallation field can then provide information about the SEU rate in the natural environment in a much shorter time-scale than can be done testing in the natural environment (thousand times or more).

In this section, three typical spallation sources are described.

4.4.2 *LANSCE (Los Alamos Neutron Science Center), LANL, New Mexico, USA*

The schematic view of the neutron beam line and irradiation facility at LANSCE is shown in **Fig. 4.19(a, b)** [4.37–4.39]. The facility uses a neutron beam generated at Target 4 of the WNR (Weapons Neutron Research) beam line at LANSCE. Neutrons are produced by bombarding a bare W target (3 cm in diameter and 7.5 cm long) with a 800 MeV proton beam accelerated by a linear accelerator. The proton beam current is around 1.5 μA. DUT and test circuit boards are placed in the ICE (Irradiation of Chips and Electronics) House, 19.7 m downstream from the neutron production target. The data acquisition (DAQ) system is also placed in the ICE House. The diameter of the neutron beam is around 8 cm at the point of irradiation.

The neutron spectrum incident on the DUT is measured over a wide energy range (from 1 to 800 MeV) using the TOF technique with a fission chamber of ^{238}U placed 60 to 320 cm upstream of the DUTs.

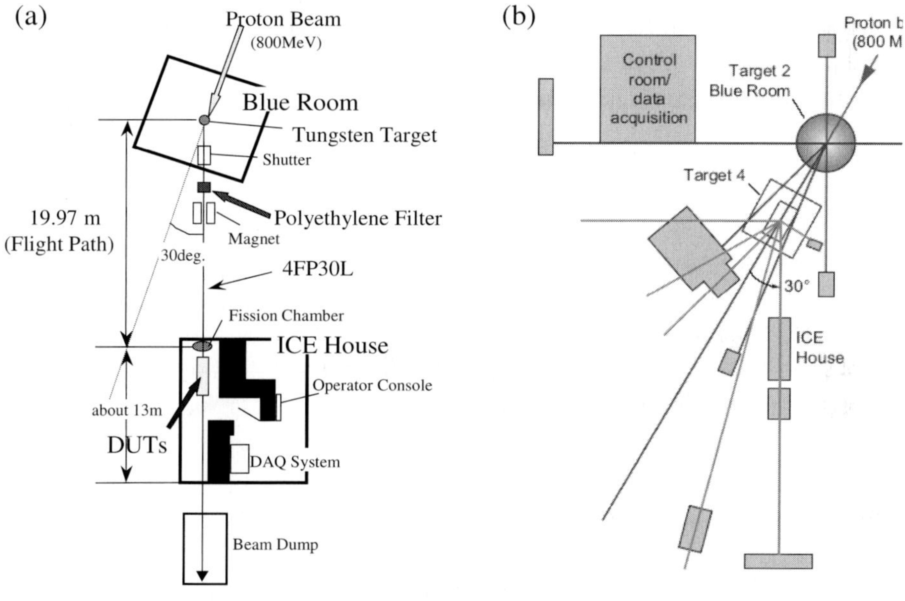

Fig. 4.19 (a) Layout of the neutron irradiation facility at Los Alamos Neutron Science Center (LANSCE) [4.39] (© 2005 IEEE). (b) Layout of the neutron beam lines at Los Alamos Neutron Science Center (LANSCE) [4.37, 4.38].

The fission chamber is also used to monitor the neutron flux during the irradiation [4.38]. The irradiation point is aligned on the basis of a center position fixed by a He-Ne Laser.

Neutron spectra incident on devices can be altered by simply placing various types of filtering material, such as polyethylene, into the neutron beam upstream from the ICE House (**Fig. 4.19(a)**). In **Fig. 4.20**, examples of spectrum shift by filtering materials are shown. From top to bottom, the spectra are without the polyethylene filter and with 5 cm-, 10 cm- or 20 cm-thick polyethylene, respectively [4.39]. (The spectra are normalized to the spectrum with no filter at 100 MeV). Neutrons with lower energy decrease drastically with increasing thickness of the polyethylene filters, whereas the spectra above 200 MeV do not change significantly. This filtering technique is useful to study the energy response of devices using a continuous spectrum of neutrons from spallation reactions.

In JESD89, the LANSCE neutron spectrum without a filter is noted as the accelerator-based spectrum that matches the natural atmospheric spectrum at New York City. It can therefore, be used to simulate SER in a natural terrestrial environment.

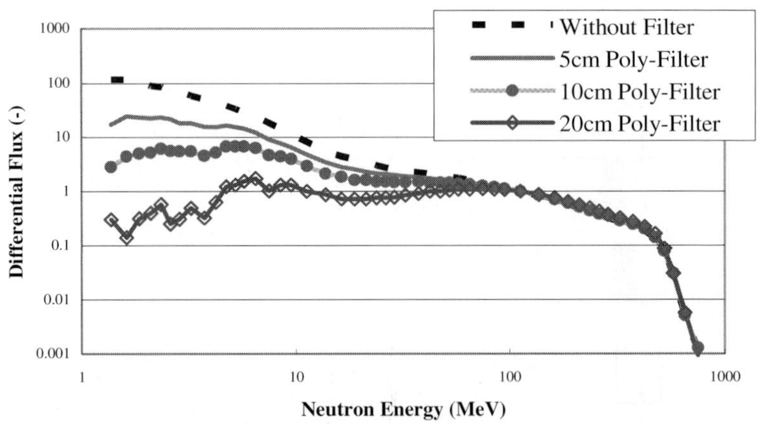

Fig. 4.20 Spectrum shift of spallation neutrons with polyethylene filters [4.39] (© 2005 IEEE).

The integrated neutron flux above 1 MeV is about 10^6 cm^2 s^{-1} at the irradiation position, which is about 10^6 greater than the flux of natural atmospheric neutrons at sea level.

4.4.3 *TRIUMF (TRI-University Meson Facility), Vancouver, BC, Canada [4.40]*

In TRIUMF, two spallation neutron sources at 450 MeV are used for SEU experiments. One is located in the NIF (Neutron Irradiation Facility) and the other in the PIF (Proton Irradiation Facility).

At the NIF, neutrons from thermal energy to high energy are generated by bombarding Al plates surrounded by water (**Fig. 4.21** [4.40]) with 400–450 MeV protons. The proton beams, which are transported from the PIF facility, typically have 100–150 nA. **Figure 4.22** shows the typical neutron energy spectrum above 0.1MeV [4.40]. The neutron flux above 0.1 MeV is 6×10^6 n cm^{-2} s^{-1}, which is about 10^7 times greater than the flux of terrestrial neutrons at sea level in New York. In the NIF, however, the physical space for devices is limited to an area of 5 cm by 20 cm. Because of the limited space, irradiation tests using a generic LSI tester is also difficult.

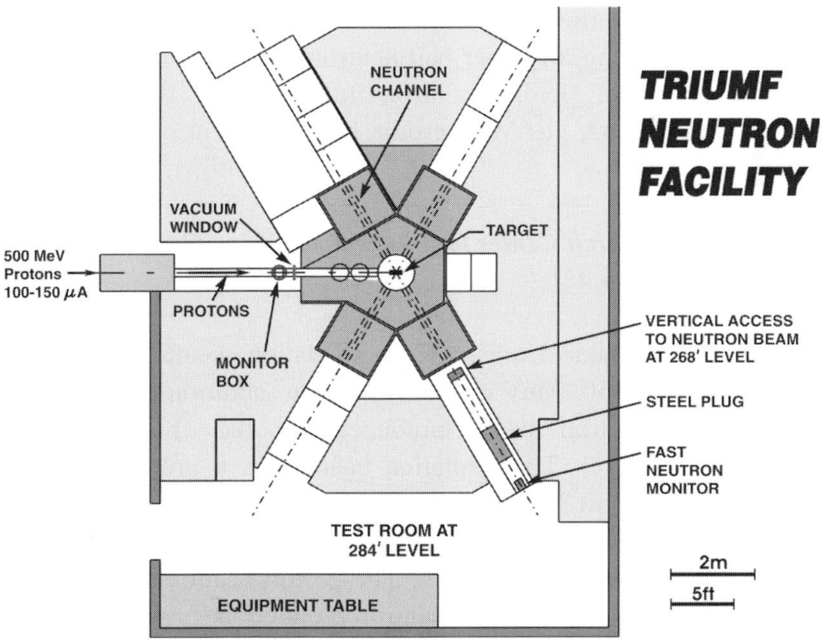

Fig. 4.21 Plan view of the experimental hall at TRIUMF [4.40] (© 2003 IEEE).

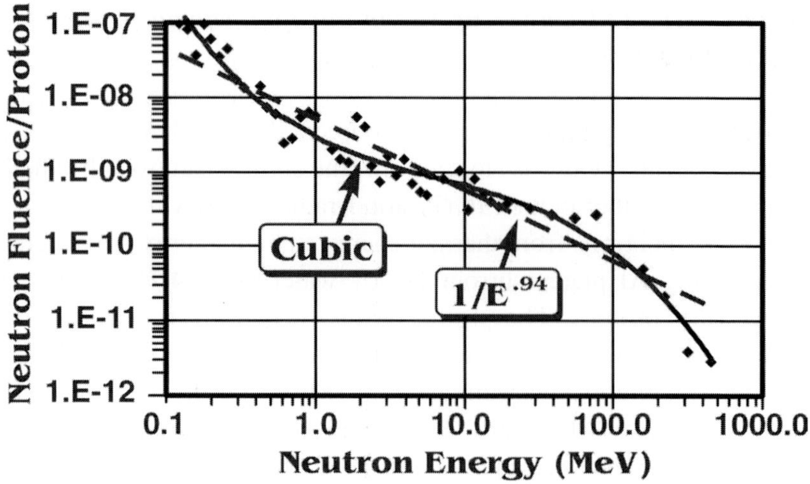

Fig. 4.22 Spallation neutron spectrum at TRIUMF; the straight line shows the spectrum of terrestrial neutrons (normalized) [4.40] (© 2003 IEEE).

A spallation neutron beam can also be obtained also at the PIF by setting up a target station and a beam dump. The neutron spectrum at the PIF is measured using a Bonner ball neutron spectrometer and a carbon activation technique. Here, a neutron flux of about 10^5 n cm^{-2} s^{-1} is available from a 3 nA 110 MeV proton beam incident on an aluminum target.

4.4.4 *RCNP (Research Center for Nuclear Physics), Osaka University, Japan [4.41, 4.42]*

At the RCNP, Osaka University, a spallation neutron beam with energies up to ~350 MeV is available, in addition to the quasi-monoenergetic neutron beam produced by the ^{7}Li(p,n) reaction (Sec. 4.3.3.(2)) [4.41]. The spallation beam uses a proton beam with incident energies up to 392 MeV. As with the case of the ^{7}Li(p,n) source, an AVF cyclotron (K = 140 MeV) and a ring cyclotron (K = 400 MeV) are used to generate a proton beam. The proton beam bombards a lead target (10 cm thick) for spallation neutron production.

Spallation neutrons emitted at 30-deg. relative to the incident beam are used for irradiation experiments. The neutron beam is collimated down to

10-cm-diamter with iron and concrete blocks. The neutron spectrum is shown in **Fig. 4.23** [4.42]. The integrated neutron flux above 10 MeV is 5.4×10^5 n cm^{-2} s^{-1} [4.41]. Currently, a tungsten target (6.5 cm thick) is used instead of lead [4.42].

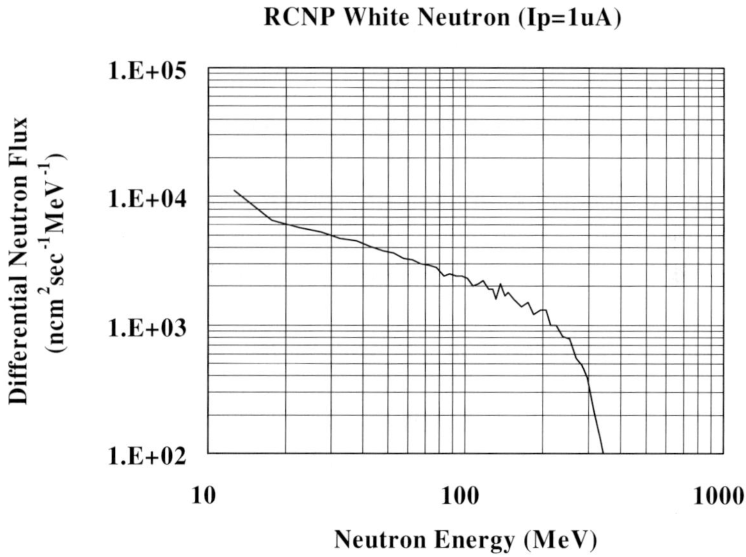

Fig. 4.23 Spallation neutron spectrum at the RCNP [4.42].

4.4.5 *Comparison of the neutron flux at various spallation neutron sources*

The main properties of three spallation neutron sources above mentioned are summarized below in **Table 4.6**.

Table 4.6 Comparison of spallation neutron source

Facility	Incident energy	Target	Typical Neutron flux (energy range)
LANSCE	800 MeV	W	6.5×10^5 n cm^{-2} s^{-1} (> 1 MeV)
TRIUMF	500 MeV	Al	6.0×10^6 n cm^{-2} s^{-1} (> 0.1 MeV)
RCNP	392 MeV	Pb	5.4×10^5 n cm^{-2} s^{-1} (> 10 MeV)

4.5 Summary

In this chapter, a review was given on accelerator-based neutron sources that are employed for irradiation testing of microelectronics devices. The sources are indispensable for testing of modern devices that are susceptible to radiation effects.

Typically, two types of neutron sources are used; monoenergetic and spallation-type. The spallation type is useful for testing conditions similar to the real natural environment, whereas the monoenergetic sources are useful for measuring the energy dependence of single event effects. Test conducted with monoenergetic sources are time consuming, but important. This is particularly true for the low energy regions close to the threshold energy. Recent progress in quasi-monoenergetic sources was also described.

The number and function of neutron sources are rather limited. However, several new sources are being developed. Improvement of these sources, particularly in intensity, is a high priority in order to facilitate faster cycles of testing and development.

Chapter 5

Review and Discussion of Experimental Data

Yasuo Yahagi

By using (quasi-) monoenergetic and spallation neutron beam facilities, a general test method is described herein. The relevant analyses techniques, including tail correction methods, Weibull-fitting, unfolding, and multi-cell upset (MCU) characterization, are also described with a variety of measured data. The authors' framework on SECIS (SElf Consistent Integrated System for neutron-induced single event) is introduced.

5.1 Monoenergetic Neutron Tests and SEU Excitation Function

The Real-Time SER test described in Chapter 3 is one of the most reliable tests on SER of a specific device, which is carried out under operational conditions of the device. The flux of terrestrial neutrons with energies above 1 MeV at sea level is, however, very low (on the order of 10 n/(cm^2 · h)). Particle accelerators can produce 10^6 times higher neutron flux than the flux at sea level. By using an accelerator, therefore, it is possible to know the susceptibility of devices to neutrons in a much shorter time than using terrestrial neutron.

One merit of accelerator-based tests that use monoenergetic neutrons is the ability to acquire essential information on the neutron energy dependence of single event upsets. Using that data, it is possible to obtain response functions of the device to neutron irradiation, and then use those functions to estimate the results of field testing.

In this chapter, SEU tests and data analysis from testing done with several energies of monoenergetic (1–15 MeV) and quasi-monoenergetic (20–400 MeV) neutrons generated by particle accelerators are described. No pure, monoenergetic neutron beam with energy more than 20 MeV exists that can be generated by particle accelerators. The quasi-monoenergetic ^{7}Li(p,n) neutron source has a continuous background of low-energy neutrons below the peak, and the neutrons are, therefore, not purely monochromatic. This is the origin of the prefix "quasi-". A systematic study done with neutron energies down to 15 MeV was carried out by our research group for the first time in the world. It was found that the threshold energy for SEUs of advanced semiconductor devices has been detected down to a few MeV, as shown in the following section.

5.1.1 *SEU cross sections in the literature*

Until now, a lot of accelerator-based irradiation tests of semiconductor devices using high energy protons or heavy ions have been carried out for applications in space technology. Similarly, neutron irradiation tests have also been done. **Figure 5.1** [5.1–5.5] shows the energy dependence of SEU cross sections of SRAMs (static random access memories) by nucleons (protons and neutrons) collected from literature. **Figure 5.2** [5.4, 5.6–5.8] shows similar information for DRAMs (dynamic random access memories). In spite of the differences between protons and neutrons, (which are discussed in Chapter 8) it is seen that for each type of memory that the SEU cross section steeply increases at low energies and saturates at high energy for both. There is little information about device structure and critical charge in the papers, except for memory density. Nevertheless, it should be assumed that SER depends on the type of device and process generation even though the SEU cross sections all exhibit the same energy dependence.

Figure 5.3 [5.9] shows the trend of SEU cross section for each memory-type as a function of the memory density. Since an increase in the memory density is usually attained by downsizing device-geometries, this figure can be roughly regarded as the relation between the susceptibility to neutrons and the process technology node. Whereas the susceptibility of DRAMs to SEU decreases as the geometries of the devices are scaled down, the SEU cross section of SRAMs steeply increases with miniaturization. The geometrical miniaturization, however, causes a decrease of the parasitic capacitance in the memory cell. Besides the aforementioned geometrical effect, a scaling down of the devices requires a decrease in the supply voltage and, as a result, the critical charge for SEU is also lowered. A reduction of the sensitive volume makes the device less sensitive to SEU, as shown by a

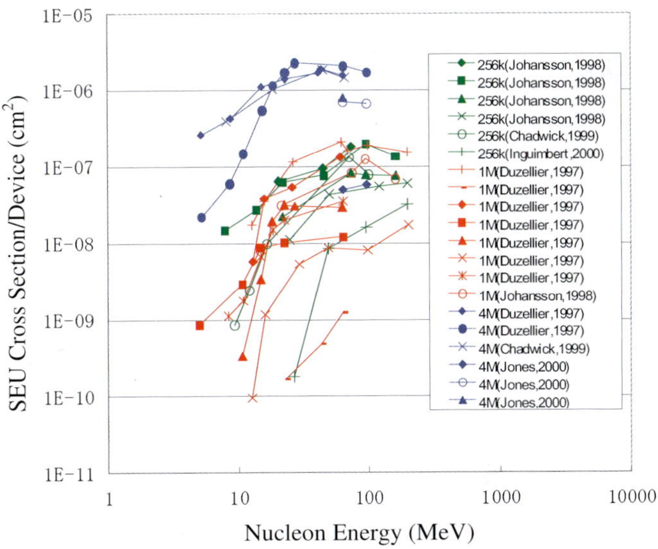

Fig. 5.1 SEU cross sections of SRAMs collected from Refs. [5.1–5.5].

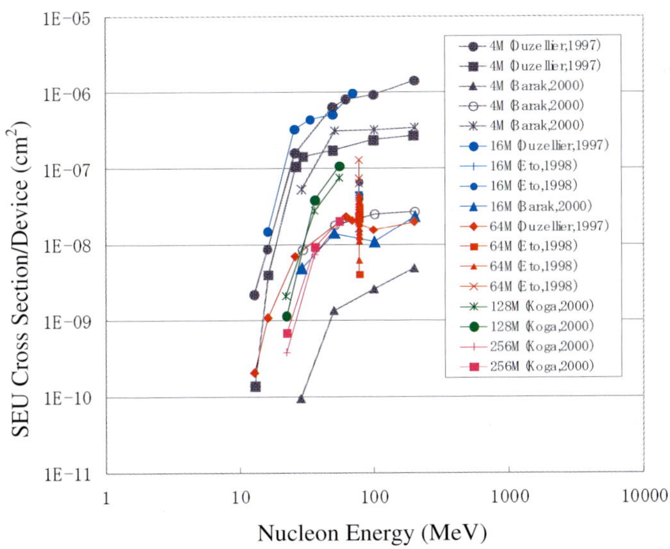

Fig. 5.2 SEU cross sections of DRAMs collected from Refs. [5.4, 5.6–5.8].

simulation [5.9], which gives a result that the SER of DRAM is proportional to the 6th power of the minimum feature size under constant critical charge. A memory cell of DRAM consists of one MOS transistor and one cell capacitor. The cell capacitance, which sustains a critical charge, makes an important contribution to reduction of SEU cross section. Unlike DRAM, SRAM is usually designed to have no cell capacitor. From the viewpoint of sensitivity to neutron-induced SEU, the reduction of the sensitive volume cannot, therefore, be compensated by the parasitic capacitance of an SRAM cell or by the supply voltage.

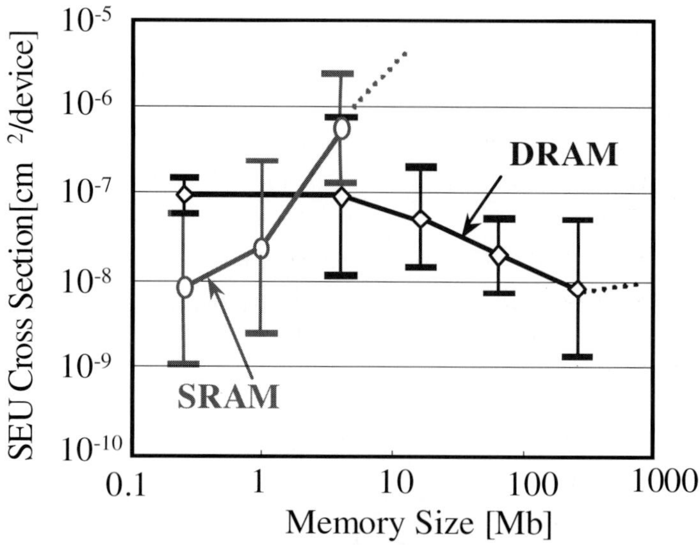

Fig. 5.3 General trend of SEU cross sections of SRAMs and DRAMs [5.9] (© 2002 IEEE).

A leading-edge SRAM cell is designed to have a capacitor connected with each storage node like the DRAM (see Sec. 3.3.1), and such an SRAM may have a low sensitivity to SEU, as does the DRAM [5.10, 5.11].

5.1.2 Irradiation test for SEU susceptibility using a mono- and a quasi-monoenergetic neutron source

Monoenergetic neutron sources are a powerful tool to obtain data on the neutron energy dependence of SEU. The energy spectrum has a narrow peak and, as such, makes it possible to obtain a set of data on the SEU susceptibility of devices at several discrete energy levels, particularly near the threshold neutron energy of SEU. A monoenergetic neutron beam is generated by specific nuclear reactions giving a peak neutron energy below 20 MeV. See, for example, the FNL of Tohoku University (Sec. 4.1.1).

A quasi-monoenergetic neutron beam is obtained by the nuclear reaction ^{7}Li(p,n) in which a proton beam of a few tens of MeV is bombarded on a Li-target. It is used to obtain data on the neutron energy dependence of SEUs. Quasi-monoenergetic neutron sources that can be used for SEU testing are described in detail in Sec. 4.1.2 and are used at facilities such as CYRIC of Tohoku University. Unlike the spectrum from a monoenergetic neutron source, the spectral shape of a quasi-monoenergetic source has a continuous tail at energies below the peak. The tail is due to many-body system interactions that accompany the ^{7}Li(p,n)^{7}Be reaction.

(1) SEU cross section and tail correction in quasi-monoenergetic neutron testing

For each irradiation test at accelerator facilities, (a) the number of SEU, (b) the neutron energy spectrum and (c) neutron fluence are obtained. An SEU cross section, σ (cm^2), is defined as number of SEU, N, per neutron fluence Φ (n/ cm^2):

$$\sigma = \frac{N}{\Phi}. \tag{5.1}$$

In the case of irradiation tests using monoenergetic and quasi-monoenergetic neutron beams, SEU cross sections should be defined for neutrons with the peak energy because the tests are aimed to obtain energy dependence of SEU. For monoenergetic neutron beams, equation (5.1) can be used to obtain the SEU cross sections. In the case of quasi-monoenergetic neutron beams, however, a tail correction (5.1) must be taken into account in order to calculate the SEU cross section, because neutrons with energy less than the peak energy are able to contribute to SEU. There are at least three methods for tail correction: the simplified iterative folding procedure proposed by Johansson *et al.* [5.1], the method using a Monte Carlo simulator CORIMS (see Chapter 6) and a method using non-zero degree neutrons [5.12].

(i) The iterative folding method

The iterative folding method needs a set of neutron irradiation tests conducted at several energy levels. In this method the tail-corrected SEU cross section for each energy level is calculated through a convergence technique. If the SEU cross section $\sigma_{SEU}(E)$ is assumed to be a *smooth* function of neutron energy, the SEU rate $R_{SEU}(E_i)$ for a quasi-monoenergetic neutron spectrum with peak energy E_i and a low-energy tail could be calculated from the following relation:

$$R_{SEU}(E_i) \propto \int_0^{E_i} \sigma_{SEU}(E_i) \frac{d\phi_i(E)}{dE}\Big|_{\theta=0^\circ} dE \tag{5.2}$$

where $\frac{d\phi_i(E)}{dE}\Big|_{\theta=0^\circ}$ is the zero-degree incident energy differential neutron spectrum with peak energy E_i. When the neutron fluence $\Phi_i(E)$ is used instead of $\phi_i(E)$ in Eq. (5.2),

$$N_{SEU}(E_i) = \int_0^{E_i} \sigma_{SEU}(E_i) \frac{d\Phi_i(E)}{dE}\Big|_{\theta=0^\circ} dE \tag{5.3}$$

where $N_{SEU}(E_i)$ is the experimentally observed number of SEU for a quasi-monoenergetic neutron spectrum, and E_i is the peak energy. First, the neutron SEU cross section $\sigma_{SEU\_l}(E_i)$ is calculated by Eq. (5.1) as well as in the case of a monoenergetic neutron spectrum, even if the neutron spectrum is quasi-monoenergetic with peak energy E_i. If the cross section is put into Eq. (5.3), then the number of SEU can be determined:

$$\boldsymbol{N^{cal}_{SEU\_0}(E_i) = \int_{E_{th}}^{E_i} \sigma_{SEU\_0}(E_i) \frac{d\Phi_i(E)}{dE}\Big|_{\theta=0^\circ} dE}\,, \tag{5.4}$$

where $\boldsymbol{E_{th}}$ is an assumed threshold energy of SEU. The next neutron SEU cross section is defined as

$$\sigma_{SEU\_1}(E_i) \equiv \sigma_{SEU\_0}(E_i) \frac{N_{SEU}^{obs}(E_i)}{N_{SEU\_0}^{cal}(E_i)}, \tag{5.5}$$

where $N_{SEU}^{obs}(E_i)$ is the experimentally observed number of SEUs. The folding procedure of putting the calculated SEU cross section into Eq. (5.3) is repeated until it converges to the point where the truncation error is small. If the convergence is reached after the (l+1)-repetition of the folding procedure, the tail correction factor C_{tail} is given by the ratio;

$$C_{tail} = \frac{N_{SEU}^{obs}(E_i)}{N_{SEU\_l}^{cal}(E_i)}, \tag{5.6}$$

where

$$\boldsymbol{N_{SEU\_l}^{cal}(E_i) = \int_{E_{th}}^{E_i} \sigma_{SEU\_l}(E_i) \frac{d\Phi_i(E)}{dE}\bigg|_{\theta=0^\circ} dE}, \tag{5.7}$$

is the calculated number of SEU for the neutron SEU cross section produced by the l-th repetition of the folding procedure. The tail-corrected SEU cross section for the neutron spectrum $\sigma_{SEU}^{corr}(E_i)$ is then given as;

$$\sigma_{SEU}^{corr}(E_i) = C_{tail}(E_i) \times \sigma_{SEU\_0}(E_i). \tag{5.8}$$

The tail correction factor for each experimented energy must be determined to insure that the SEU cross section $\sigma_{SEU}(E)$ is a smooth function of neutron energy.

In a simplified model of this method described in Ref. [5.1], the tail part of the quasi-monoenergetic neutron spectrum is treated as constant in the equations above, as graphically shown in **Fig. 5.4 (A)**. The simplified spectrum is shown in **Fig. 5.4 (B)**, and it is mathematically

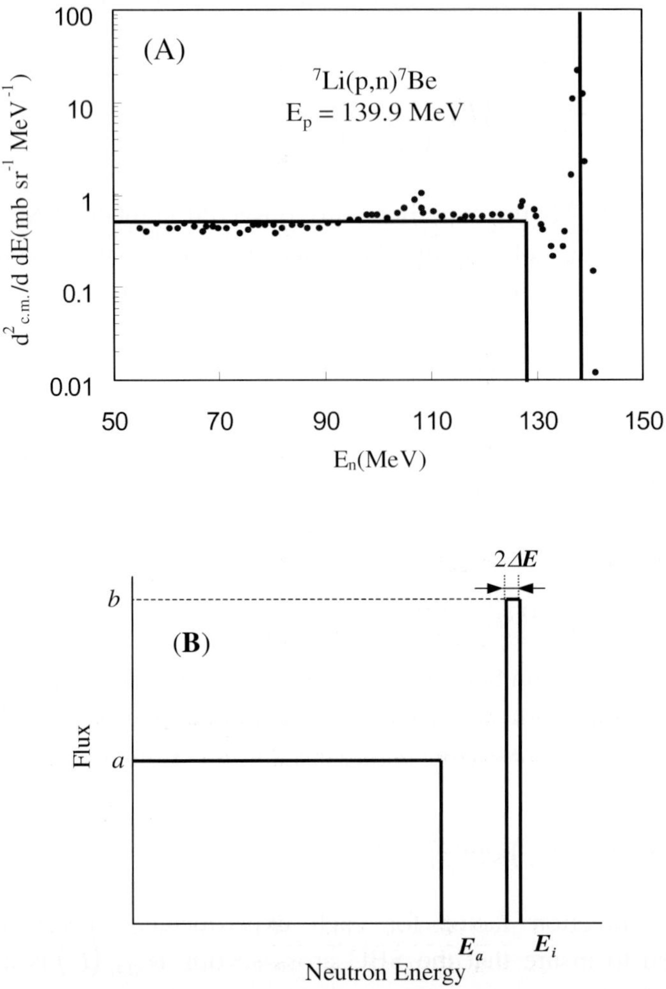

Fig. 5.4 (A) Experimental neutron spectrum from ^{7}Li(p,n) reaction at 0 deg. with $E_p =$ 139.9 MeV. The solid line represents the approximation of the spectrum [5.1] (© 1998 IEEE). (B) The schematically depicted simplified neutron spectrum.

expressed as:

$$\left.\frac{d\phi_i(E)}{dE}\right|_{\theta=0^\circ} = \begin{cases} a & (0 < \boldsymbol{E} < \boldsymbol{E}_a) \\ b & (\boldsymbol{E}_i - \boldsymbol{\Delta E} \le \boldsymbol{E} \le \boldsymbol{E}_i + \boldsymbol{\Delta E}) \\ 0 & \textbf{(except above)} \end{cases}. \quad (5.9)$$

where a is a constant corresponding the tail part of the quasi-monoenergetic neutron spectrum, E_i is the peak energy and $\boldsymbol{\Delta E}$ is the half width of the peak.

(ii) Method using the Monte Carlo simulator CORIMS

On the other hand, a tail correction can be applied for each test by using the Monte Carlo code CORIMS. In CORIMS, the SEU of the device is simulated for each energy bin in the neutron spectrum. The width of the energy bin used in the simulation is set to about 1 MeV, according to a previously measured or calculated neutron spectrum. As schematically depicted in **Fig. 5.5**, the correction coefficient C_{peak} is obtained as the ratio of the numbers of events calculated by CORIMS:

$$\mathbf{C}_{\mathbf{peak}} = \frac{\textbf{Number of error events caused by peak energy neutrons}}{\textbf{Number of error events caused by all the neutrons}}. \quad (5.10)$$

The experimentally obtained SEU cross section,

$$\sigma_{SEU\,(Exp)} = \frac{\textbf{Number of error events caused by all the neutrons}}{\textbf{Fluence of neutrons with peak energy}}, \quad (5.11)$$

is multiplied by the correction coefficient C_{peak}, to determine the corrected SEU cross section:

$$\sigma_{SEU}^{Corr} = C_{peak} \times \sigma_{SEU\,(Exp)}. \quad (5.12)$$

In **Fig. 5.6**, the SEU cross sections per device corrected by CORIMS are shown for each peak energy of the quasi-monoenergetic neutron beam: at CYRIC (26–70 MeV), TSL (90, 180 MeV) and RCNP (above 20 MeV

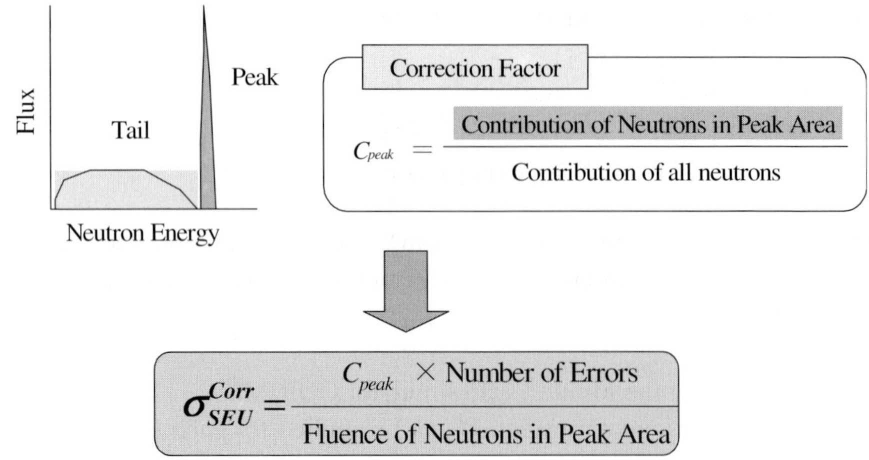

Fig. 5.5 A schematic of the tail correction procedure by the SEU simulator package CORIMS.

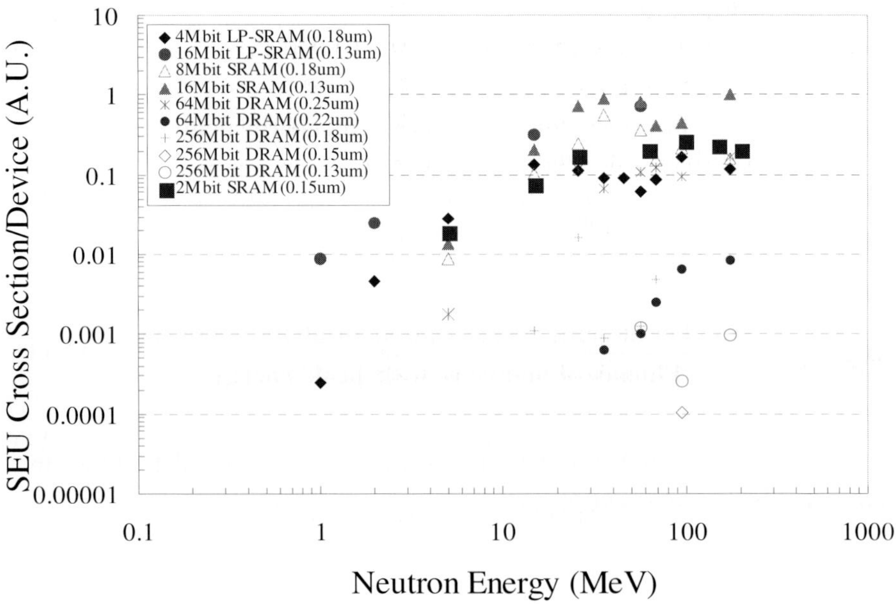

Fig. 5.6 SEU cross sections corrected by a SEU simulator package CORIMS.

for 150 nm SRAM). In this figure, the results using monoenergetic neutron beams at FNL (1, 2, 5 and 15 MeV) are also shown. As described in Section 5.1, the SEU cross section steeply increases with neutron energy and is saturated at higher energies.

The correctness of the model used in CORIMS determines the validity of the method described above. This is validated, as described later in Section 5.2.3, by checking the consistency of the neutron energy dependence of the corrected SEU cross sections with the results obtained by the unfolding method using spallation neutron beams, since CORIMS is not used in the unfolding procedure at all. This is also validated from the fact that the measured SERs are in good agreement both with the estimated results and with the simulated results.

(iii) Method using non-zero degree neutrons

The contribution of peak-energy neutrons to SEU can be experimentally separated from the contribution from the tail energy by using Be(p,n) neutron beams at 0 and 16 deg. Sisterson *et al.* [5.12] used Be(p,n) neutron beams to measure the energy dependence of neutron-induced reaction cross sections in copper. For a particular quasi-monoenergetic neutron beam, the total number of SEU, $N_{Total}^{(0)}$, and the number of SEU from the tail, $N_{Tail}^{(1)}$, are measured at 0 deg. and 16 deg. respectively. The same is true for the neutron fluence $\Phi_{Total}^{(0)}$ and $\Phi_{Tail}^{(1)}$. The total fluence at 0 degrees, $\Phi_{Total}^{(0)}$, is experimentally given by:

$$\Phi_{Total}^{(0)} = \Phi_{Tail}^{(0)} + \Phi_{Peak}^{(0)} . \tag{5.13}$$

The total number of SEU at 0 degrees is expressed by:

$$N_{Total}^{(0)} = N_{Tail}^{(0)} + N_{Peak}^{(0)} . \tag{5.14}$$

The SEU cross sections are calculated by the equations

$$\sigma_{Total} = \frac{N_{Total}^{(0)}}{\Phi_{Total}^{(0)}} \tag{5.15}$$

and

$$\sigma_{Tail} = \frac{N_{Tail}^{(0)}}{\Phi_{Tail}^{(0)}}. \tag{5.16}$$

Eq. (5.16) is also given by

$$\sigma_{Tail} = \frac{N_{Tail}^{(1)}}{\Phi_{Tail}^{(1)}}, \tag{5.17}$$

which is experimentally measured using Be(p,n)-neutrons at 16 degrees. The SEU cross section for peak energy neutrons is expressed by

$$\sigma_{Peak} = \frac{N_{Peak}^{(0)}}{\Phi_{Peak}^{(0)}}. \tag{5.18}$$

Using equations (5.15) to (5.17) to modify Eq. (5.14), the tail-corrected SEU cross section is given by;

$$\frac{N_{Total}^{(0)}}{\Phi_{Total}^{(0)}} \times \Phi_{Total}^{(0)} = \frac{N_{Tail}^{(0)}}{\Phi_{Tail}^{(0)}} \times \Phi_{Tail}^{(0)} + \frac{N_{Peak}^{(0)}}{\Phi_{Peak}^{(0)}} \times \Phi_{Peak}^{(0)} \tag{5.19}$$

$$\sigma_{Total} \times \Phi_{Total}^{(0)} = \sigma_{Tail} \times \Phi_{Tail}^{(0)} + \sigma_{Peak} \times \Phi_{Peak}^{(0)} \tag{5.20}$$

$$\sigma_{Peak} = \frac{\sigma_{Total}\Phi_{Total}^{(0)} - \sigma_{Tail}\Phi_{Tail}^{(0)}}{\Phi_{Peak}^{(0)}}. \tag{5.21}$$

(2) Energy-dependent function of neutron SEU cross sections

As described above, it is very convenient to have a mathematical expression of the energy dependence of neutron SEU cross sections. That expression is refered to as an SEU function. The following five expressions of the SEU function have been proposed from prior results of proton irradiation and heavy-ion irradiation tests:

a) The one-parameter Bendel's expression [5.13],

$$\sigma(E) = \left(\frac{24}{E_{th}}\right)^{14} \left\{1 - \exp(-0.18\sqrt{Y})\right\}^4 \tag{5.22}$$

where

$$Y = \sqrt{\frac{18}{E_{th}}}(E - E_{th}) \tag{5.23}$$

and E_{th}(MeV) is the threshold energy of SEU.

b) The two-parameter Bendel's expression [5.14],

$$\sigma(E) = \left(\frac{B}{E_{th}}\right)^{14} \{1 - \exp(-0.18\sqrt{Y})\}^4 \tag{5.24}$$

where

$$Y = \sqrt{\frac{18}{E_{th}}}(E - E_{th}), \tag{5.25}$$

B is the Bendel parameter, and E_{th}(MeV), threshold energy of SEU.

c) The integration form of the Weibull function [5.15]

$$\sigma(E) = \sigma_{\infty}\left[1 - \exp\left\{-\left(\frac{E - E_{th}}{W}\right)^S\right\}\right] \tag{5.26}$$

where σ_{∞} (cm^2) is a saturated value of the SEU cross section, E_{th}(MeV) is the threshold energy of SEU, and W(MeV) and S are the Weibull parameters.

d) The modified Weibull function [5.16]

$$\boldsymbol{\sigma(E) = A\left[1 - \exp\left\{-\left(\frac{E - E_{th}}{W}\right)\right\}\right] \times \exp\left(-\frac{E}{E_0}\right)} \tag{5.27}$$

where A(cm^2) is a cross-section magnitude parameter, E_{th}(MeV) is the threshold energy of SEU, and W(MeV) and E_0 (MeV) are shapefactors of the fitting function.

e) A polynomial function of energy [5.17]

$$\sigma(E) = \sum_{i=0}^{M} a_i E^i \tag{5.28}$$

or, in the expression of a logarithm,

$$\log(\sigma(E)) = \sum_{i=0}^{M} a_i (\log E)^i \, . \tag{5.29}$$

In order to make it clear which function better represents the SEU cross sections obtained by monoenergetic and quasi-monoenergetic neutron irradiation tests, the fitted results for 4 Mbit SRAM with 180 nm technology are compared. A curve fitting procedure using a non-linear function was carried out using the commercial softwear Origin® (Ver.6.1) by Microcal Software, Inc. The algorithm used in the program for least-square fitting is the Levenberg-Marquardt method. **Figure 5.7** shows the fitting results for the 4 Mbit SRAM using the indicated functions. The saturated value of each function is fit to the measured SEU cross section.

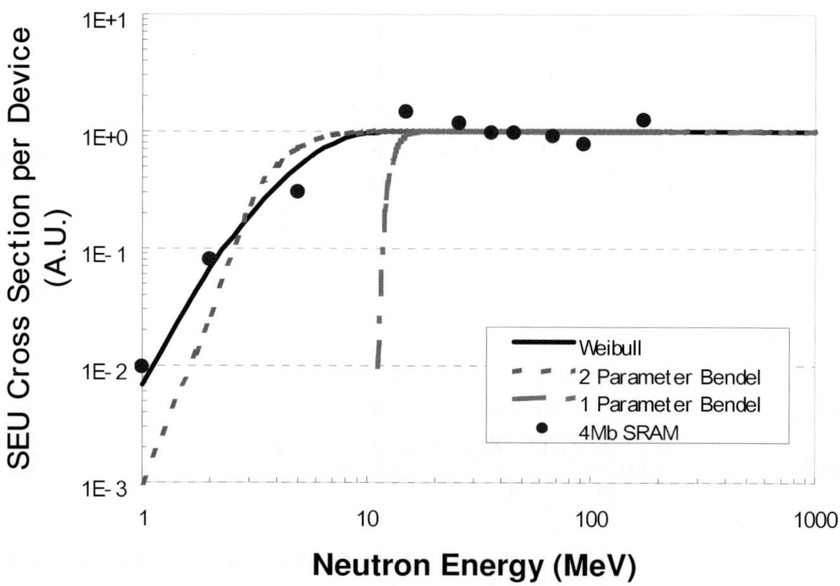

Fig. 5.7 A comparison of fitting functions for the neutron energy dependence of SEU cross sections of 4 Mbit low power consumption SRAM; Weibull-function, 1- and 2-parameter Bendel function.

For proton irradiations, the one-parameter Bendel's expression poorly represents the measured data in the low energy region, as pointed out by Stapor *et al.* [5.18]. Also, it can not attain a proper value of the threshold energy of the calculated SEU if the saturated value of the calculated SEU cross section is fit to the data. If, on the contrary, the threshold energy of SEU is fit to the measured value, the saturated value of SEU cross section is 1×10^{12} cm^2, which is clearly too large. In the case of the two-parameter Bendel's expression, the function gives too steep of an increase in the low-energy region if the saturated value of the SEU cross section is properly fit to the experimental results. As a result, the two-parameter Bendel's expression poorly fits the measured data. The modified Weibull expression has difficulties in fitting the data of 4 Mbit low power consumption SRAMs, because the data has no steep dip in the SEU cross section below 180 MeV.

In the case of the unmodified Weibull function, the demerits of both Bendel's expressions are avoided and the Weibull function expresses the entire range of SEU cross sections very well, including the threshold energy for SEU. **Figure 5.8** shows the result of curve-fitting with the Weibull function for each device, and **Table 5.1** summarizes the parameters of the function; E_{th}, W, and S. Since only a few SEU in the low energy region have been observed so far on DRAMs except for 64 Mbit with 250 and 220 nm process, curve-fitting was carried out only for these two DRAMs. Because of the limitation in capacity for each test board used in the experiment, only four chips for 256 Mbit DRAM were set on the test board. As such, statistically meaningful data on 256 Mbit DRAMs have not yet been obtained owing to their high immunity to SEU and limited number of DRAMs tested.

Table 5.1 Weibull-parameters of each device.

	SRAM					DRAM	
	Low Power		High Speed		Embedded		
Memory Capacity (Mb)	4	16	8	16	2	64	64
Process (nm)	180	130	180	130	150	250	220
E_{th} (MeV)	0.7	0.2	4.0	3.5	3.5	3.0	12.0
W (MeV)	5.3	18.9	16.0	16.0	17.0	20.0	60.0
S	2.1	1.4	1.6	1.6	1.1	2.0	3.0

It can be seen in **Fig. 5.8** and **Table 5.1** that devices with the larger value of W have a wider energy range from E_{th} to the value where SEU cross section reaches to the saturated value, and that a larger value of S leads to a steeper increase in the curve at lower energies. There seems to be no relation between the critical charge and W or S. In the case of the lower power-consumption SRAMs, the threshold energy E_{th} is below 1 MeV, which is much less than other types of devices. Regarding the two DRAMs, an increase of E_{th} from 3 MeV up to 12 MeV is brought about by the reduction of the process technology from 250 to 220 nm. Since these two devices are operated at almost same voltage, the miniaturization of the device has a large influence on the increase in SEU immunity. It should be also noted that SEU immunity depends on the size of diffusion layer, the 3-dimensional distribution of impurity concentration, the structure of memory cell, and the cell layout in a chip.

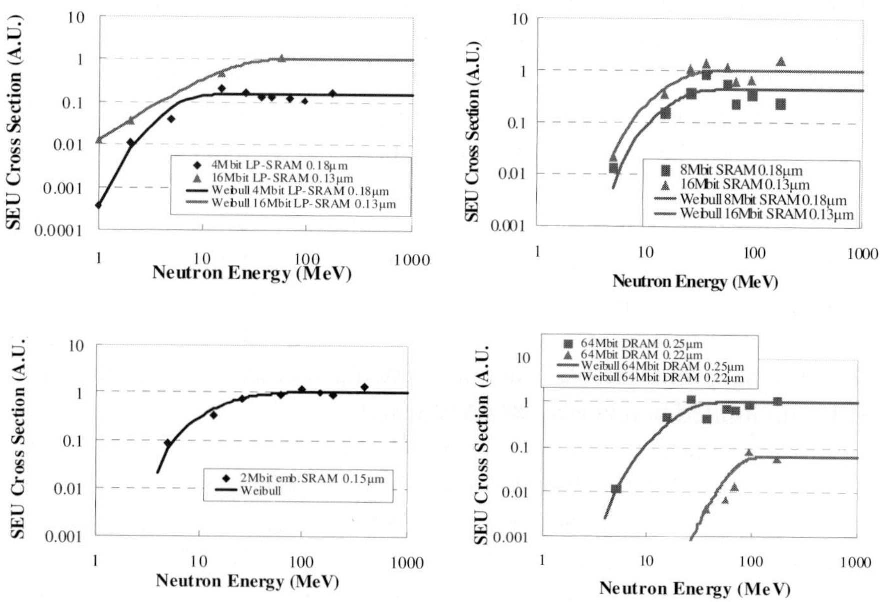

Fig. 5.8 The solid symbols are tail-corrected SEU cross sections for each device and the solid lines are results of Weibull-fitting. The Weibull-parameters are summarized in Table 5.1.

5.1.3 *Measurement of the threshold energy for SEU and its importance*

In the last section, the threshold energy E_{th} for the SEU of a device is introduced by expressing the energy dependence of SEU cross section as a function. The SER of a device under irradiation from a neutron differential flux $d\phi(E)/dE$ is given as an integral form

$$SER(\mathrm{FIT}) \propto \int_{E_{th}}^{\infty} \sigma_{SEU}(E) \frac{d\phi}{dE} dE\,, \tag{5.30}$$

where $\sigma_{SEU}(E)$ is the SEU function. The threshold energy E_{th} for SEUs of the device is required in order to estimate SER.

(1) E_{th} for neutron SEU in previous studies

The handling of E_{th} in previous studies is reviewed here. Ziegler and Lanford [5.19], who first predicted SEUs caused by cosmic rays at ground level, estimated the E_{th} for neutron SEU was about 8 MeV by investigating the kinetics of recoil nuclei of Si, Mg or alpha particles produced by interactions between neutrons and Si nuclei, using a Burst Generation Rate (BGR) calculation. Normand *et al.* [5.20] estimated that E_{th} for neutron SEU lies down between 4 and 5 MeV by investigating recoil nuclei of Si from the ^{28}Si(n, n') ^{28}Si reaction, also by using a BGR calculation. Eto *et al.* [5.7] carried out a series of irradiation experiments using spallation neutron beams with varying spectral shapes at LANSCE, and they concluded that E_{th} for neutron SEU is about 5 MeV. Agosteo *et al.* [5.21] stated that since the threshold energy of the nuclear reaction ^{28}Si(n, α) ^{25}Mg is 2.7 MeV, E_{th} for neutron SEU is about 5 MeV. Dyer *et al.* [5.22] carried out a series of irradiation experiments using monoenergetic and quasi-monoenergetic neutron beams with energies above 14 MeV, and they concluded E_{th} for neutron SEU is about 3 MeV, as determined from the results of Weibull fitting and simulation. They, however, carried out the Weibull fitting to the data without a tail correction for the quasi-monoenergetic neutron beam.

It can be summarized that most of the previous works attributed E_{th} for neutron SEU to nuclear reaction ^{28}Si(n, x) to be about 5 MeV or ^{28}Si(n, α) ^{25}Mg to be about 3 MeV.

(2) Importance of experimentally measuring E_{th} for neutron SEU

Because the current trend in downsizing the scaling of devices, especially CMOS devices, brings a reduction of the critical charge for SEU, it is expected that contribution from terrestrial, low-energy neutrons to SEU will increase because of the higher flux of low-energy terrestrial neutrons. In this section, the influence of E_{th} on the estimation of SER without measuring E_{th} or without performing SEU tests with low-energy neutrons is discussed.

In this example, a Weibull-type SEU function is applied to test results on low power consumption SRAM with 180 nm technology node. If we determine the SEU function without measurements using monoenergetic neutrons of less than 15 MeV, for example, almost all of the Weibull-curves with E_{th} < 15 MeV could be acceptable, as shown in **Fig. 5.9**. **Figure 5.10** shows the calculated value of *SER*(FIT) for each E_{th} according to equation (5.30). The calculated SER for E_{th} = 10 MeV, for example, is about 50 % smaller than that for E_{th} = 2 MeV. It is, therefore, very important to carry out SEU tests for low energy neutrons and to determine E_{th} for neutron SEU, in order to accurately evaluate the SER of each device.

(3) Measured results of E_{th} for neutron SEU

Irradiation tests for neutron SEU play a very important role both for acquiring the SEU function and for determining E_{th}. **Table 5.2** gives a summary of the experimentally obtained neutron E_{th} for the SEU of each device. As described in Section 5.4.1, the E_{th} for neutron SEU is assumed to lie around 5 MeV owing to the nuclear reaction of ^{28}Si(n,α) ^{25}Mg. The experimental result at this time reveals, however, that $E_{th} \leqq 1$ MeV for the two low power consumption SRAMs. This is the first measurement of E_{th} below 5 MeV in the world.

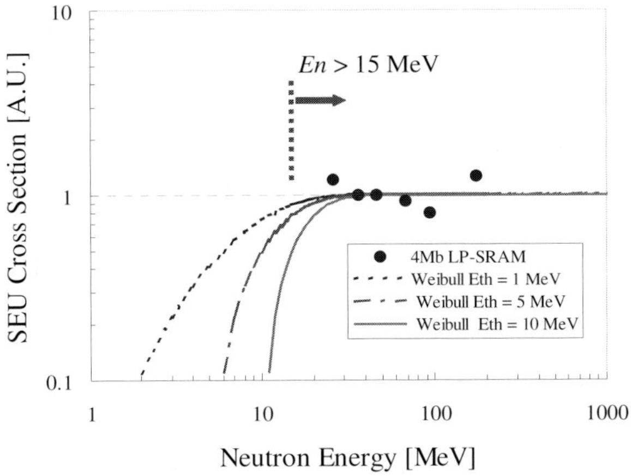

Fig. 5.9 Weibull-fitting for 4 Mbit SRAM with variation in the SEU threshold energy E_{th}. For SEU cross section data above 15 MeV, each of the curves could be acceptable.

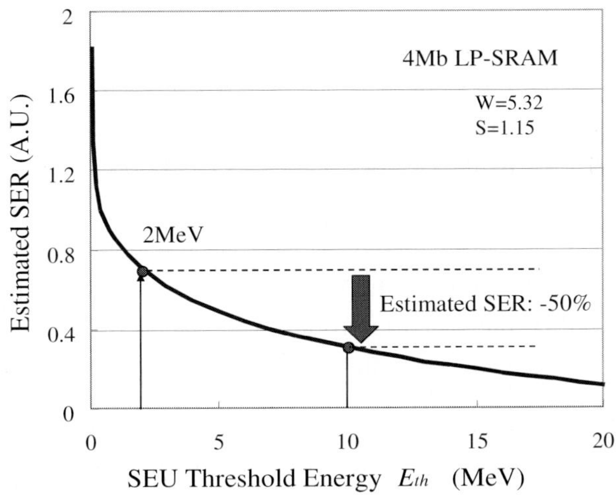

Fig. 5.10 The relative SER for 4 Mbit SRAM as a function of E_{th}, estimated by Eq. (5.30). The lower the threshold energy of SEU becomes, the greater influence it has on the estimation of SER.

Table 5.2 Experimentally obtained E_{th} for each device.

Type		Capacity (Mb)	Process (nm)	E_{th} (MeV)
SRAM	Low Power	4	180	$E_{th} \leqq 1$
		16	130	$E_{th} \leqq 1$
	High Speed	8	180	$2 < E_{th} \leqq 5$
		16	130	$2 < E_{th} \leqq 5$
	Embedded	2	150	$2 < E_{th} \leqq 5$
DRAM		64	250	$2 < E_{th} \leqq 5$
		64	220	$(E_{th} \leqq 26)$
		256	180	$5 < E_{th} \leqq 15$
		256	150	$(E_{th} \leqq 94)$
		256	130	$(E_{th} \leqq 56)$

According to the nuclear data base JENDL 3.3 [5.23] (**Fig. 5.11**), it is impossible to explain the reason why $E_{th} \leqq 1$ MeV for low-power consumption SRAMs by Si(n,α) reactions. Even if one includes ^{29}Si (natural abundance of 4.67 atomic %), the threshold energy of ^{29}Si(n,α) reaction is above 1 MeV.

The device's structure and the materials used in device were carefully scrutinized in order to explain such a low E_{th}. A model device was created using Cu as wire material, SiO_2 as the interlayer, WSi_2 as the contact plug which connects the diffusion layer with the wire, and Si_3N_4 as the side wall of the gate electrode on the Si substrate. This structure imitates the low power consumption SRAMs only to examine the effect of heavy elements such as W and Cu, and of light elements such as O and N on SEU. Simulation was performed by using SEALER (Single Event Adverse and Local Effects Reliever) [5.24], the newest version of CORIMS expanded to multiple elements, including isotopes. In Chapter 6 it is shown that the contribution of a light element such as oxygen to SEUs is extremely large.

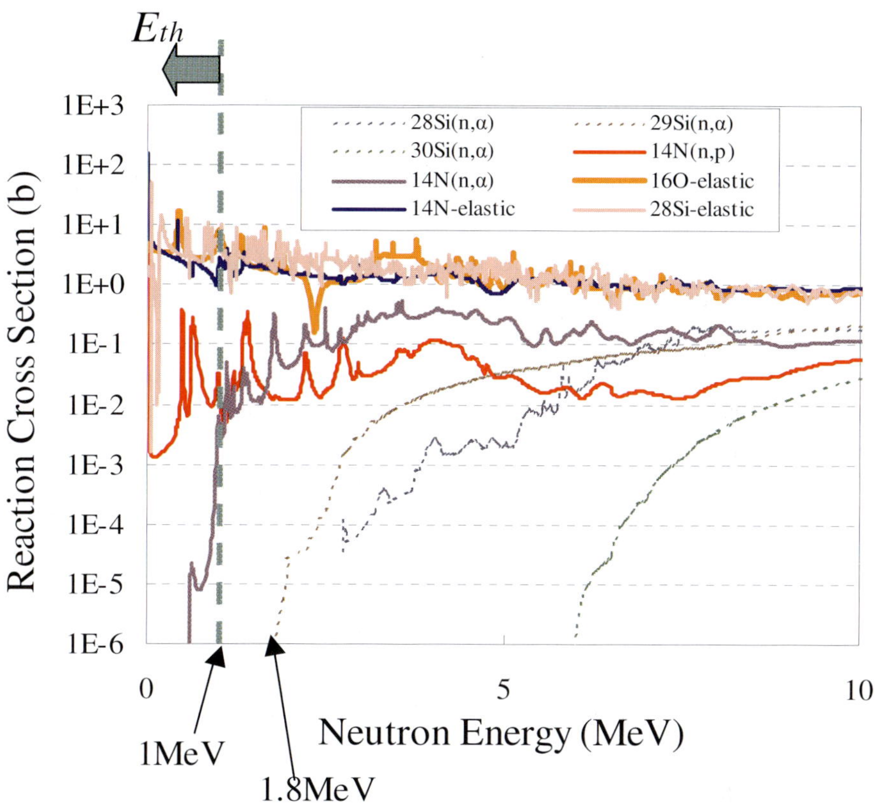

Fig. 5.11 Nuclear reaction cross sections of the light elements used in semiconductor devices [5.23]. SEU induceed by 1 MeV neutron can be attributed to α by $^{14}N(n,\alpha)$ reactions, or by recoil-nuclei of elastic scattering.

Regarding the threshold energy for SEU, only ^{28}Si has been considered. The simulated results in Chapter 6, however, reveal that other nuclides, in particular light elements, dominantly contribute to the SEU caused by low energy neutron. Recoil nuclei of light elements produced by elastic scattering can also bring about SEUs in the low energy region. This means that the reduction of critical charge with further miniaturization of the devices leads to a lower threshold energy for SEU. From that view point, introducing new materials composed of light elements may help to raise the SEU susceptibility; for example, using carbon nano-tubes as wire material [5.25] or in the channel region [5.26].

5.2 Application of SEU Excitation Functions

5.2.1 *Spallation neutron irradiation tests and the unfolding method*

In the previous section, the SEU testing method and SER (soft-error rate) estimation method using monoenergetic and quasi-monoenergetic neutron sources is described in detail. The neutron energy dependence of SEU cross sections is determined using the testing method, and it is expressed by a specific function, the SEU function. Integration over energy of the SEU function of a specific device gives the SER of the device, as does testing using the terrestrial neutron spectrum at a location on the ground. The SER estimation method can be validated by comparison with results from field tests.

There is a wide variation in the spectral shape and flux, as shown in **Fig. 5.12** [5.9, 5.27–5.33], if the observed location is fixed and field testing needs a long testing period, and the tests are restricted in number. Moreover, there are quite a few papers on temporal variation of terrestrial neutron spectrum, such as the study by Nakamura *et al.* [5.32]. These reasons make it difficult to compare the SER values deduced by the estimation method with those by field tests in order to validate the estimation method using monoenergetic and quasi-monoenergetic neutron beams. The temporal variation of the neutron spectrum can be neglected by using a well-defined neutron spectrum generated by a particle accelerator. The spallation neutron source with neutron energies range up to 800 MeV at LANSCE (Los Alamos Neutron Science Center) in LANL (Los Alamos National Laboratory) is one of the useful sources for just that purpose. LANSCE is referred in JEDEC Standard/JESD89 [5.28] as the preferred facility to carry out accelerator SEU tests of semiconductor devices. The irradiation hatch on the beam line FP30L was reconstructed in 2001 as ICE House (Irradiation of Components and Electronics House) and it has been utilized for SEU testing of semiconductor devices [5.34].

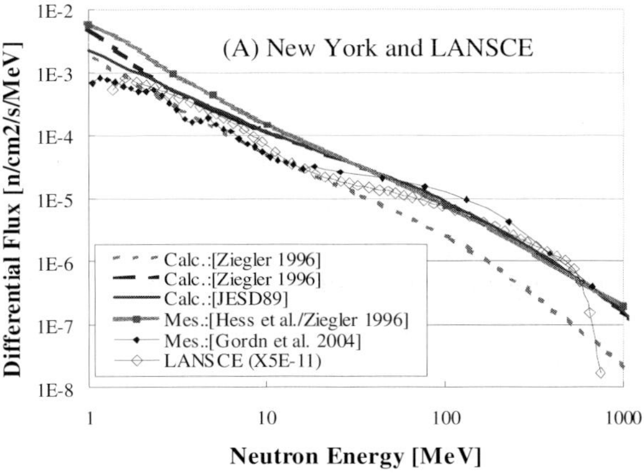

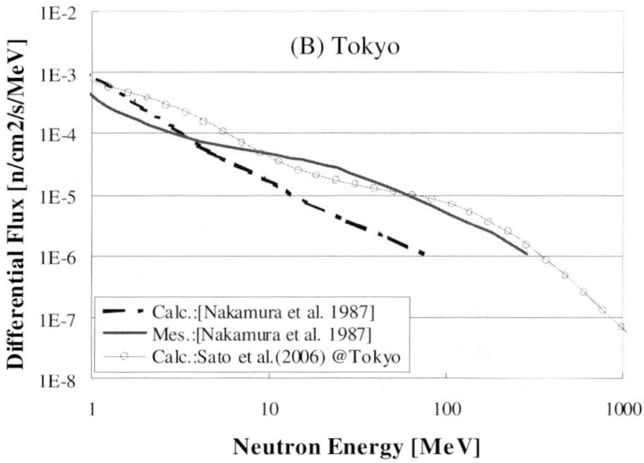

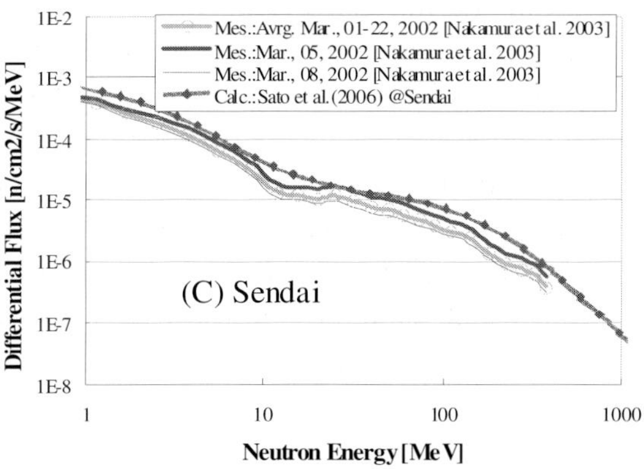

Fig. 5.12 Terrestrial neutron spectra at (A) New York [5.27–5.30] in the USA, (B) Tokyo [5.31–33] and (C) Sendai [5.32, 5.33] both in Japan. Spallation neutron spectrum at LANSCE [5.9] is shown in (A) as a reference (© 2005 IEEE).

There are two reasons why LANSCE is recommended in JEDEC Standard/JESD89 as a suitable test facility for the estimation of SER at the ground level: (a) the spallation (or white) neutron spectrum generated at LANSCE approximately matches the terrestrial neutron spectrum of New York City in JESD89, and (b) the high flux of neutrons leads to shorter SER test times. For neutrons at LANSCE with energy of more than 1 MeV, the average spectral energy is 77 MeV. On the other hand, the average energy for terrestrial neutrons in JESD89 is 46 MeV; the difference is about factor 2. Therefore, it must be noted that the SER will be overestimated if the spectrum at LANSCE is regarded as the terrestrial neutron spectrum [5.35].

5.2.2 *Experiments using the spallation neutron beams at LANSCE*

The schema of the neutron flight path FP30L at LANSCE is shown in **Fig. 5.13** [5.36]. The spallation neutron beam is generated as follows: protons accelerated by the linac are injected into a storage ring, and then a pulsed proton beam of approximately 1.5 μA with energy of 800 MeV is bombarded on a tungsten target (Target 2: 3 cm in diameter and 7.5 cm long). A fission chamber using ^{238}U is placed at the uppermost stream in the ICE House, at 19.78 m from the target, and is used to measure the neutron spectrum over a wide energy range (from 1 to 800 MeV) using a TOF (time of flight) technique. It also monitors the neutron flux during the irradiation test [5.37]. Different types of energy spectra can be obtained by simply setting various types of filter material, such as polyethylene, in the path of the neutron beam outside the ICE House (**Fig. 5.13**). **Figure 5.14** shows four LANSCE neutron spectra. From top to bottom, they were generated without polyethylene filters and with 5 cm-, 10 cm- or 20 cm-thick polyethylene filters (denoted by $\Psi^{[i]}$ (I = 1,2,3,4), respectively) [5.36]. The spectra are normalized to the flux at 100 MeV. Low energy neutrons drastically decrease with increasing thickness of the polyethylene filters. The polyethylene filters with the thicknesses mentioned above do not vary the spectral shape above 100 MeV.

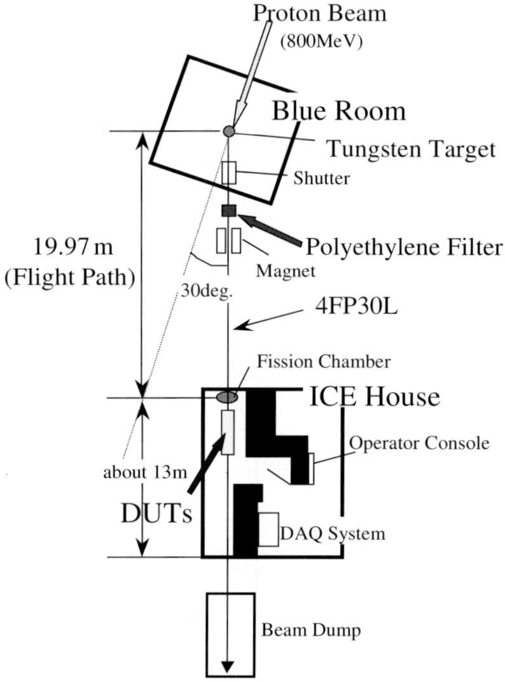

Fig. 5.13 Schematic view of the beam line FP30L of spallation neutron source at LANSCE [5.36] (© 2005 IEEE).

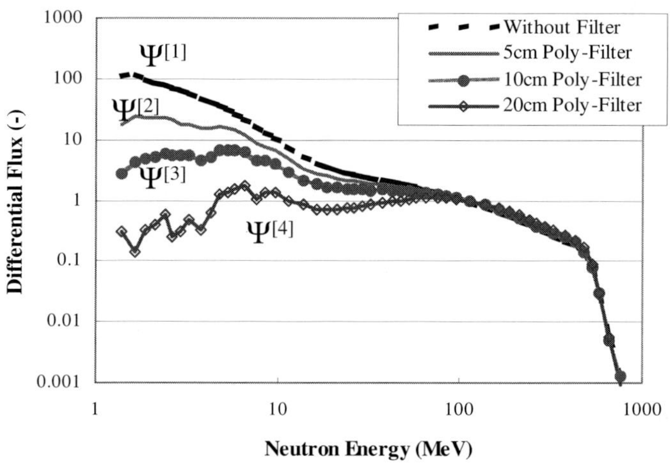

Fig. 5.14 Variation of spallation neutron spectra at LANSCE using polyethylene filters [5.36] (© 2005 IEEE).

Test circuit boards were placed 60 to 320 cm from the fission chamber in the ICE House. The diameter of the neutron beam there is about 8 cm. The irradiated part was placed in the beam on the basis of the center position fixed by a He-Ne laser. The change in the average neutron flux with changing polyethylene thickness is summarized in **Table 5.3**. Averaged neutron flux from 1.5 to 800 MeV is approximately 6.5×10^5 $(n/(cm^2 \cdot s))$ without a filter, and about 4×10^5 $(n/(cm^2 \cdot s))$, 2.5×10^5 $(n/(cm^2 \cdot s))$ and 2×10^5 $(n/(cm^2 \cdot s))$ with 5cm-, 10 cm- and 20 cm-thick polyethylene filters, respectively. **Figure 5.15** shows the filtering effect

Table 5.3 Filter thickness dependence of averaged neutron flux.

Filter Thickness (cm)	Averaged Flux $(n/(cm^2 \cdot s))$
0	6.5×10^5
5	4.0×10^5
10	2.5×10^5
20	2.0×10^5

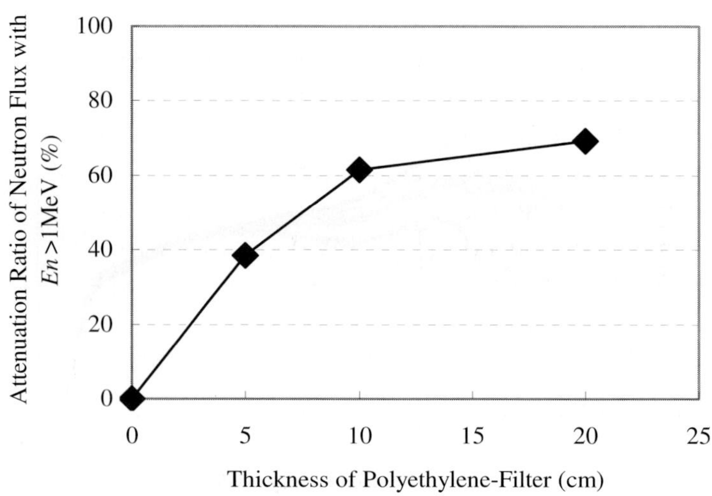

Fig. 5.15 Attenuation effect of polyethylene filter on the averaged neutron flux.

of the polyethylene blocks. It saturates above a thickness of 10 cm. In case of devices with high susceptibility to neutrons, an appropriate number of SEUs to determine that susceptibility to neutrons can be obtained within 2 hours. In order to obtain a greater number of SEUs and reduce the statistical error, the duration of the irradiation can reach up to 12 hours.

5.2.3 *Validation of the SER estimation method using monoenergetic and quasi-monoenergetic neutron beams*

The first purpose of the experiment using spallation neutron source at LANSCE is the validation of the SER estimation method using monoenergetic and quasi-monoenergetic neutron beams. For an experiment using a spallation neutron spectrum $\Psi^{[i]}$, two values can be acquired: the number of SEUs of the device, $N^{[i]}_{SEU\_obs}$ and the neutron fluence $\Phi^{[i]}_{E_{\min}=E_0}$ (n/cm^2) above E_0 (MeV). From these values, the averaged SEU cross section under a spallation neutron spectrum $\Psi^{[i]}$,

$$\sigma^{[i]} = \frac{N^{[i]}_{SEU\_obs}}{\Phi^{[i]}_{E_{\min}=E_0}} \tag{5.31}$$

can be calculated. The following is a procedure to validate the SER estimation method using monoenergetic and quasi-monoenergetic neutron beams: an SEU function, which expresses the response function of a device to neutrons, is processed to a neutron spectrum $\Psi^{[i]}$ at LANSCE, and the averaged SEU cross section is calculated. Then the calculated SEU cross section is compared with the experimental averaged SEU cross section $\sigma^{[i]}$. A Weibull-type SEU function derived from the irradiation tests with monoenergetic and quasi-monoenergetic neutron beams is expressed as

$$\sigma_{Weibull}(E_n) = \sigma_\infty\left[1-\exp\left\{-\left(\frac{E_n - E_{th}}{W}\right)^S\right\}\right], \tag{5.32}$$

where σ_∞ (cm^2) is the saturation value of SEU cross section, E_{th}(MeV) the threshold energy for SEU, and W (MeV) and S, are the Weibull parameters. The estimated SEU cross section under a spallation neutron spectrum $\Psi^{[i]}$,

$$\sigma_{SEU}^{[i]} = \frac{\int_{E_{th}}^{800MeV} \sigma_{Weibull}(E_n) \frac{d\phi^{[i]}(E_n)}{dE_n} dE_n}{\int_{E_{th}}^{800MeV} \frac{d\phi^{[i]}(E_n)}{dE_n} dE_n}, \tag{5.33}$$

where $\phi^{[i]}$ (n/(cm^2 · s)) is the neutron flux of the spallation neutron spectrum $\Psi^{[i]}$ with energy greater than E_0 (MeV), can be obtained using a folding technique with the spallation neutron spectrum $\Psi^{[i]}$. The numerator in the right hand side of Eq. (5.33) expresses the number of errors per unit of time and its denominator expresses the number of neutrons per unit time, i.e. the error rate and the neutron flux, respectively. Consequently, the validation of the SER estimation method using monoenergetic and quasi-monoenergetic neutron beams is done by comparing Eq. (5.31) with Eq. (5.33). It is enough to compare just the results for the spectrum without a filter, but the results with filters are also compared each other. At the same time, the dependence on E_{min} is also examined. **Figure 5.16** shows the results [5.36]. The results are normalized to the value of the averaged SEU cross section for 16 Mbit SRAMs irradiated by a spallation neutron spectrum behind a 20 cm polyethylene filter.

Both the estimated SEU cross section for $E_{min} = 10$ (MeV) and the experimental averaged SEU cross section of each device increase with the increase of the thickness of the filter. This trend is due to neutrons with energy less than the threshold energy of device E_{th}. As shown in **Fig. 5.14**, the neutrons in the low energy part decreases with the increase of the thickness of the filter, and the proportion of neutrons, especially those with energy less than the threshold energy of device E_{th} that does not cause SEU, decreases. This means, therefore, that in the case of devices with higher E_{th}, the neutron fluence is overestimated. The experimental results for $E_{min} = 10$ (MeV) slightly increase or are almost constant with the increase of the thickness of the filter. Since E_{th} of devices can not be directly determined by the experiment using a spallation neutron source, the value of E_{min} is a completely arbitrary selection.

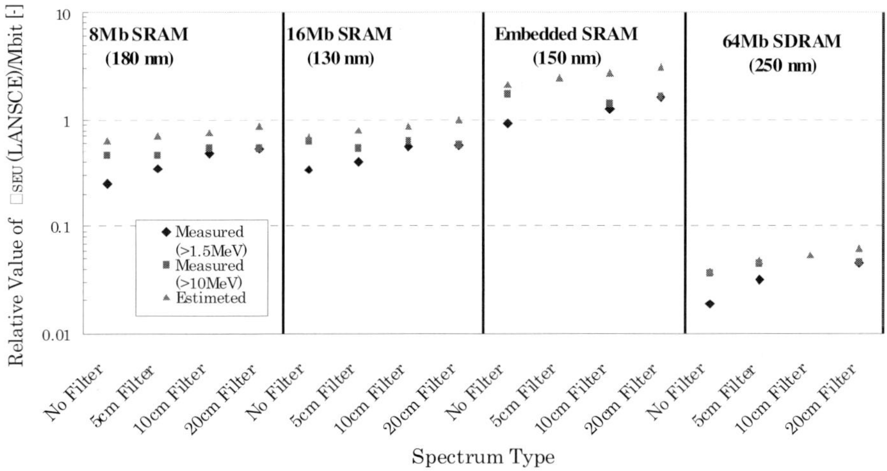

Fig. 5.16 Comparison of SEU cross sections from estimation using monoenergetic and quasi-monoenergetic neutron irradiation tests (solid triangles) and from tests at LANSCE [5.36] (© 2005 IEEE).

As shown in the Section 5.1, all the examined devices have a threshold energy $E_{th} > 1.5$ (MeV). This means that for $E_{min} = 1.5$ (MeV), the neutron fluence is over estimated, as described above. Each measured value of the SEU cross section for $E_{min} = 10$ (MeV) without a polyethylene filter agrees with the estimated value to within 20%. This result means that the SER estimation method using monoenergetic and quasi-monoenergetic neutron beams is validated.

5.2.4 *Derivation of the SEU function from spallation neutron tests*

The unfolding technique described in this section is based on the inverse of the procedure in Section 5.2.3. Here the method derives the neutron energy dependence of the SEU cross section from a set of spallation neutron tests. For an experiment using spallation neutron spectrum $\Psi^{[i]}$, the number of SEUs of device, $N_{SEU\_obs}^{[i]}$ and neutron fluence $\Phi^{[i]}_{E_{\min}=E_0}$ (n/cm^2) greater than E_0 (MeV) can be determined, as described in Section 5.2.3. The relation between them is expressed by the following equation:

$$N^{[i]}_{SEU\_obs} = \int_{E_{th}}^{800MeV} \sigma(E_n) \frac{\partial \Phi^{[i]}(E_n)}{\partial E_n} dE_n \ . \tag{5.34}$$

On the basis of the experiment using monoenergetic and quasi-monoenergetic neutron sources, the SEU function $\sigma(E_n)$ is well described as a Weibull type function $\sigma_{Weibull}(E_n)$ (see Equation (5.32)) and the $\sigma_{Weibull}(E_n)$ can be determined by solving that relation based on the results spallation neutron experiments. The mathematical procedure is called unfolding. The Weibull-type SEU function includes 4 parameters (the saturated value of SEU cross section σ_∞, the threshold energy for SEU E_{th}, a width factor W and a shape factor S), and they are determined only by the experimental results of spallation neutron irradiation tests, independently from the fitting results described in Sec. 5.1. Substituting Equation (5.32) into Equation (5.34),

$$\begin{aligned} N^{[i]}_{SEU\_obs} &= \int_{E_{th}}^{800MeV} \sigma_\infty \left[1 - \exp\left\{-\left(\frac{E_n - E_{th}}{W}\right)^S\right\}\right] \frac{\partial \Phi^{[i]}(E_n)}{\partial E_n} dE_n \\ &= \sigma_\infty \int_{E_{th}}^{800MeV} \left[1 - \exp\left\{-\left(\frac{E_n - E_{th}}{W}\right)^S\right\}\right] \frac{\partial \Phi^{[i]}(E_n)}{\partial E_n} dE_n \ . \end{aligned} \tag{5.35}$$

In order to reduce the number of parameters by one, the spectra $\Psi^{[i]}$ ($i=2,3,4$) are treated on the basis of the spectrum without filter $\Psi^{[1]}$: the ratio experimentally obtained from the observed number of SEUs

$$R^{[i]} \equiv \frac{N^{[i]}_{err\_obs}}{N^{[1]}_{err\_obs}} \tag{5.36}$$

and the function $\Gamma^{[i]}(E_{th}, W, S)$ with three parameters

$$\Gamma^{[i]}(E_{th},W,S) \equiv \frac{\int_{E_{th}}^{E_{\max}} \left(\frac{\partial \Phi^{[i]}}{\partial E_n}\right)\left[1-\exp\left\{-\left(\frac{E_n - E_{th}}{W}\right)^S\right\}\right] dE_n}{\int_{E_{th}}^{E_{\max}} \left(\frac{\partial \Phi^{[1]}}{\partial E_n}\right)\left[1-\exp\left\{-\left(\frac{E_n - E_{th}}{W}\right)^S\right\}\right] dE_n} \tag{5.37}$$

is defined. Equations (5.36) and (5.37) are derived from Equation (5.34). The three parameters E_{th}, W and S are determined so as that the sum of the square of the difference between the calculated ratio $\Gamma^{[i]}$ and the observed ratio $R^{[i]}$

$$\Sigma \equiv \sum_{i=2}^{4} \{\Gamma^{[i]}(E_{th},W,S) - R^{[i]}\}^2 \tag{5.38}$$

is minimized. The calculation proceeds as follows: the initial values of E_{th}, W and S are $E_{th}^{\min}$, W^{min} and S^{min}, the minimum values of the three parameters. The two parameters W and S are fixed to W^{min} and S^{min}, respectively, and a temporary value of E_{th}, $E^*_{th\_l}$, is increased by a small increment Δh, that is,

$$E^*_{th\_l+1} = E^*_{th\_l} + \Delta h . \tag{5.39}$$

Then E_{th} is obtained such that the residual sum of the squares Σ is smaller than a limiting value, ε :

$$\Sigma \leq \varepsilon . \tag{5.40}$$

At this time E_{th} can be temporarily fixed. The limiting value ε is then set at a lower value than before, and the same procedure is performed for W and S. Iteration of this calculation provides the optimum parameter values when the value of ε approaches zero. Then σ_∞ is calculated from Equation (5.35)

$$\sigma_\infty = \frac{N^{[i]}_{err\_obs}}{\int_{E_{th}}^{E_{\max}} \left(\frac{\partial \Phi^{[i]}}{\partial E_n}\right)\left[1-\exp\left\{-\left(\frac{E_n - E_{th}}{W}\right)^S\right\}\right] dE_n} \tag{5.41}$$

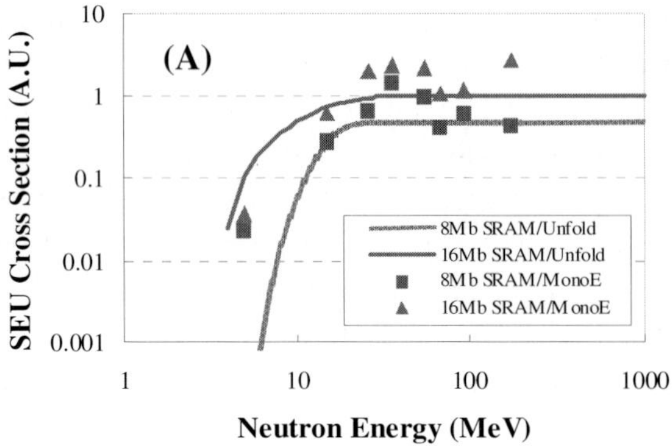

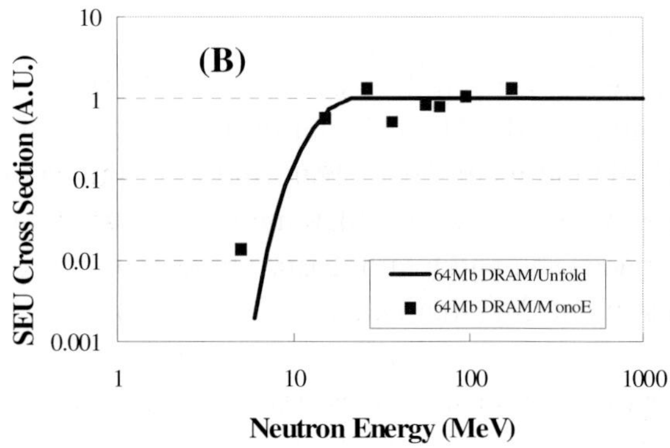

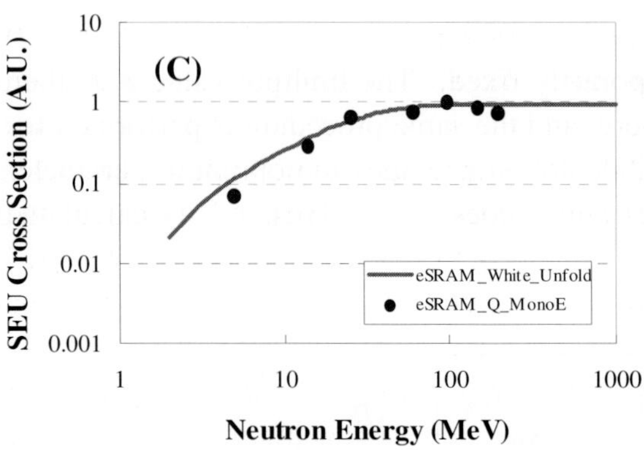

Fig. 5.17 The Weibull functions (solid or dotted lines) derived by the unfolding technique, and the SEU cross sections obtained by experiments with monoenergetic and quasi-monoenergetic neutron beams (solid symbols) [5.36] (© 2005 IEEE).

and the full expression of the Weibull-type SEU function is obtained. Thus, the unfolding technique enables us to determine the energy dependence of the SEU cross sections by just using a spallation neutron source without using monoenergetic and quasi-monoenergetic neutron sources. In **Fig. 5.17**, the acquired Weibull-type SEU function for each device is shown as a line, and the experimentally obtained SEU cross sections using monoenergetic and quasi-monoenergetic neutron sources are shown as solid symbols [5.36]. The Weibull-type SEU function acquired by the unfolding technique reproduces the experimental results using monoenergetic and quasi-monoenergetic neutron sources very well. This means that the unfolding technique is a very useful method to determine the energy dependence of the SEU cross sections of devices.

5.2.5 *A framework on our SER evaluation system – SECIS*

The SER evaluation techniques using high-energy particle accelerators are integrated as an SER evaluation system, SECIS (SElf-Consistent Integrated System for SER evaluation system), combined with field-testing and measurement of the terrestrial neutron spectrum. SECIS consists of five closely interlinked key techniques (**Fig. 5.18**): (i) field-testing of typical devices (see Sec. 3.3 and 3.5), (ii) measuring SEU cross section as a function of neutron energy (E_n) using mainly quasi-mono-energetic neutron beams along with a necessary correction using the numerical simulation package CORIMS (developed for nuclear spallation and charge collection physics in the device [5.9], see Sec. 6.2), (iii) measuring the terrestrial neutron spectrum at a specific location (see Sec. 2.4), (iv) measuring geographic coordinates and terrestrial neutron dose in the field (see Sec. 3.4), (v) a numerical simulation by CORIMS of field- and accelerator-testing of memory devices. The goal of this system is to evaluate SER of devices directly by the simulator CORIMS. The handling of the SEU cross section data acquired by accelerator-testing and the estimation method of SER on the basis of the accelerator-testing data are described in this section. It is expected that repetition of the procedure from (i) to (v) converges the evaluated value of SER obtained by SECIS with a high degree of accuracy. In order to confirm

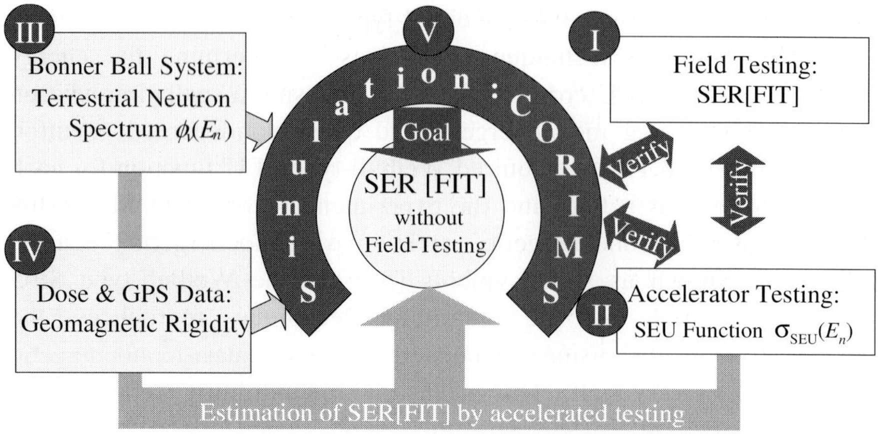

Fig. 5.18 Overview of the SER evaluation system SECIS (modified © 2002 IEEE).

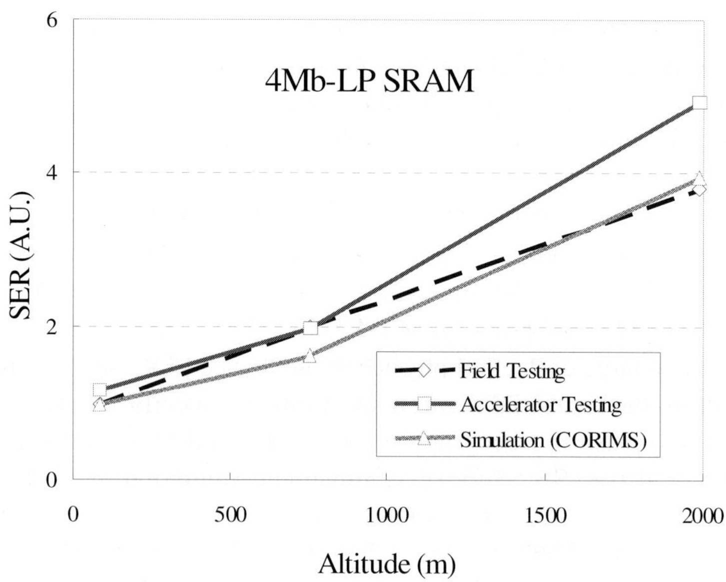

Fig. 5.19 Comparison of terrestrial neutron-induced SER results evaluated by Real-Time SER testing (field testing), accelerator testing using mono- and quasi-monoenergetic neutron beams and simulator CORIMS simulations.

the usefulness of SECIS, a comparison among SER values obtained by field-testing, accelerator-testing and simulation is carried out. **Figure 5.19** shows the series of the SER values of low-power consumption CMOS SRAM with 180 nm process technology at three different locations in Japan at altitudes of 86 m, 755 m and 1988 m. Overall, a good agreement between the measured field-testing results and the estimated or simulated SER from SECIS is obtained, as shown in **Fig. 5.19**. The accuracy of the accelerator test was < 30%, and the accuracy of the simulated results was < 20%. The results indicate that the SER evaluation system SECIS works well.

5.3 Analysis of Multi-Cell Upsets (MCUs)

In recent years, "multi-cell upset (MCU)" has been in a topic of interest [5.38–5.42]. MCU is defined as simultaneous errors in more than one memory cell that are induced by a single event. The concept of MCU, therefore, contains both the upset that can be corrected by error detection/correction code (EDC/ECC) and the upset that cannot. Though the latter case can be avoided by the interleave technique [5.39], the former is still a problematic issue in high performance devices, like content addressable memories (CAMs) [e.g. 5.43] used in network processors and routers. Since a CAM performs the search of information downloaded into the memory upon a system startup such as LUT (look-up table), the presence of MCU in memory cells has a significant impact on the search time, due to error correction. For system design, therefore, it is very important to evaluate the MCU as well as the soft-error rate (SER) of the device in advance.

5.3.1 *MCU ratio and its neutron peak-energy dependence*

The MCU ratio is defined as the ratio of the number of MCU over the total number of SEU. The neutron peak-energy dependence of MCU ratio is shown in **Fig. 5.20**, where the value is normalized to the ratio for a 130 nm device at 400 MeV. The data were obtained at FNL and CYRIC of Tohoku University, TSL of Uppsala University, and RCNP of

Osaka University. For energies above 20 MeV, where the tests were carried out using quasi-monoenergetic neutron beams, the contribution of neutrons in the low-energy tail part to MCU is included. The MCU ratio for the 130nm device steeply increases with the peak energy of the monoenergetic or the quasi-monoenergetic neutron spectrum. The MCU ratio of the 130 nm device is saturated around an energy of 180 MeV. For the 180 nm device, MCUs are produced by neutrons with an energy greater than 40 MeV. In other words, the threshold energy of MCU is between 40 and 60 MeV, where the MCU ratio is almost at a constant value. The threshold energy of MCUs for a 180 nm device is much higher than that of SEU, which lies between 2 and 5 MeV. For the 130 nm device, the threshold energy of MCU lies in the energy range between 2 and 5 MeV, which is almost the same as its SEU threshold energy [5.42, 5.44]. As in the case of the energy dependence of SEU cross sections, the neutron peak-energy dependence of MCU ratio *R(E)* is expressed by a Weibull-type function:

$$R(E) = R_\infty \left[1 - \exp\left\{ -\left(\frac{E - E_{th}}{W} \right)^S \right\} \right], \tag{5.42}$$

where E (MeV) is the neutron peak-energy, E_{th} (MeV) is the MCU threshold energy, R_∞ is the saturated value of the MCU ratio, and W and S are the fitting parameters [5.46]. The fitted results are shown as solid lines in **Fig. 5.20**. The results are normalized to the value of the MCU ratio for 180 nm at 180MeV.

Figure 5.21 depicts a typical failure bit map of MCUs. The typical error pattern for all "1" or all "0", which is called ALX or ALL0/ALL1, is the ranged one along the bit line. On the other hand, for checkerboard or checkerboard compliment which is called CBX or CHB/CHBc, the cluster like pattern is dominant. They are categorized by the specially designed program MUCEAC [5.45].

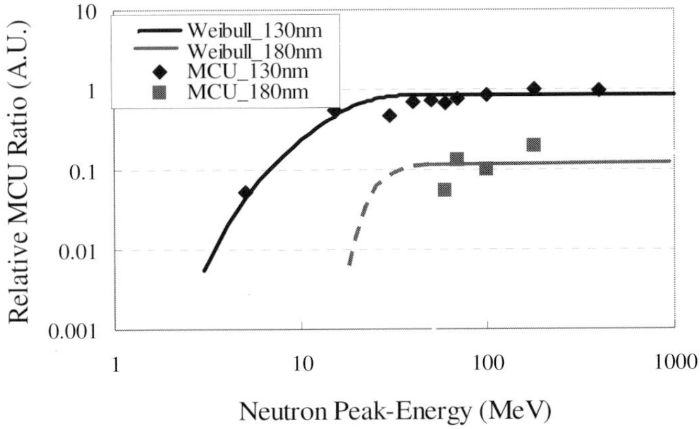

Fig. 5.20 Approximation function of the MCU ratio for an all "1" or all "0" data pattern, given by the Weibull-type function [5.46] (© 2007 IEEE).

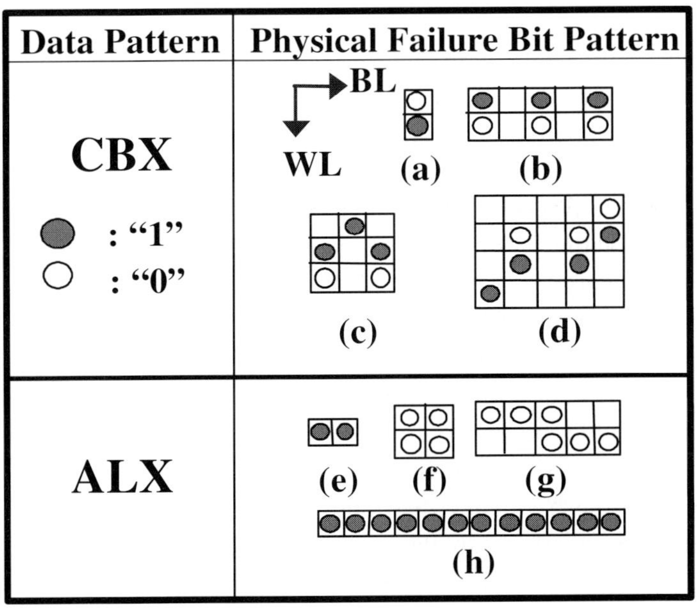

Fig. 5.21 Typical failure bit map of MCUs depending on the stored data pattern [5.45] (© 2006 IEEE).

5.3.2 *MCU ratio and frequency distribution function of MCU*

In **Fig. 5.22** [5.46] the frequency distributions of MCU with an ALX data pattern obtained by monoenergetic and quasi-monoenergetic neutron sources for the two SRAMs are shown. For the 180 nm device the ratio decreases exponentially as the multiplicity increases. For the 130 nm device, the ratio of double MCU lies between 25 and 42%, triple MCU at several % and the others at about 1%, except the frequency distribution for 5 and 15 MeV. The number of double-bit error in the 130 nm device increases with the neutron peak-energy and exceeds the single-bit error with energies greater than 100 MeV. In **Fig. 5.23** [5.45] the frequency distribution of MCU for 130 nm SRAM in another series of systematic experiments at TSL is shown. **Figure 5.24** [5.45] shows the dependency of the three MCU categories of "cluster", "MCU on word line" and "MCU on bit line" on the neutron peak energy and stored data pattern. MCUs for the CBX data pattern are categorized, in most cases, into the three types of the categories. On the other hand, the ALX data pattern gives almost only MCU on the bit line.

Figure 5.25 shows the frequency distributions of the two SRAMs using the two spallation neutron sources of LANSCE and RCNP. Despite the difference of a factor of two in the highest neutron energy, the frequency distributions of the devices with 180 nm technology match each other. For the 130 nm device, it is also the same, which suggests that neutron energy up to 400 MeV may be enough to obtain the frequency distribution of MCU; especially considering the neutron peak-energy dependence of the MCU ratio, which is saturated at an energy less than 392 MeV, as shown in **Fig. 5.20**. Unlike the 180 nm device, the number of double bit errors in a 130 nm device is comparable in frequency to that of single bit errors.

In **Fig. 5.26** the frequency distribution result of MCUs with the CBX data pattern obtained by spallation neutron tests is summarized. The function of the frequency distribution, or the probability function, for SRAM with 180 nm technology can be represented by an exponential function (5.43) or simply by the geometric function (5.44):

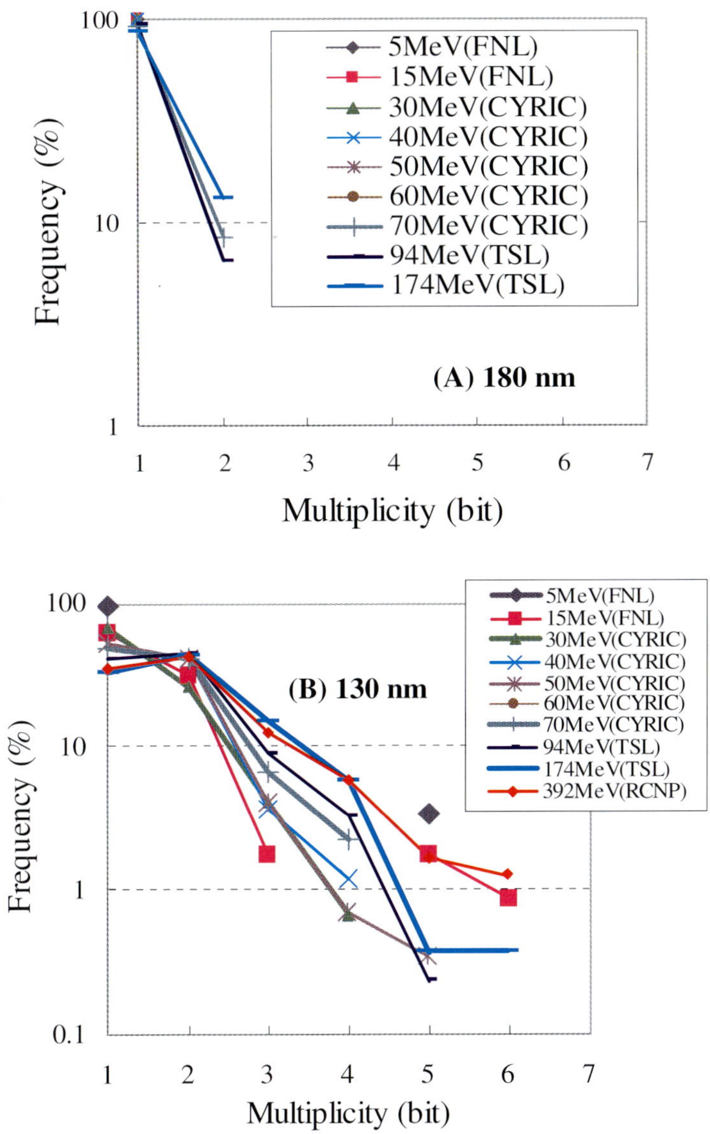

Fig. 5.22 Neutron peak-energy dependence of frequency distributions of MCU for (A) 180 and (B) 130 nm SRAMs [5.46] (© 2007 IEEE).

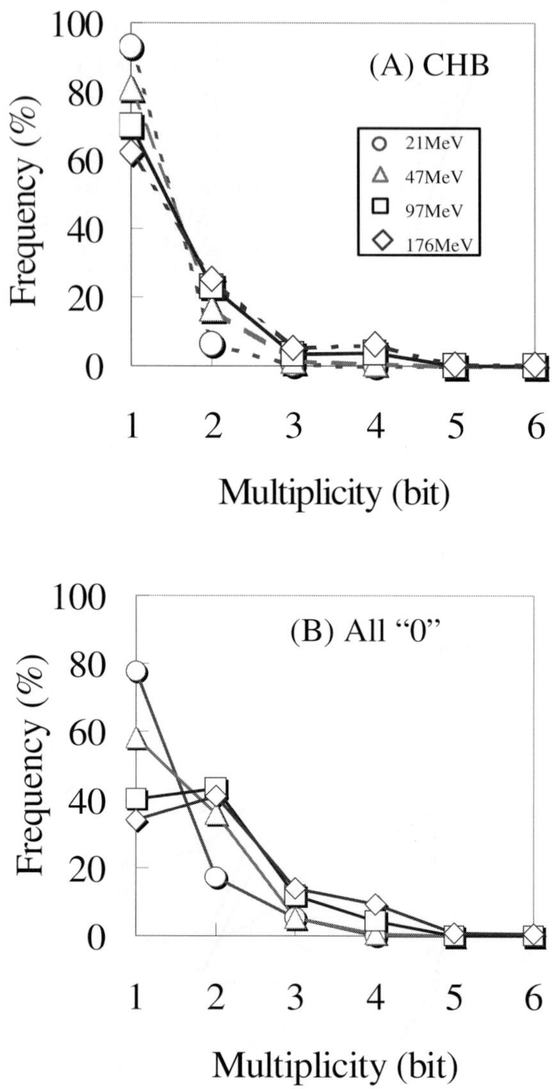

Fig. 5.23 Dependency of the MCU distribution on neutron peak-energy and data pattern for 130 nm SRAM obtained at TSL: (A) checkerboard pattern and (B) all "0" pattern [5.45] (© 2006 IEEE).

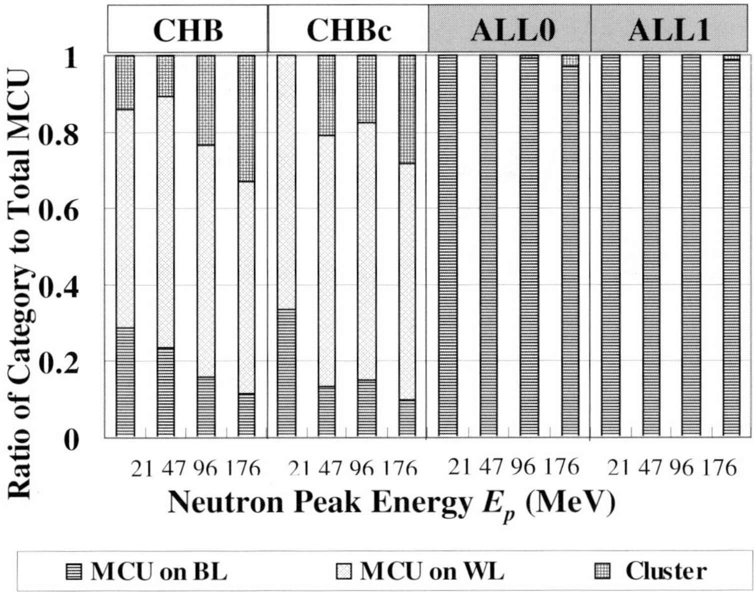

Fig. 5.24 Dependency of the three MCU categories on neutron peak-energy and data pattern for 130 nm SRAM obtained at TSL [5.45] (© 2006 IEEE).

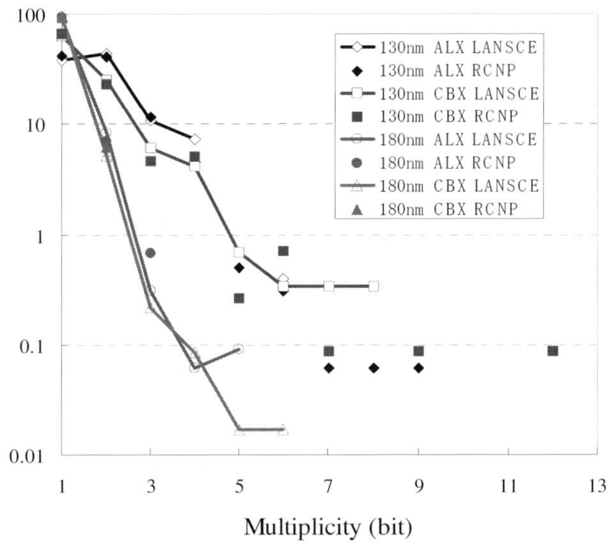

Fig. 5.25 Frequency distributions of MCU and their dependency on data pattern for 130 and 180 nm SRAMs using two spallation neutron sources [5.46] (© 2007 IEEE).

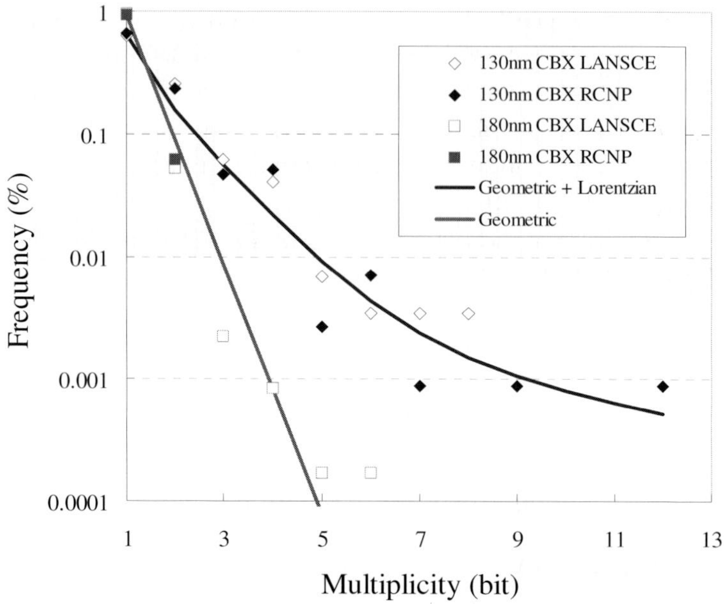

Fig. 5.26 Fitting results of the frequency distributions of MCU [5.46] (© 2007 IEEE).

$$p(n) = a \exp\{b(1-n)\} \tag{5.43}$$

$$p(n) = a(1-a)^{n-1} \tag{5.44}$$

where a is the frequency of single bit error, b is a fitting parameter and n is the number of multiplicity. Multiplicity $n=1$ means a single bit upset. On the other hand, the frequency distribution function for the 130 nm device needs a function with a tail over the higher multiplicity, such as the Lorentz distribution:

$$p(n) = c \times a(1-a)^{n-1} + d \times \frac{\Gamma/2}{(n-m)^2 + (\Gamma/2)^2} \tag{5.45}$$

where a is the frequency of single bit error, m is the location parameter, Γ is the scale parameter, c and d are fitting parameters and n is a number of multiplicity. The fitting results are shown in the figure as solid lines [5.46].

The Monte Carlo SEE simulator CORIMS was employed in order to comprehend the MCU feature in the 130 nm device. **Figure 5.27** shows the simulated results of the MCU distribution under several neutron conditions. The device model embedded in CORIMS produces only a frequency distribution, which is equivalent to that of the 180 nm device. This is probably due to the lack of bipolar models in CORIMS. The most plausible explanation for the prominent frequency distribution of MCU emerged in the 130 nm device can be provided by the mechanism of the Multi-Coupled Bipolar Interaction (MCBI) in the 3D device simulation by DAVINCI™ [5.47].

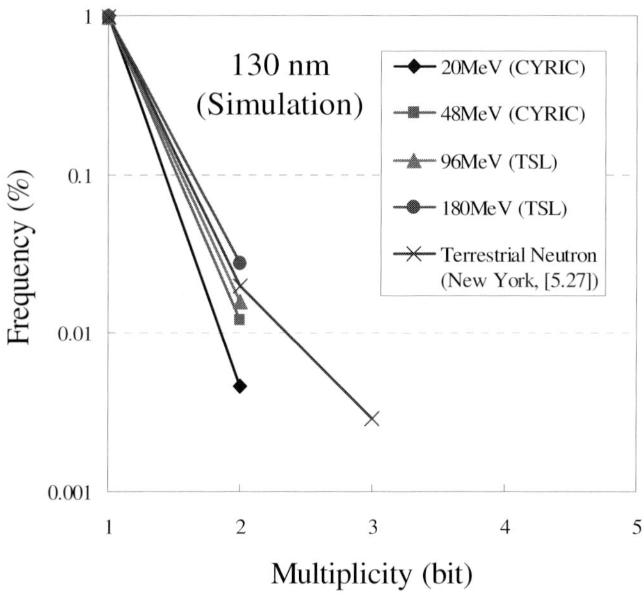

Fig. 5.27 Simulated results of the frequency distributions of MCU using CORIMS in which *no* model of bipolar action is implemented [5.46] (© 2007 IEEE).

5.4 Summary

In this Chapter, the SEE test method using neutron beams generated by high-energy particle accelerators is described. Several important results listed below result from the tests:

(1) by using accelerator-based monoenergetic and quasi-monoenergetic neutron irradiation tests, a systematic understanding of neutron-induced SEU is achieved, and the dependence of SEU cross section on neutron energy, i.e. an SEU excitation function, is well reproduced using the Weibull-type function, rather than the Bendel-type one,
(2) monoenergetic neutron tests below 15 MeV for advanced downsized devices revealed that they have a low threshold energy of a few MeV,
(3) tests using the spallation neutron beam help to validate the SER estimation method based on monoenergetic neutron tests,
(4) an SEU excitation function (Weibull-type function) of a device is also derived by using spallation neutron beams,
(5) the relation among the accelerator-based tests, the field measurements and the soft-error simulation by CORIMS on the basis of SECIS is confirmed,
(6) the dependence of MCU ratio on neutron energy is well expressed by the Weibull-type function,
(7) failure bit pattern of MCU depends strongly on the stored data pattern, and
(8) for the function type of the MCU frequency distribution in advanced CMOS SRAMs, a Lorentz component due to the effect of parasitic bipolar action can be added to an exponential or geometric function.

Chapter 6

Monte-Carlo Simulation Methods

Eishi Ibe

A Monte-Carlo simulation model, which is implemented in the program packages CORIMS/SEALER, is described over the entire physical process related to neutron soft-errors. The model can be applied to any nucleon (neutron/proton) radiation fields, including accelerator and terrestrial radiation fields.

6.1 Nuclear Reaction Model

The simulation packages CORIMS [6.1–6.4]/SEALER [6.5, 6.6] developed by Ibe are equipped with vigorous numerical algorithms for nuclear spallation reactions, ion track analysis in an infinite cell-matrix, and charge collection to the storage nodes. A simplified device layout, using 3D CAD parameters, including storage nodes, channels, wells, and isolation oxide are incorporated in the device model with an effective and systematic ion-track analyses.

The Monte-Carlo method is applied to a number of steps in the physical process, from nuclear spallation reactions, to charge collection to storage node, or MCU/MBU analyses, as partly illustrated in **Fig. 6.1**.

The soft-error rate of any MOSFET device is calculated in any given radiation environments, such as the terrestrial neutron field at any location on the earth and neutron fields at accelerator facilities (See Chapter 4). The spectra used by the code are differential neutron energy spectra (neutron flux as a function of energy), as illustrated in **Fig. 6.2**. The minimum and maximum energies of each energy bin are selected to give an equal flux over all the energy bins. The flux in each energy group may not be necessarily equal, but it is recommended to make the weight of each energy bin as constant as possible in order to minimize the error in the total calculation.

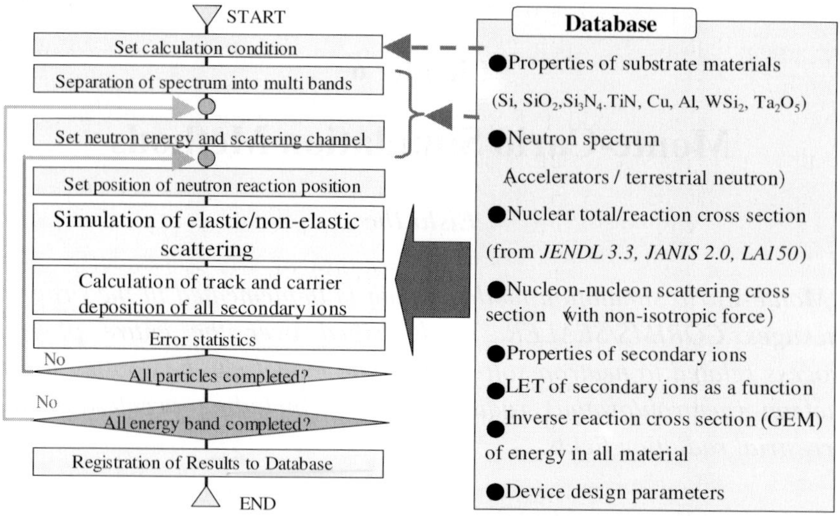

Fig. 6.1 Overall CORIMS/SEALER basic structure and procedures.

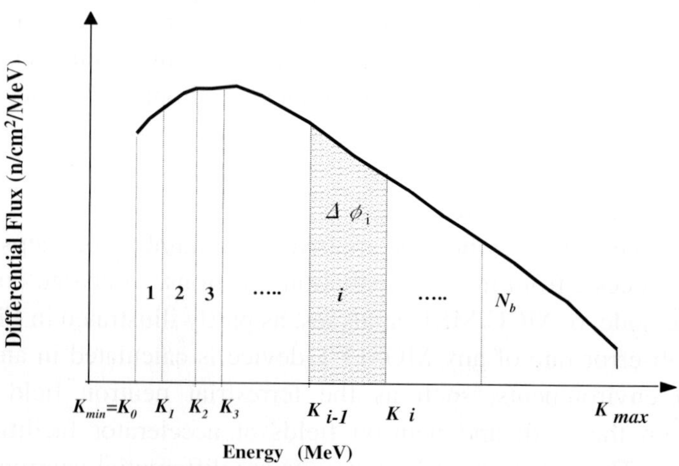

Fig. 6.2 Grouping of the energy bins for an arbitrary neutron spectrum.

Charge production from direct irradiation by ions can also be treated, for example, alpha source (^{258}U, ^{232}Th, ^{210}Po, etc) contamination is implemented although it is not the major point of this book.

The Intra-Nuclear Cascade (INC) model [6.7] is applied to the prompt collision process, where the many-body collision among nucleons (neutrons and protons) is treated numerically as a cascade of relativistic binary collisions between two nucleons in the nucleus, as shown in **Fig. 6.3**.

The evaporation model [6.8] of light particles from an excited nucleus is also applied for the delayed nuclear reaction process, where nucleons (n and p), deuterons (^{2}H or D), tritons (^{3}H or T), helium and residual nucleus are released, as illustrated in **Fig. 6.4**. The type, energy, and direction of each secondary ion are automatically determined in the Monte-Carlo nuclear reaction simulation.

In the evaporation model, the inverse reaction cross section has to be calculated between the residual nuclei and light particles. This is done by either a look-up table/interpolation [6.7] or by using the GEM model [6.9].

In the INC and evaporation models, each binary collision is analysed on the basis of a relativistic Lorentz transformation. Since the Lorentz transformation can only be applied to the direction of movement of the particle, the z-axis is first aligned to the direction of the particle (aligned-labo-system).

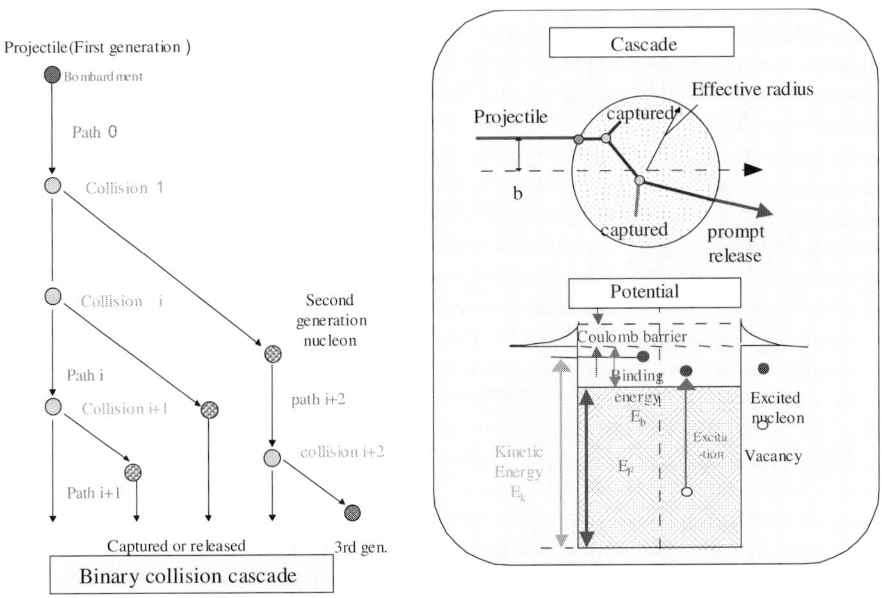

Fig. 6.3 Intra-nuclear cascade and the relevant physical model.

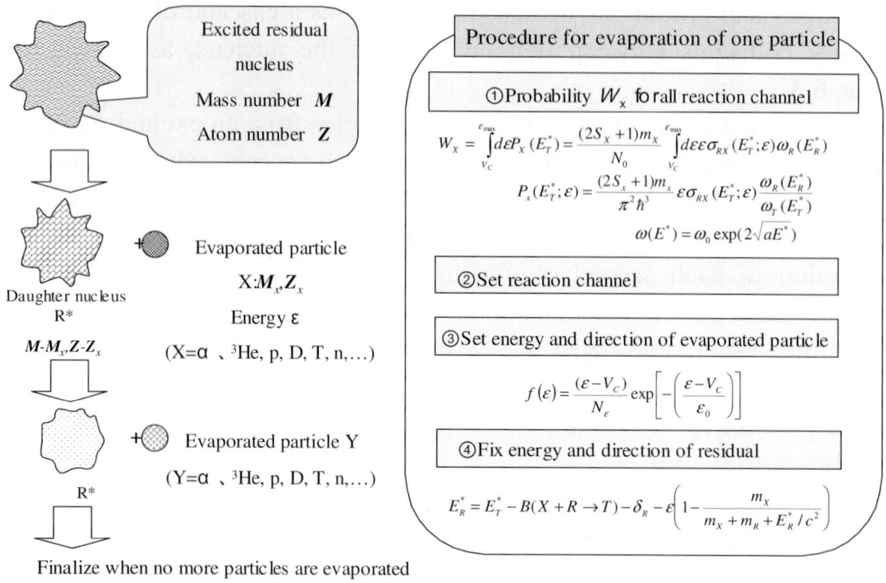

Fig. 6.4 Evaporation model of light nuclei.

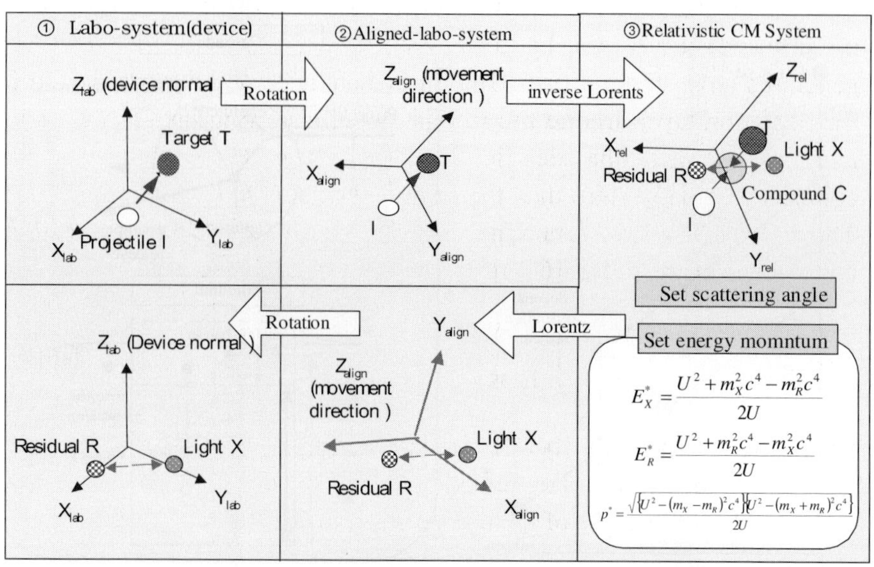

Fig. 6.5 Transformation of labo-CM-labo system for relativistic binary collisions.

Next, the inverse Lorentz transformation is applied to determine the relativistic CM system, and then, the direction and energies of scattered particles are determined. The direction of the scattered particles in the labo-system are calculated after Lorentz transformation is applied to the relativistic CM system. The overview of the procedures is shown in **Fig. 6.5**.

6.2 The Device Model

6.2.1 *The single bit model and basic charge collection mechanism*

The model layout of a MOSFET SRAM cell is illustrated in **Fig. 6.6**. Since the active regions are isolated by an oxide in the lateral direction, and since the wells line up across the word lines, charge collection in the lateral direction is tightly limited in this device. Bits in a word are aligned along the word line. As such, MCUs along that word line are critically important in this device from a viewpoint of MBU.

The charge collection mechanisms (funneling and drift-diffusion mechanisms) are implemented into CORIMS/SEALER. When the n^+ or p^+ storage node is hit by a secondary ion, electron-hole pairs are generated along the ion path. They move along with the electric field in the depletion layer around the storage node, distorting the electric field, and thus elongate the electric field. As a result, the amount of the collected charge is more than the charge generated in the depletion layer. This effect is called "funneling" and its length L_f in which the charge collected is expressed as [6.10]:

$$L_f = \left(1 + \frac{\mu_n}{\mu_e}\right)\frac{W}{|\cos\theta|} \tag{6.1}$$

where, μ_n, μ_e : the mobility of hole and electron, respectively;
W : the thickness of depletion layer;
θ : is the angle of the ion track to normal.

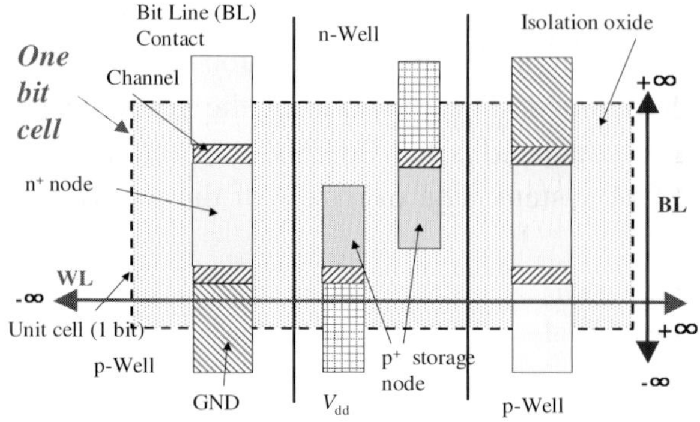

Fig. 6.6 Layout of a SRAM single bit model [6.3].

The secondary ion does not necessarily pass through the depletion layer in the drift-diffusion mechanism. The hole and electrons drifts or diffuse into the storage node. In CORIMS/SEALER, this effect is implemented simply by a drift-diffusion "box" beneath the storage node.

6.2.2 *The MCU model*

As described as in Chapter 5, features of MCUs depend significantly on the data patterns CHB, CHBc, All0 and All1. The layout of bits in a word is of primary importance in designing the ECC and interleaving parameters. In the cell matrix model of CORIMS/SEALER, these parameters can be implemented and their sensitivity can be surveyed. As summarized in **Fig. 6.7**, there are five possible MCU mechanisms: (A) A successive hit by one secondary ion; (B) Multi-hits by multi-ions; (C) Charge sharing by adjacent nodes; (D) parasitic bipolar action between a S/D channel; and (E) Fake MCU caused by failure in the peripheral circuits (e.g. decoder/encoder). The bipolar mode including (D) and the peripheral circuit mode (E) in **Fig. 6.7** are very crucial modes, but they are not treated in this book because of their intrinsic difficulty in simple modelling. Together with 3D multi-cell device simulations, the in-depth study will be introduced in another publication in the near future.

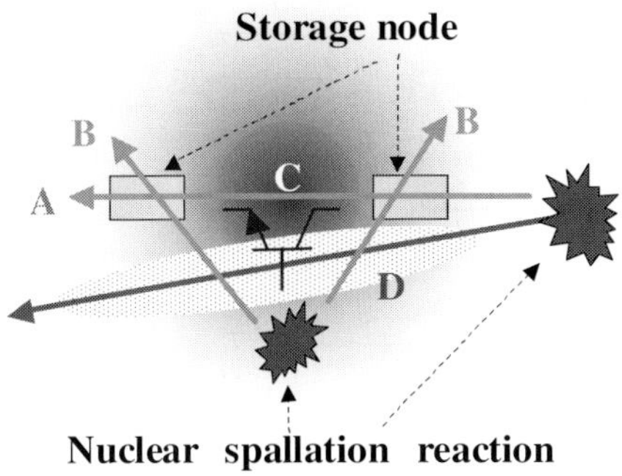

Symbol	Name
A	Successive hits by one ion
B	Multi hits by multi-ions
C	Diffusion/drift (charge share)
D	Bi-polar action
E	Peripheral circuit (fake)

Fig. 6.7 Mechanisms of MCU [6.3].

6.2.3 *Dynamic cell-shift method to simulate an infinite cell matrix*

Naturally, a model with a number of fixed cells may be applied to investigate MCU effects. Such a method, however, has inherent limitations on the memory and speed of the simulations. We have developed a dynamic cell-shift method to overcome such limitations.

Figure 6.8 illustrates the basic idea of the dynamic cell-shift method. When an ion crosses a memory cell matrix along the line A-B-C-D, the track of the ion may be traced as long as the ion has a possibility to hit sensitive volumes. This method requires a cell-matrix that is wide enough when compared to the ion's range. The proposed method does not need an actual cell matrix. Instead, it utilizes just one physical cell. When an ion reaches the boundary at B, the track is virtually shifted to B'-C' by using an one-bit-shift in the Y-coordinates. The proposed method does not trace the track C-D but, instead, traces the track C"-D", using an increment or decrement in the X and/or Y addresses. In this way, any ion track can be traced until it stops, regardless of the length of the ion's range.

6.2.4 *The method to implement data pattern into cell matrix*

Any cyclic data pattern in a rectangular zone can be implemented in CORIMS/SEALER. The basic idea is illustrated in **Fig. 6.9**. Once the data "1" and "0" pattern is set in a unit rectangular zone with a certain

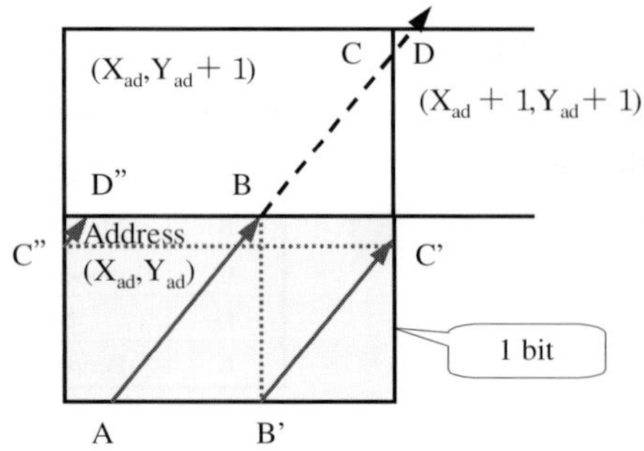

Fig. 6.8 Dynamic cell-shift method to quantify MCU features.

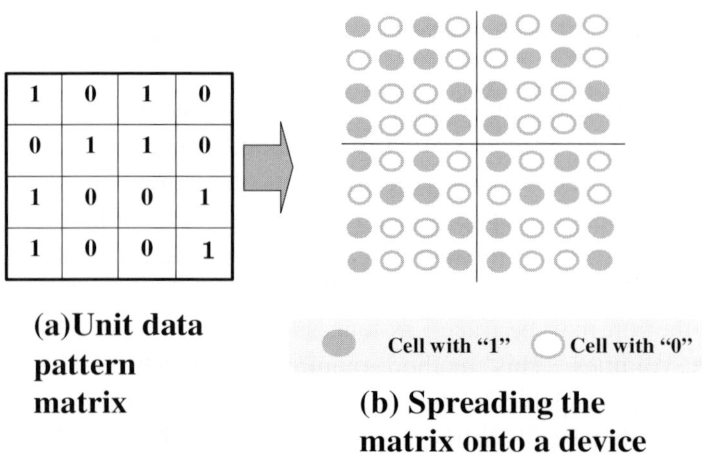

Fig. 6.9 Method to spread a data pattern onto a cell array [6.3].

number of memory bits, the unit is "painted" infinitely in the WL and BL directions.

6.2.5 *The method to implement bit patterns in a word*

Bits in a word are naturally aligned in the WL direction since they are accessed by a "high" condition of the WL. Corresponding bits in a word are selected among the BLs. Again, the bits in a word naturally align in one single WL. If necessary, however, other layouts like "zig-zag" or "tilted" may be applied, as illustrated in **Fig. 6.10** (although the algorithm in the decoder would be very sophisticated in order to enable such layouts). Such "unrealistic" layouts can also be implemented in the simulator by giving intervals along WL and BL with positive (tilted) or negative (zig-zag) signs.

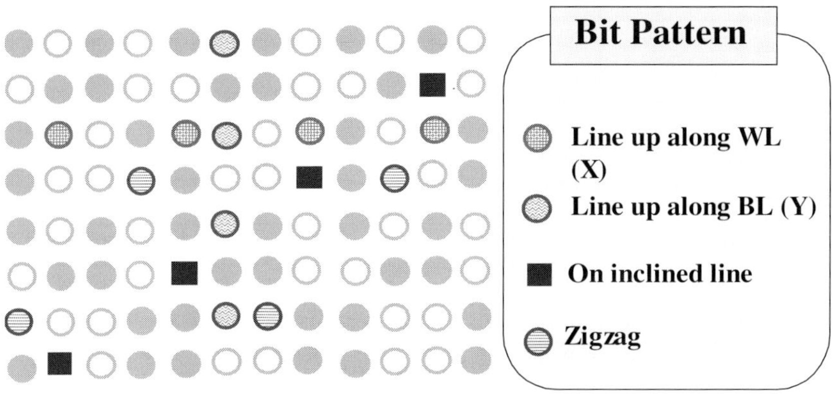

Fig. 6.10 Definition of bit patterns in the same word [6.3].

6.3 Numerical Data for SER Simulations

The properties of each element and materilas related to nuclear reactions and LET are required for reliable SEU/MCU simulations. While most of them are known with sufficient accuracy, some important data are still missing. In this section, the reliable data are tabulated, and the methods used to approximate the specific values are presented. It is noted that even if the original data is accurate, an approximations may be used in the Monte-Carlo simulations to save CPU time.

6.3.1 *Materials in silicon semiconductors*

Table 6.1 summarizes typical 10 materials currently used in silicon semiconductors with their relevant specific parameters, such as moleculer weight, density, W-value (energy necessary to create one electron-hole pair), calculated total scattering cross section, and the calculated reaction rate with terrestrial neutron at sea level in Tokyo.

Table 6.1 List of materials in silicon semiconductors and relevant features.

Symbol	Major component	Molecular weight	Density ($g\ cm^{-3}$)	W-value (eV/pair)	σ at 100MeV[1]	Reaction rate in Tokyo (molecules × $h^{-1}cm^{-3}$)
Al	Inter-connect	26.98	2.7	-	0.44	0.097
Si	Substrate	28.09	2.33	3.6	0.44	2.3
Cu	Inter-connect	63.55	8.96	-	0.86	0.16
W	contact	183.84	19.3	-	1.84	0.52
SiO_2	Interlayer oxide	60.09	2.33	-	1.05	0.085
Si_3N_4	Buffer	140.28	3.17	-	4.17	0.18
TiN	Electrode	61.87	5.21	-	1.91	0.42
WSi_2	Contact	240.01	9.3	-	2.72	0.94
Al_2O_3	Gate oxide	101.96	3.97	-	1.79	0.19
Ta_2O_5	Capacitor oxide	441.89	8.2	-	6.57	1.05

[1]Total cross section at 100 MeV(barn/molecule)

6.3.2 *Elements in silicon semiconductors*

Table 6.2 summarizes the properties of the elements in the materials listed in **Table 6.1**. Charcteristic values relevant to nuclear physics are shown, such as the Fermi energies for neutrons and protons, binding energy, coulomb barrier, nucleus radius, total scattering cross section for 100MeV neutrons, and the reaction rate at sea level in Tokyo.

6.3.3 *Total cross section*

Total cross sections for all materials used in the devices are necessary to determine which material reacts with the incident high energy neutrons.

The total scattering cross section of a device Σ^{total} is the sum of the total cross sections of all of the materials the device is made of. Namely,

$$\Sigma^{total} \doteq N\Sigma P_i^{total} = \Sigma(N_i\Sigma_i^{total}) \tag{6.2}$$

Table 6.2 List of elements in silicon semiconductor and their relevant features.

Symbol	Z	M	Natural abundance (%)	Average Mass	Fermi energy (MeV)		Average binding energy (MeV)	Coulomb barrier (MeV)	Static mass (MeV)	Radius (fm)	Spin	Total cross section (barn)[1]	Reaction rate in Tokyo[2]
					n	p							
N	7	14	99.53	14.01	33.4	33.4	7.476	1.84	13040.3	2.892	1	0.71	5.18e-28
		15	0.37				7.7		13969	2.959	0.5		
O	8	16	99.76	16.00	33.4	33.4	7.977	2.05	14895.2	3.024	0	0.3	2.3e-28
		18	0.2				7.708		16762.1	3.145	0		
Al	13	27	100	26.98	34.2	32.6	8.332	3.02	25126.6	3.6	2.5	0.44	7.5e-28
Si	14	28	92.22	28.09	33.4	33.4	8.449	3.23	26053.3	3.644	0	0.44	5.4e-28
		29	4.69				8.449		26984.4	3.687	0.5		
		30	3.09				8.521		27913.4	3.729	0		
Ti	22	46	8.25	47.87	35.2	31.6	8.657	4.56	42793.6	4.3	0	1.2	1.8e-27
		47	7.44				8.662		43724.3	4.331	2.5		
		48	73.72				8.724		44652.2	4.361	0		
		49	5.41				8.712		45583.7	4.391	3.5		
		50	5.18				8.756		46512.3	4.421	0		
Cu	29	63	69.17	63.55	35.4	31.3	8.753	5.64	58604.1	4.775	1.5	0.86	1.7e-27
		65	30.83				8.758		60465.4	4.825	1.5		
Ta	73	181	100	180.9	37.6	29.0	8.024	11.2	168615.7	6.788	3.5	2.5	1.3e-26
W	74	180	0.12	183.8	28.9	34.1	8.026	11.3	167582.5	6.775	0	1.8	1.0e-26
		182	26.5				8.019		169446.9	6.8	0		
		183	14.31				8.009		170380.2	6.813	0.5		
		184	30.64				8.006		171312.4	6.825	0		
		186	28.43				7.989		173178.6	6.85	0		

1 For most abundant isotope at 100 MeV), 2 For most abundant isotope (nuclei(/s)

where, N: Total number of molecules in a device;

N_i: Total number of molecules of i-th material in a device;

Σ_i^{total}: Total cross section of the i-th material.

The material that undergo an interaction with a high energy neutron is selected by a common Monte-Carlo method based on the reaction probility P_i.

The total scattering cross section $\Sigma^{\text{total}}(\mathbf{A}_x\mathbf{B}_y)$ of a specific material $\mathbf{A}_x\mathbf{B}_y$ is expressed as:

$$\Sigma^{\text{total}}(\mathbf{A}_x\mathbf{B}_y)=x\sigma_A^{\text{total}}+y\sigma_B^{\text{total}} \tag{6.3}$$

where, σ_A^{total}, σ_B^{total} are the total scattering cross sections of the elements A and B at neutron energy E_n, respectively. The element that interacts with a high energy neutron is determined based on Eq. (6.3).

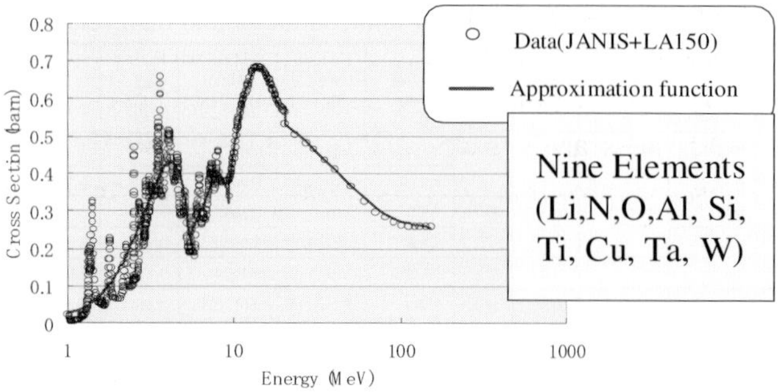

$$\sigma_{nonela} = \begin{cases} -0.0202\times E_n^3 + 0.15144\times E_n^2 - 0.1999\times E_n + 0.095 & (0 \le E_n < 5.53) \\ -0.0092\times E_n^3 + 0.1799\times E_n^2 - 1.0975\times E_n + 2.369 & (5.53 \le E_n < 9.352) \\ 0.001133\times E_n^3 - 0.0572\times E_n^2 + 0.9343\times E_n - 4.2959 & (9.352 \le E_n < 20) \\ -2.13\times 10^{-7}\times E_n^3 - 8.0\times 10^{-5}\times E_n^2 + 0.0102\times E_n + 0.7052 & (20 \le E_n < 150) \\ 0.256 & (150 \le E_n) \end{cases}$$

Fig. 6.11 Total cross section [6.11, 6.12] and approximation function for ^{14}N.

The total scattering cross section is a function of neutron energy so that the approximation formulae

$$\sigma^{total} = c_0 + c_1 E_n + c_2 E_n^2 + c_3 E_n^3 + \cdots \tag{6.4}$$

is applied. The coefficients c_0, c_1, c_2, c_3,,, are given in Appendix A.2 for the specified energy regions. The order in the fomula is adjusted according to the complexity of the energy dependency. The original data are obtained from JANIS [6.11] and LA150 [6.12]. A typical example of the original total scattering cross section and the approximated results are shown for ^{14}N in **Fig. 6.11**.

6.3.4 *Non-elastic reaction cross section*

The element that interacts with the incoming neutron is determined by the procedure mentioned in section 6.3.1. The next step is to determine the reaction channel, namely non-elastic or elastic. Since

$$\sigma^{total} = \sigma^{ela} + \sigma^{nonela}, \tag{6.5}$$

the channel to be used can be determined because the non-elastic reaction cross-section can be obtained in a manner similar to Eq. (6.4). An example of the original non-elastic reaction cross section and the approximated results are shown in **Fig. 6.12** for ^{14}N [6.6]. Coefficients and the original/approximated non-elastic reaction cross sections are summarized in A.3.

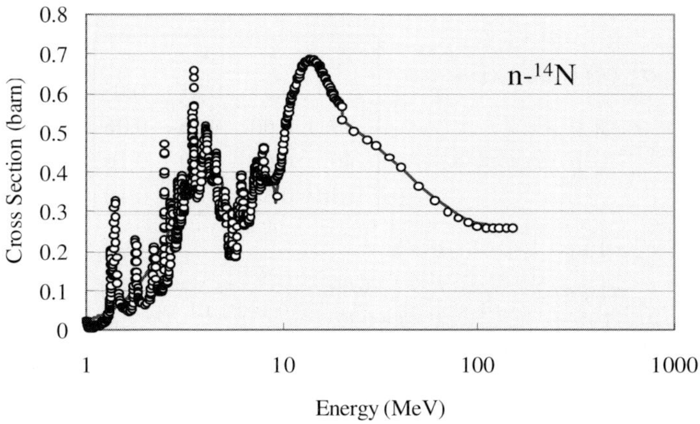

Fig. 6.12 Example of the original data [6.11, 6.12] of the non-elastic reaction cross section (circles) and approximation (solid line) for ^{14}N [6.6].

6.3.5 *Inverse binary reaction cross section*

In order to determine the evaporation reaction cross section from an excited nucleus R, the binary reaction cross section corresponds to the reaction in the laboratory system given by:

$$\mathbf{T} + \mathbf{P} \Rightarrow \mathbf{R} \tag{6.6}$$

where, **T**: target nucleus of the binary system;

P: projectile nucleus of the binary system.

In the real physical situation, **P** is equivalent to a light nucleus that has evaporated from the residual excited nucleus **R**.

In CORIMS, where the only substarate material is Si, the number of evaporation channels is not so large (on the order of 100), so the look-up table method [6.7] was appllied. In SEALER, where all of the materials used in the silicon semiconductor are implemented, the look-up method cannot be applied due to sparseness of the data base. Instead of using the look-up method, so the GEM(Generalized Evaporation Model) [6.9] was incorporated into SEALER. The inverse reaction cross section can be calculated with a relatvely small number of parameters, as shown in **Fig. 6.13**.

$$\sigma_{inv}(\varepsilon) = \pi R_b^2 c_j \left(1 - \frac{k_j V}{\varepsilon}\right)$$

$$V = \frac{Z_j Z_d e^2}{R_c}$$

Z_d	k	k_α	c
≦20	0.51	0.81	0.0
30	0.60	0.85	-0.06
40	0.66	0.89	-0.10
≧50	0.68	0.93	-0.10

$$c_p = 1 + c \qquad k_p = k$$

$$c_d = 1 + c/2 \qquad k_d = k + 0.06$$

$$c_t = 1 + c/3 \qquad k_t = k + 0.12$$

$$c_{^3He} = c_\alpha = 0 \qquad k_{^3He} = k_\alpha - 0.06$$

$$R_b = \begin{cases} 1.5 A_d^{1/3} & \text{(for Proton)} \\ 1.5\left(A_d^{1/3} + A_j^{1/3}\right) \end{cases}$$

$$R_c = \begin{cases} 1.7 A_d^{1/3} & \text{(for Proton)} \\ 1.7 A_d^{1/3} + 1.2 \end{cases}$$

Fig. 6.13 Calculation of the inverse reaction cross section based on GEM [6.9].

A typical example of the calculated inverse reaction cross section and data from the literature [6.13] are shown in **Fig. 6.14** for the target atom Si.

Other examples are shown in A.4. Since the data from the literature are still poor, it is difficult to evaluate the accuracy of the GEM results, but the trends and the values seem to be roughly consistent.

6.3.6 *LET calculation for a composite material*

The element with the maximum atomic number is 78 (Tungsten) and the number of relevant materials is 8. Therefore, the maximum number of combinations of secondary ions and substrates is 8 × 78 = 624. The LET for each combination can be calculated with SRIM [6.14], and the results are summarized in polynomial form,

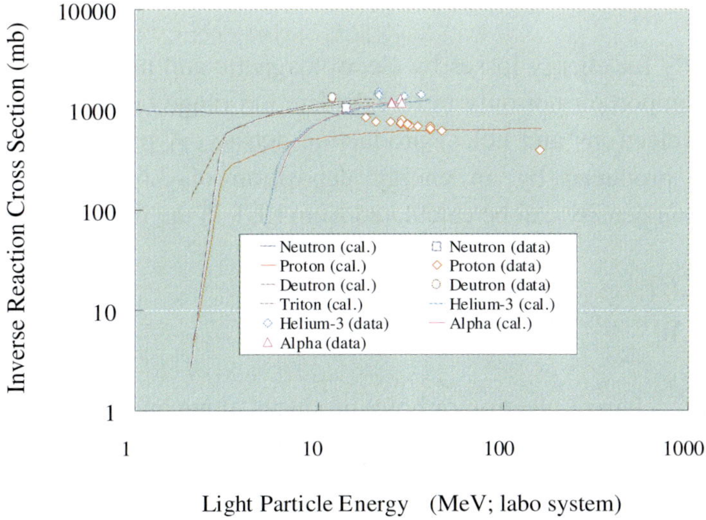

Fig. 6.14 Comparison between the calculated inverse reaction cross section and literature data [6.6].

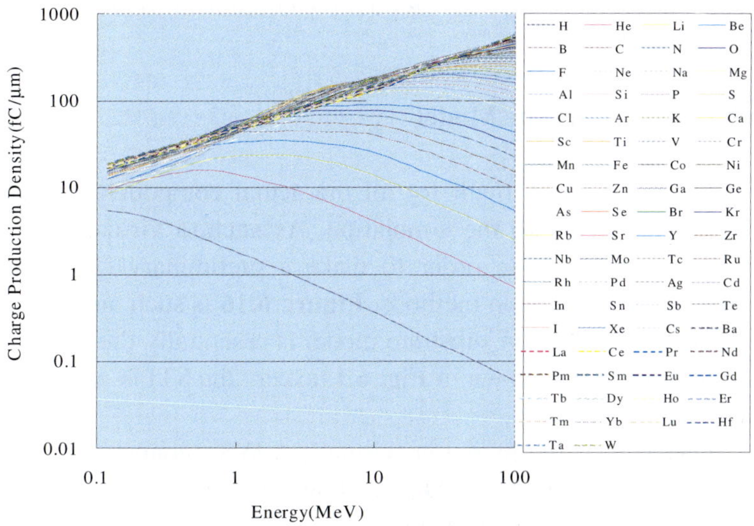

Fig. 6.15 Calculated charge deposition density in Si substrate for various ions.

$$\text{LET}=a_0 + a_1\text{Log}(E_n) + a_2\{\text{Log}(E_n)\}^2 + a_3\{\text{Log}(E_n)\}^3 + \cdots \quad (6.7)$$

separately for energy losses by electromagnetic and nuclear interactions. LET is important not only to to calculate ion range but also to calculate charge (electrons and holes) production density. A pair of electron and hole is produced by an energy deposition of 3.6 eV. The charge deposition density can be calculated from LET in the following manner:

$$\frac{dq}{dx} = \frac{LET_e}{W} \times \rho, \cdots \quad (6.8)$$

where, q : charge(electron or hole) produced along an ion path;
x: length of an ion track along with the ion path;
LET_e: linear energy transfer due to electronic loss;
W: energy required for production of a pair of hole and electron;
ρ : density of substrate.

Calculated results for the charge deposition density of various ions in Si substrate are shown in **Fig. 6.15**.

Calculated results of LET for all possible combinations are shown in A.5.

6.4 A Virtual Composite Model

At present, it is very difficult for all the actual component layout and sizes to be included into the simulation. As such, a virtual composite SRAM device is made in order to make a preliminary check on the feasibility of the simulation methods. **Figure 6.16** is such an example of such a device. The Silicon substrate model is essentially the same as the original SRAM model shown in **Fig. 6.1**, except the STI is made of SiO_2 instead of Si.

The upper part of the model is 'virtual'. A WSi_2 contact is placed on the diffusion layer. A Si_3N_4 LDD placed on the side of the gate. A copper metal line is placed in the interlayer oxides.

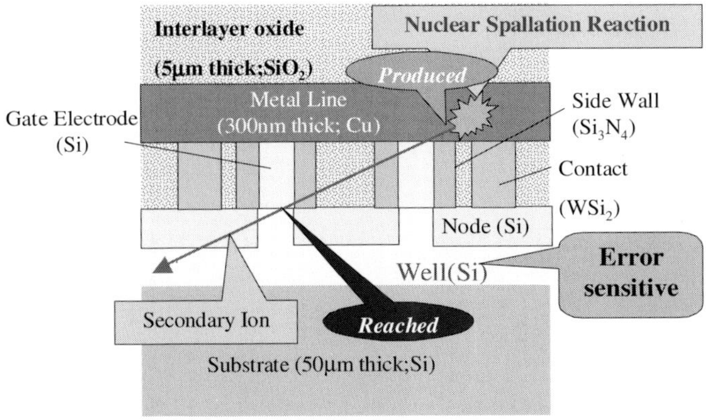

Fig. 6.16 A virtual SRAM model with composite materials [6.6].

Chapter 7

Simulation Results and Their Implications

Eishi Ibe

The model is validated with terrestrial field and accelerator test results. Unique results are given for asymmetrical multi-cell upset, material effects in the components, and threshold energies using realistic 3D device models.

7.1 Validation of the Model

Validation of SER simulations has several steps through nuclear reactions, accelerator testing and testing results in the field.

7.1.1 *Nuclear reaction model*

For validation of the nuclear reaction model, nuclear reaction data of protons on ^{27}Al [7.1], as well as some other reactions, are used as demonstrated in Tang's work [7.2]. Typical simulation results are summarized in **Fig. 7.1** where the energy spectra of the secondary protons and alphas are shown with the literature data. The spectra of secondary particles were shown at 15, 30, 90 degrees in the laboratory system. The agreements between literature data and calculation results seem acceptable.

7.1.2 *Accelerator test results*

For validation of the combined models for nuclear reactions and charge collection/diffusion to the storage, results from accelerator-based testing are used. The simulated results are shown in **Fig. 7.2** for 4, 8, 18 Mbit SRAM using (quasi-) monoenergetic neutron tests [7.3], and in **Fig. 7.3** for 16 Mbit SRAM for the spallation neutron tests at LANSCE, together with the experimental results as a function of the filter thickness (polyethelene filter thickness = 0 (full), 5, 10, 20 cm). In the (quasi-) monoenergetic neutron experiment, the errors between the simulations and the data are around 20–30%, on average. In **Fig. 7.3**, experimental results are obtained with the minimum energies E_{th} of 1.5 and 10 MeV in the fluence calculation. The estimated results in **Fig. 7.3** are also obtained from the folding method with the Weibull excitation function (See Chapter 5) by using the quasi-monoenergetic test results shown in **Fig. 7.2** [7.4]. The measured values vary with the minimum energy used in the fluence calculation. When the filter thickness is thin (0 or 5 cm), the discrepancy between both results is large by around factor 2. By contrast, when the filter thickness is thick (10 or 20 cm), both results become closer. This results implicate that the the portion of the lower energy neutron can be attenuated easily by a relatively thin filter whereas the portion of higher energy neutrons cannot be attenuated so easily even by thicker filters as previously shown in **Fig. 5.14**.

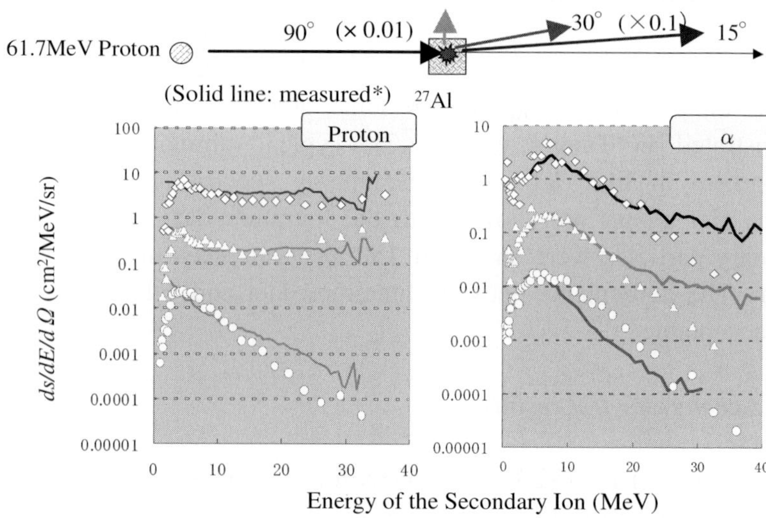

Fig. 7.1 Comparison between experimental [7.2] and calculated results of nuclear spallation.

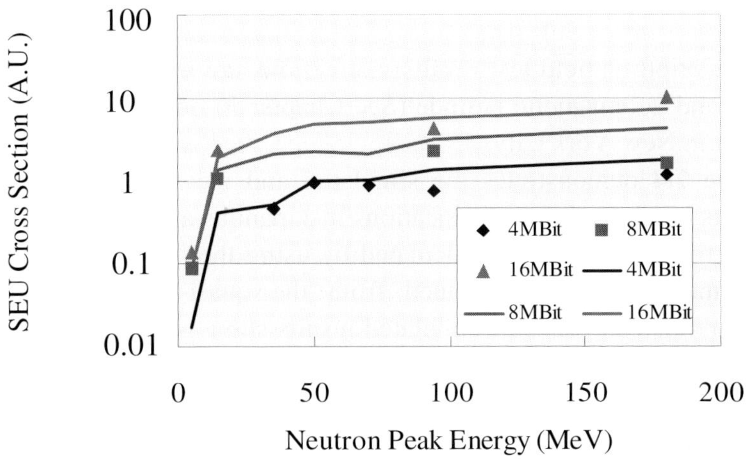

Fig. 7.2 Comparison between measured (solid symbol) and simulated (solid line SEU cross section of SRAMs using a (Quasi-) monoenergetic neutron beam) [7.3].

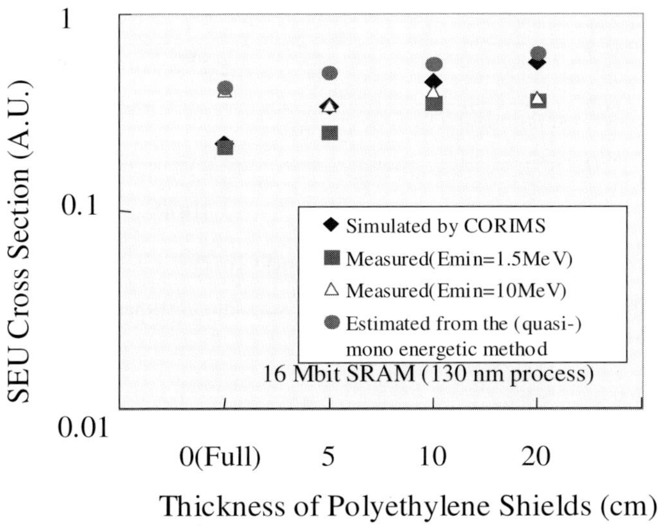

Fig. 7.3 Comparison of simulated, estimated (based on SEU cross section data from (quasi-) monoenergetic test results) and measured results from LANSCE tests [7.4].

7.1.3 *Field test results*

For simulation of field test results, corrections are made with respect to altitude and geomagnetic latitude (See Chapter 3) on the neutron flux at sea level in New York City.

Figure 7.4 demonstrates the simulated and measured results in the field tests conducted in three locations in Japan, as a function of altitude [7.5]. Estimated results are calculated by using the folding method with the excitation function obtained from the quasi-monoenergetic test results are also shown. The simulated results show good agreement with the measured results, within ±20%.

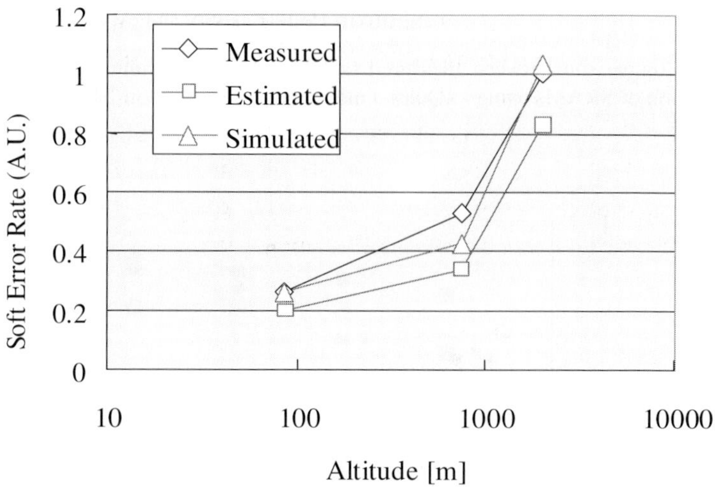

Fig. 7.4 Comparison of soft-error rate data and simulated/estimated results for 180 nm process SRAM from field testing [7.4].

7.2 Impact of Scaling

Increasing the critical charge is known to be a very effective method for reducing the alpha-ray induced SER. For the neutron SER, however, it has been pointed out that the slope is not so steep as the alpha SER, as shown in **Fig. 7.5** [7.5] (Note that reducing V_{dd} is equivalent to reducing the critical charge.), indicating increasing the critical charge is not an effective countermeasure for the neutron-induced soft-error.

The lower box of **Fig. 7.6** shows the simulated differential cross sections of the frequency of charge collection to the storage node through which protons or alphas or heavier or total ions are passed [7.6]. For the present device, the maximum collected charge is about 100 fC. Two bending points for the total ion can be seen in the curve at 2 fC and at about 10 fC, although the bending point at the colledted charge of about 10 fC is rather faint.

In the upper box of **Fig. 7.6**, the SEU cross section obtained by integration below a certain collected charge (= critical chage) of the differential cross section for the total ion in the lower box is shown. Consistently with **Fig. 7.5**, simulation results from CORIMS reveal that the SEU cross section does not change so much in the range of the critical charge 10–100 fC, where heavier secondary ions dominate charge collection. In the mid-range (2–10 fC), the slope is slightly steeper than the higher region, where alpha particles dominate. In the lower region, below 2 fC, the slope is steeper than the mid-range, where protons dominate. This trend does not appear in **Fig. 7.5**, but the simulation implicates that the curve should be much steeper at lower V_{dd} below 0.5 V if the device can be operated at such a lower voltage.

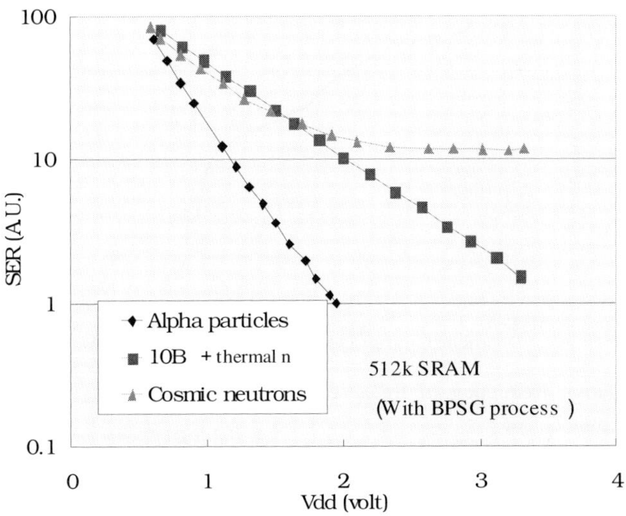

Fig. 7.5 Dependency of SER upon Vdd [7.5].

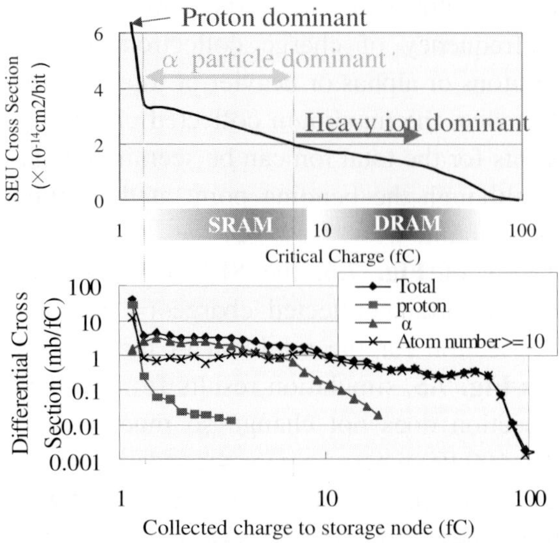

Fig. 7.6 Paradigm shift with a decrease in critical charge: As Q_C decreases, the dominant ion over SER changes from heavy ions to alphas, and then to protons [7.6] (© 2001 IEEE).

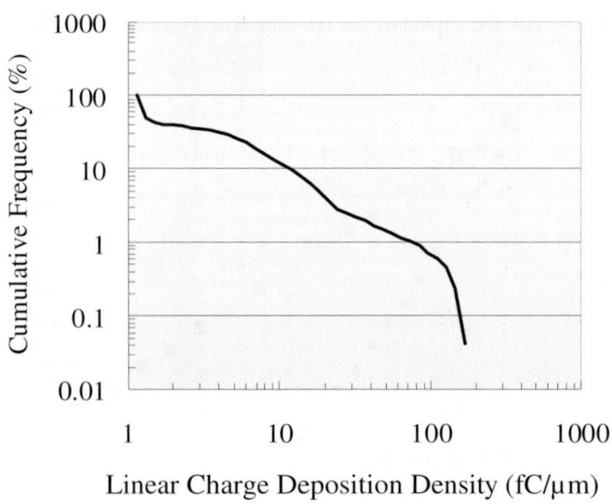

Fig. 7.7 Cumulative frequency of linear charge deposition density for the terrestrial neutron spectrum in Tokyo.

It is noteworthy that the absolute values in the curves, for instance the maximum collected charge, are subject to change with changing the device architectures.

Figure 7.7 may be useful for more general purpose. The figure shows the cumulative frequency of the charge deposition density in a Si substrate at the boundary of the sensitive volume. As long as the material of a device is made of Si, the curve holds true under a terrestrial neutron environment. As a result, **Fig. 7.7** can be used for the device hardening design. For instance, if a device is tolerant to a charge density of 170 fC/μm, the device is absolutely immune to terrestrial neutron soft-errors.

Figure 7.8 shows simulated results from the effects of spatial scaling and reduction of critical charge on the SEU cross-section per bit [7.6], together with experimental results measured at LANSCE [7.7, 7.8]. It can be said that spatial scaling has a significantly beneficial impact on SEU susceptibility: 1-D spatial scaling with a factor k results in roughly k^2 scaling of SEU susceptibility. As for 3-D scaling, k^6 scaling of SEU susceptibility is expected. Physically, this can be interpreted that the SEU susceptibility is in proportion to the product of two volumes (6-D in total) in the device. Namely, the storage node volume which determines the probability to be hit by the secondary ions, and the surrounding volume which determines the amount of charge collected to the storage node. It is also shown that if the critical charge Q_c is scaled simultaneously, the beneficial effect of spatial scaling deteriorates significantly, as exemplified in the 3-D & Qc case.

These simulated results give a clear explanation for the conflicting trends observed in **Fig. 1.1**: For DRAMs, where the critical charge has been strenuously kept high by engineering of the capacitance, the beneficial effect of spatial scaling has prevailed. By contrast, the critical charges of SRAMs are generally reduced with scaling due to reduction in parasitic capacitance. As a result, the beneficial effect is eventually blocked. Further decrease in the critical charge is an inevitable direction in the road map of ULSI devices as a consequence of reducing the supply voltage, area of the (parasitic) capacitors, and dielectric constant of the interlayer dielectrics so that increase in SEU susceptibility of memory devices is also inevitable in general.

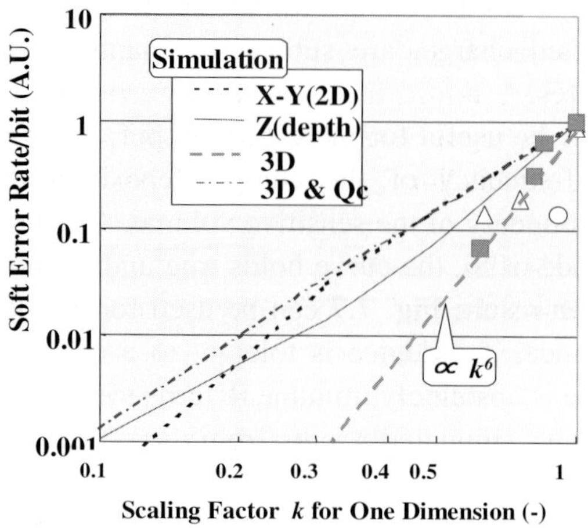

Fig. 7.8 Effect of scaling on soft error rate (Δ, O: measured at LANSCE by Eto (1998) [7.7] ■: Sato *et al.* (1998) [7.8]; *k* is calculated from the square root of the dimensionless area of the device) [7.6] (© 2001 IEEE).

7.3 Asymmetry in Multi-Cell Errors

7.3.1 *Dispersed and nearest neighbor MCUs*

Any data patterns of "1"s and "0"s in the cell matrix can be expressed in the form of a unit 2D block, as shown in **Fig. 7.9**. In general, the

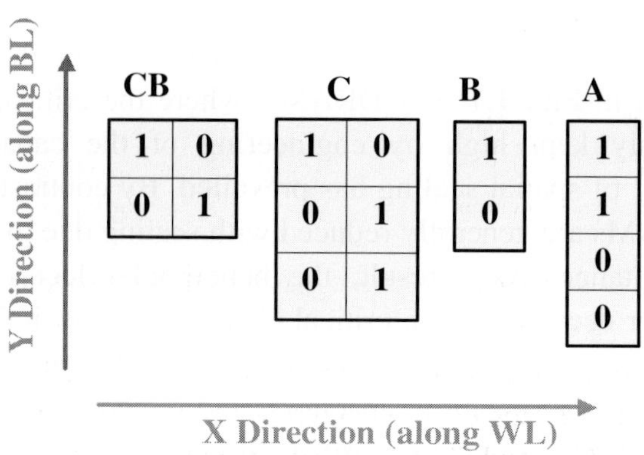

Fig. 7.9 Definition of the basic data pattern [7.4].

patterns #FF, #00, and CB ("checker board") are applied in experiments. The patterns A, B, and C are added for wider options. As explained in section 6.2, any bit pattern in the same word in the memory matrix can also be treated by giving a bit interval in the WL (along the word line) and/or BL (along the bit line) direction. SER simulation in Tokyo was carried out with as many as 10 million reactions to obtain a statistically enough number of MBUs.

Figure 7.10 shows the impact of reducing the critical charge on the spread and the number of SBUs [7.9].

The simulation is carried out for 130 nm SRAM under terrestrial neutrons in Tokyo, using a data pattern C. With a critical charge of 7.4 fC (right), the spread and number of error bits are not large because heavier secondary ions with short ranges dominate. Meanwhile, with a critical charge of 2 fC, the spread and number of error bits are much larger, suggesting an increase in MCU.

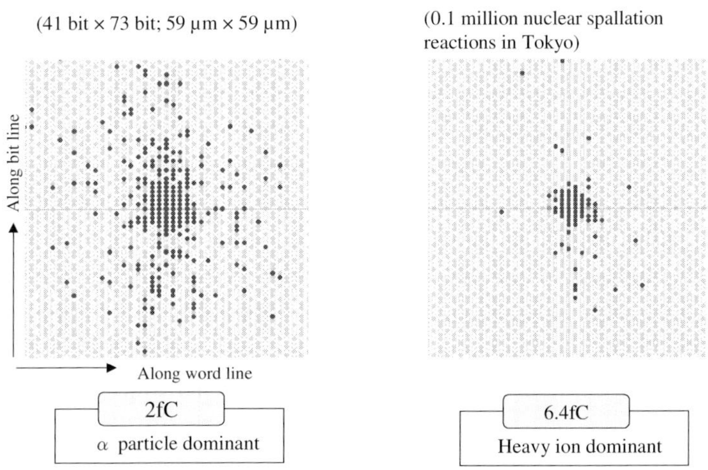

Fig. 7.10 Effects of a decrease in the critical charge [7.3] on the spread of error bits.

Figure 7.11 summarizes the nearest neighbor (NN; in WL, BL, or diagonal direction), dispersed (non-NN), and multi-ion errors [7.10]. It is seen that (a) WL-NN and diagonal NN depend significantly upon the data pattern, (b) for patterns #FF and #00, most MCUs are in the NN

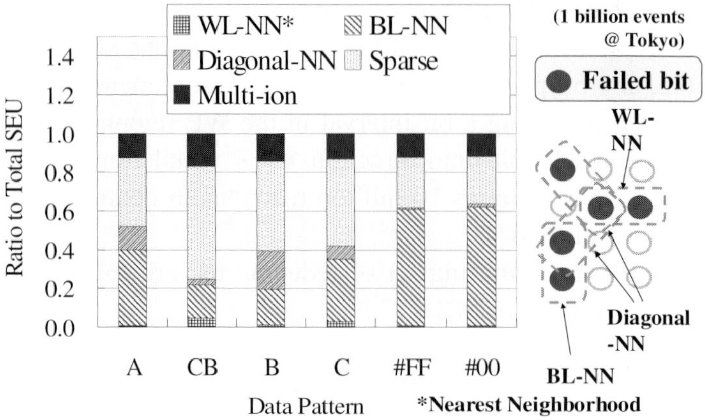

Fig. 7.11 Dependency of the basic categories on the data pattern [7.10].

position (70%) but substantial errors are not (dispersed; 30%), and (c) inversely, for patterns CB, B, A, a dispersed MCU is the main pattern. This boils down to the fact that the common experimental method in which only NN errors (or plus next NN) are counted as MCUs leads to a substantial underestimation of MCUs and, inevitably, MBUs. In addition, multi-ion MCUs appear to be not so common (about 10%). The result that WL-NN MCU has the highest probability with a CB pattern was validated with a 70 MeV quasi-mono neutron accelerator-based test at CYRIC, as demonstrated in **Fig. 7.12**. The data patterns #FF, ♯00, and CB were applied to the SRAMs, and only X-NN MCUs with the CB pattern were found, as evidenced by the double-bit errors circled in the failed bit map. The origin of the enhanced susceptibility of the CB pattern can be explained in terms of the data in each storage node.

7.3.2 *MBU sensitive to architecture*

Figure 7.13 shows the normalized SER for SEU, MCU, and MBU (the interval is zero bit between bits in the same word) for each data pattern. While simulated SEUs vary only slightly with data pattern, MCUs and, in particular, MBUs, depend surprisingly strongly upon the data pattern: Data pattern C or CB showed a much larger MBU rate than the pattern #00, by a factor as large as 8 in some cases.

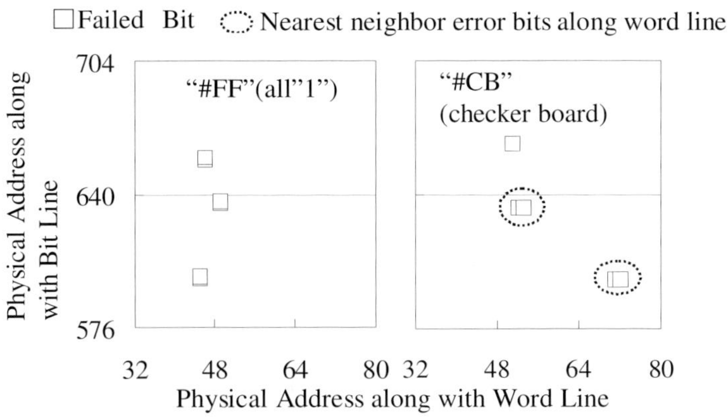

Fig. 7.12 MCU pattern appeared in a 130nm SRAM for the 70MeV (quasi-) monoenergetic neutron test at CYRIC [7.10].

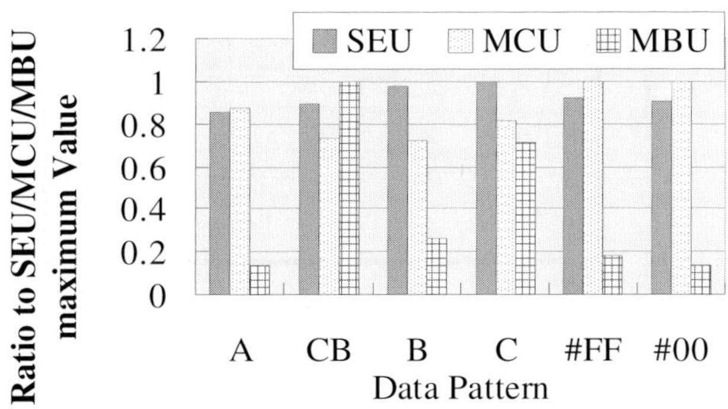

Fig. 7.13 SEU, MCU, MBU dependency on data pattern [7.10].

7.3.3. *The margin-of-interleave technique to suppress MBUs*

Figure 7.14 shows simulated results for the effects of the separation distance (defined as 1 for NN) along a word line or bit line. It is seen that only a 3 or 4 bit separation distance (<8 μm) along a word line will be enough to suppress MBU below 0.01 FIT/Mbit. In fact, a low power SRAM with the same process but with a lower Q_c and a bit separation distance of 2 bits in the WL direction showed no MBUs in accelerator-based tests [7.12]. It is concluded that a separation distance much shorter than the one commonly applied in interleave techniques can be applied to suppress MBUs. These results are supported by recent experimental results [7.11]. The interleave techniques, however, may not work if Q_c decreases below 1 fC. In this Q_c range, secondary alpha particles and protons play a major role, and as a result, SEU ranges well beyond 100 μm.

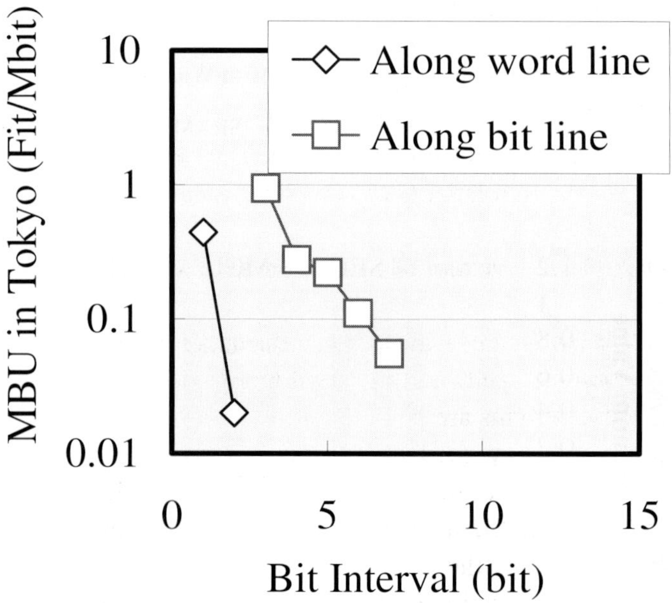

Fig. 7.14 Effect of bit separation distance (1: nearest neighbor) to mitigate MBUs [7.3].

7.4 Material Effects

In this section, some preliminary results are introduced from a calculation done with a virtual SRAM model with composite materials (see **Fig. 6.16)** [7.12, 7.13].

7.4.1 *Secondary ions from different materials*

Figure 7.15 shows the production yields of secondary particles from three different Si-based materials. Among three Si-based materials (Si, SiO_2, Si_3N_4), the Si target has heavy ions (Al, Mg) and light ions (p, α) as the dominant secondary ions when bombarded by 100 MeV neutrons. Meanwhile, it is found that medium weight secondary ions (C, N) are produced from SiO_2 and Si_3N_4 substrates on the same order of Mg and Al production from a Si substrate.

Figure 7.16 shows the energy spectra of secondary ions *produced (Later, we will see that elements with high production yields do not necessarily affect SEU) in* the WSi_2 substrate. It is seen that W (through an inelastic reaction) and Ta with low energy are produced from a W target in WSi_2 with a relatively high probability.

7.4.2 *Results from a virtual composite material device*

Figure 7.17 shows the energy spectra of secondary ions that *reached the sensitive region* (Si well) of the virtual composite-material device. The following obeservations are pointed out:

(A) Among the heavy secondary ions, Na has the highest probability to reach the sensitive region while Al has the highest probability of production from Si, as seen in **Fig. 7.16**. This implies that the probability to reach the sensitive region is determined from the production rate and location (range) relative to the sensitive region.

(B) The probabilities of medium weight ions (C, N) to reach the sensitive region are not so high. This is due to the assumption of a thickness of SiO_2 (5 μm) that is much thinner than the Si substrate (50 μm). This implies that the total amount of target nuclei in the device is another key parameter.

(C) The probabilities of Ta and W to reach the sensitive region are negligibly small, even though WSi_2 is in contact with the sensitive region. This may be due to the very short range of these secondary ions, since the LET is relatively high and the energy is very low, as seen in **Fig. 7.16**.

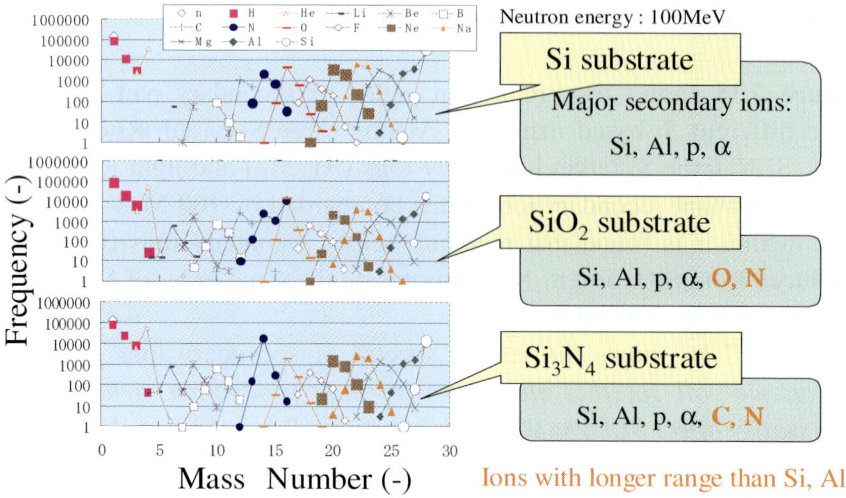

Fig. 7.15 Difference in the secondary ions *produced in* the indicated substrates [7.13] (© 2005 IEEE).

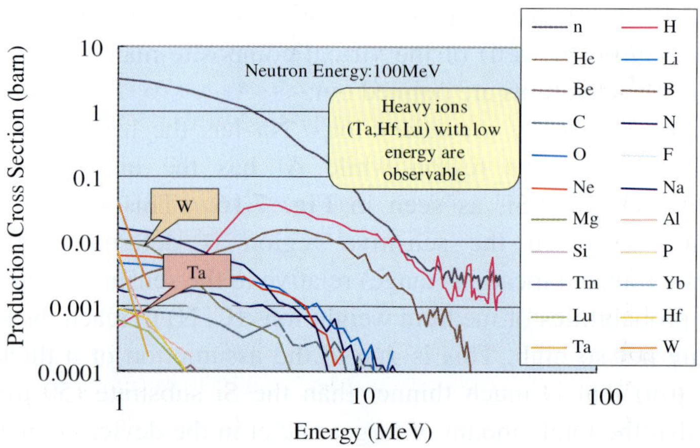

Fig. 7.16 Energy spectra of secondary ions *produced* in WSi_2 substrate [7.13] (© 2005 IEEE).

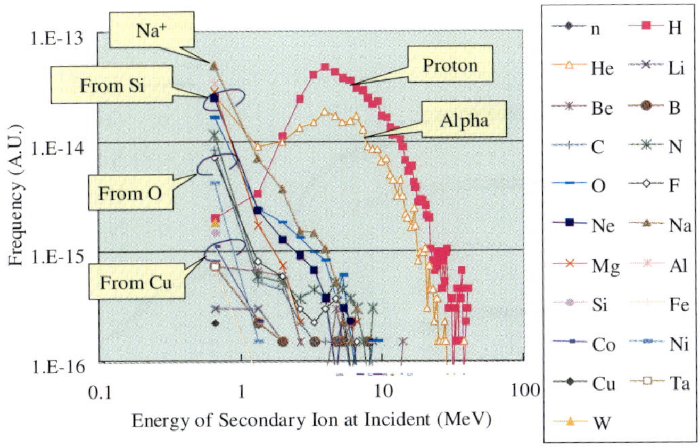

Fig. 7.17 Example of simulated results for a virtual device: Shown are the secondary ions which *reach* the sensitive volume [7.13] (© 2005 IEEE).

7.5 Threshold Energy of SEU Excitation Function

Figure 7.18 shows the calculated charge density at the inlet of the sensitive region (Well) created by irradiation from 2 MeV neutrons [7.12]. The cumulative frequency over a specific charge density is plotted for oxygen, silicon, elastic scattering, non-elastic scattering, and total scattering. The major findings are:

(A) Although a slight charge deposition is created by non-elastic (actually inelastic n-n') scattering from 2 MeV neutrons, the charge density from elastic scattering is as high as 40 fC/μm. This density can cause SEUs, depending on the device structure/size and critical charge. Thus, it can be major mechanism behind the low SEU threshold energy that is below the threshold energy of nuclear interactions with a Si nucleus [7.14].

(B) For low-energy neutron irradiations, as in Gasiot's work (14 MeV) [7.15], the contribution from elastic scattering is relatively high, and as such, the dependency of SEUs on the neutron beam direction may be quantitatively explained in the future.

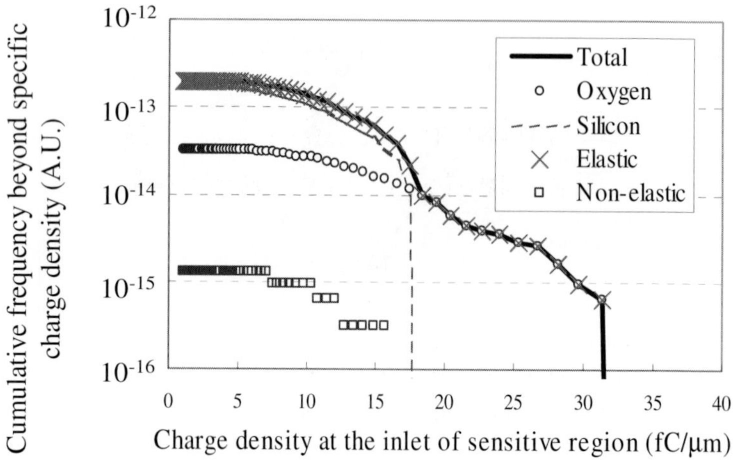

Fig. 7.18 Charge density spectra *at the surface* of the well for lower-energy ($E_n = 2$ MeV) neutrons [7.12].

Chapter 8

International Standardization of the Neutron Test Method

Eishi Ibe

The most recent developments with respect to the international standardization of testing methods are introduced, covering JESD89A, 89-1, 89-3, and EDR4705 (Japanese guidelines).

8.1 The Current Status of Standardization

As for SEUs, particularly SBU (Single Bit Upset), co-operative international research has been made for the standardization of the test methods. The JEDEC (Joint Electron Device Engineering Council) issued JESD89 as a standard method for alpha and neutron induced soft-errors in 2001 [8.1]. The JESD89 Task Group with about 70 members in Europe/USA/Asia was organized in 2004 to improve JESD89 and issue the revised version JESD89A [8.2]. The JEITA (Japan Electronics and Information Technology Industry Association) organized a SER committee in 2002 to evaluate JESD89 and issue a Japanese guideline for the test methods. The draft of the guideline was introduced at the 8th JEDEC/JEITA joint meeting in 2004 to promote an international agreement on the test standards. The guideline was issued in December, 2005 as EDR4705 [8.3].

Field testing (including high altitude tests), alpha-ray test methods, monoenergetic proton, (quasi-) monoenergetic neutron, spallation neutron and thermal neutron test methods are commonly refered to in the standards. Field testing requires hundreds or thousands of devices undergoing months of testing. It is so time consuming and costly that its application is quite limited, and is used mainly for verification. Since the alpha test method is not in the scope of this book, and the thermal neutron test is required only for devices with high ^{10}B concentration, only accelerator-based methods are introduced here, with an enphasis on their own merits and demerits, as summarized below [8.4].

8.2 Monoenergetic Proton Method

Monoenergetic protons produced at cyclotron facilities are used as test probes in lieu of neutron. The merits (M) and demerits (D) of using proton beams are:

> The flux is high: suitable to obtain good statistics and get data for rare error modes. (M)
> The number of facilities is relatively large. (M)
> Proton dosimetry is easier than neutron dosimetry. (M)
> Heating and charging of DUT must be considered. (D)
> It is difficult to obtain the threshold energy E_{th} of the cross section (normally more than 10 to 20 MeV is necessary to get enough range to reach the sensitive volume. Unliding of the DUT or an accurate energy calculation at the sensitive volume is necessary (D).
> Total reaction cross-section is not the same as that of neutron, particularly, below 50 MeV (D).

The probabilities for the nuclear reaction channels ((n,α), (n,p), ...) are markedly different from those by neutrons.

This leads to an assumption that we see somehow physically different phenomena from neutrons by proton irradiation, so care must be taken when proton SEU cross section data is used for FIT estimation in the field.

8.3 (Quasi-) Monoenergetic Neutron Method

Neutron beams with a flux peaked at a specific energy can be obtained by specific nuclear reactions such as Li(p,n)B reaction. In the Li(p,n)B reaction, a thin Li plate is used for the target. Typical neutron spectra and shapes of SEU cross sections are introduced in Chapter 4. The merits and demerits of this approach are:

> Neutron cross-section data is obtained directly as a function of energy. Threshold energy data is the most reliable. (M)

> An approximation (excitation) function for the SEU cross section vs. neutron energy can be used to obtain the FIT number for any kind of neutron environments, including building interiors and accelerators. (M)

A Weibull fit is recommended for the approximation function because it explicitly contains the threshold energy E_{th} [8.5] and saturation level σ_∞ as illustrated in **Fig. 8.1**:

$$\sigma_{approx}(E) = \sigma_\infty\left[1-\exp\left\{-\left(\frac{E-E_{th}}{W}\right)^S\right\}\right] \tag{8.1}$$

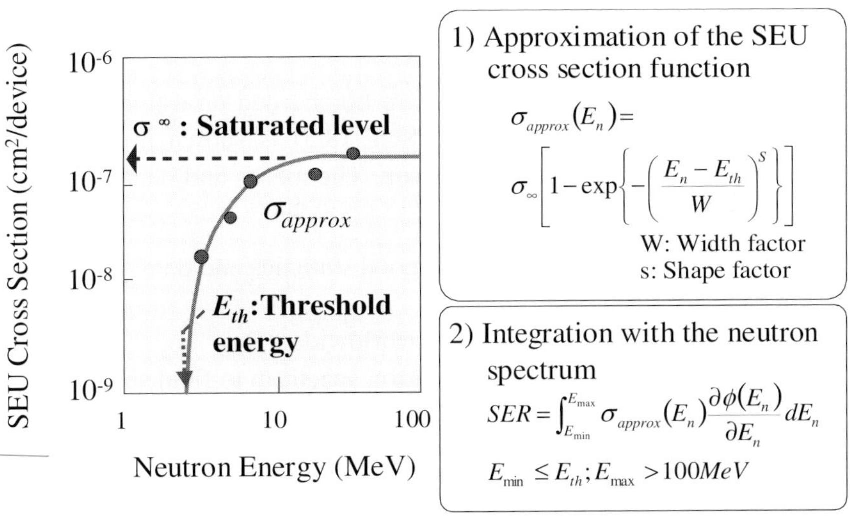

Fig. 8.1 Basic procedures of the (quasi-) monoenergetic test method used to make an estimation of the soft-error rate at ground level.

The characteristics of SEU responses can be tabulated and compared by using the Weibull fit parameters for any semiconductor devices. (M)

- Elimination of the contribution from neutrons outside the peak (tail correction) must be carried out for quasi-monoenergetic neutron tests. The common correction methods as well as the folding method in Eq. (8.2) should be incorporated into any data analysis programs. (D)
- Dosimetry and spectrometry are rather difficult. (D)
- The available facilities are limited worldwide. (D)

For (quasi-) monoenergetic neutron/proton tests, it is recommended that the SEU excitation function and differential flux are used to obtain FIT values for SER, as follows:

$$SER(FIT) = 3.6\times 10^{9}\int_{E_{th}}^{E_{\max}} \frac{\partial\phi(E)}{\partial E}\sigma_{appx}(E)dE \qquad (8.2)$$

8.4 Spallation Neutron Method

In this method, high energy protons are bombarded on a thick W or Pb target. A typical spectrum at LANSCE [8.5] is shown in Chapter 4, along with a terrestrial neutron spectrum. The merits and demerits of this approach are:

- The flux is high, which allows for good statistics and data on rare error modes. (M)
- The available facilities are limited worldwide. (D)
- The similarity of the spallation neutron spectrum to field spectrum is critical. (D)
- The flux of higher energy above around 100 MeV is relatively low, so that this method is not suitable for quantification of rare but fatal error mode. (D)

On the condition that the white neutron spectrum is close enough to the terrestrial neutron spectrum, the FIT value can be obtained by the following procedure:

First, the effective cross section is calculated from the number of events (N_{err}) and neutron fluence ($\Phi(E_{\text{min}},E_{\text{max}})$) in a energy region from E_{min} to E_{max};

$$\sigma_{eff} = \frac{N_{err}}{\Phi(E_{\min}, E_{\max})} \tag{8.3}$$

Then, the FIT number is calculated from the flux in the field (ϕ_{field}) by using the following equation.

$$SER(FIT) = 3.6\times10^{9}\sigma_{eff}\phi_{field}(E_{\min}, E_{\max}) \tag{8.4}$$

It is noteworthy that the estimated SER may vary with the energy range because of difference in neutron spectrum shape between the spallation neutron source and terrestrial neutron.(D)

Considering the merits and demerits from all of the methods listed above, it may be reasonable to conclude that a combination of these methods can be used on a case-by-case basis. When using a combination of methods, it is critical to keep consistency between the data obtained from different methods, as exemplified in SECIS (SElf-Consistent Integrated System for terrestrial neutron soft-error) [8.7].

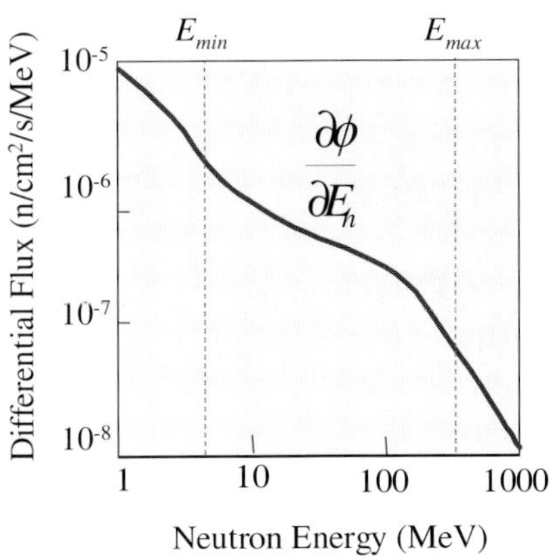

1) Determine energy range and calculate fluence

$$\Phi(E_{\min}, E_{\max}) = \int_{E_{\min}}^{E_{\max}} \frac{\partial\phi}{\partial E_n} T_{irrad} dE_n$$

T_{irrad}: *Irradiation* time

2) Calculate effective cross section

$$\sigma_{eff} = \frac{N_{err}}{\Phi(E_{\min}, E_{\max})}$$

3) Estimate SER

$$SER = \sigma_{eff}\phi_{field}(E_{\min}, E_{\max})$$

Φ_{field} :Flux in field

Fig. 8.2 Basic procedures of the spallation neutron test method to make an estimation of soft-error rate at ground level

For other SEE types, including MCU/MBU, reliable characterization and quantification methods are not fully established. This remains as a challenge for the future.

Chapter 9

Summary and Challenges

Eishi Ibe

This chapter briefly summarizes the contents of the book. The topics which are not included in this book, and items left for action in the future, are also discussed.

9.1 Standard Test Methods for Multi-Cell Upset

Standard testing methods are introduced in this book with a focus on the (quasi-) monoenergetic and spallation neutron testing methods. The present international testing methods [9.1, 9.2] mainly focus on the testing methods for SBUs. Testing methods and algorithms should be expanded for other important failure modes; in particular, MCUs in which MCBI [9.3], SEL [9.4], and SEFI [9.5] are included. As these failure modes take place simultaneously even in a single chip, classification rules and algorithms for failure modes, which take place should no doubt be established on an interdisciplinary basis in order to make valid comparisons over different generations and architectures of semiconductor devices [9.6, 9.7]. More accurate and effective neutron quantification techniques in neutron spectrometry, dosimetry, and topology for lateral uniformity are required to make such classifications more meaningful.

9.2 Neutron-Induced Errors in Logic Devices

As the down-scaling of devics proceeds, soft-errors of combinational/ sequential logic devices are being a significant concern over device/system reliability because ECC cannot be applied to logic devices. It is a common approach to quantify three derating/masking effects, as summarized in **Fig. 9.1**. Namely:

(1) Logic masking/derating

The error signal cannot pass a gate if the status of the other input does not allow the signal to go forward.

(2) Electric masking/derating

During propagation in a circuit, if the error signal decays down to the high threshold level of a gate, the error signal cannot affect the output of the gate.

(3) Window masking/derating (Timing derating)

When the error signal arrives, if the gate input is not active due to a lack of the clock signal, the error signal cannot propagate.

SER of logic device =

Nominal FIT× Logic derating× Timinig derating× Electric masking

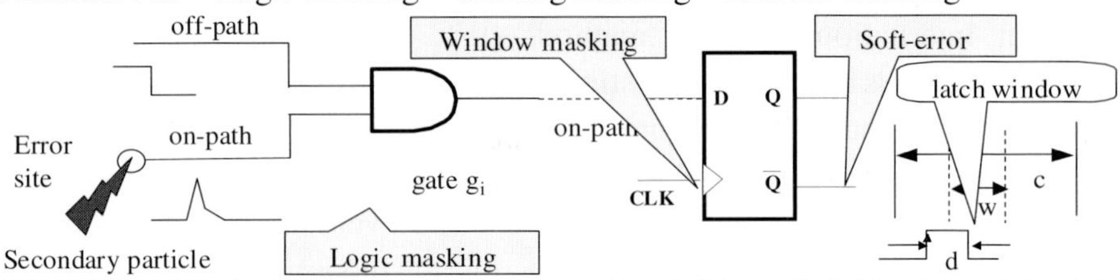

Concept	Definition	Remarks
Logic derating/ masking	Blockage by logical condition at the inputs of a gate (ex:AND gate with a "0" input.)	
Timing derating/ Windowmasking	Blockage by the time window of F/F input.	$P=0$ $(d< w)$ $=(d-w)/c$ $(w<d<c+w)$ $=1$ $(d>c+w)$
Electric masking/ masking	Attenuation of the voltage and duration of a noise pulse below critical level	

Fig. 9.1 Basic concept of masking of logic circuits.

One of the ultimate goals of soft-error research on logic devices is to estimate SER in any neutron environment. A circuit level simulation technique, such as HSPICE [9.8], is one of the options that can be used, but it is very time consuming in computation [9.9, 9.10]. A more effective methodology needs to be established.

9.3 In-depth Study on Material Effects

Effects of composite materials can be one of the most crucial points in the significance of failure modes and countermeasures. Gasiot *et al.* [9.11] revealed that the error rate (cross section) depends on the direction of a 14 MeV neutron beam, as shown in **Fig. 9.2**. From front side irradiation, where the neutron beam is facing SiO_2, the error rate is higher by a factor of about two than from the rear side, where the beam faces a Si substrate.

It is pointed out that the difference in the SEU cross section between the front and rear side is assumed to be due to the difference in the materials Si and SiO_2.

Schwank [9.2] has pointed out that the range of SEL cross sections of SRAM is too wide, by as much as an order of five. They attribute such a wide range to material effects, particularly heavy component elements such as tungsten.

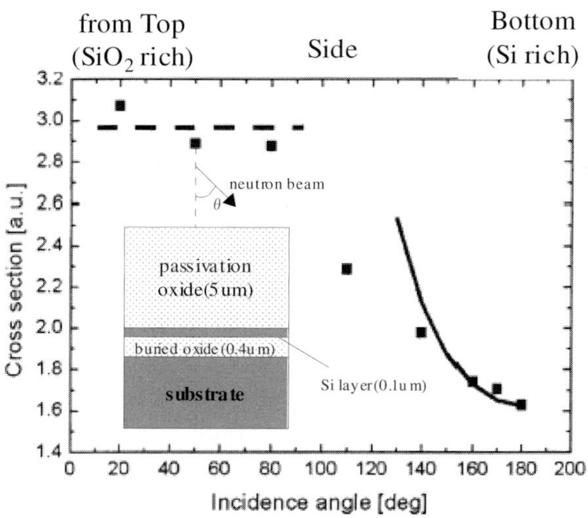

Fig. 9.2 Possible effects of materials: susceptibility depends on the front material facing the beam [9.6].

9.4 Possible Feedbacks from the System Side

The standard techniques for devices introduced here are quite important and invaluable. However, validation for their accuracy and usefulness for systems/components is very difficult at present because of their complexities as illustrated in **Fig. 9.3**. We have to buildup our quantitative knowledge on the neutron-induced soft-error from the device level via the component level up to the system level to estimate real impact on the infrastructures of the society. Any challenges from the system side to obtain direct information on which error mode is fatal/crucial are quite important and encouraged. For example, as Bradley and Normand pioneered in their work for defibrillators [9.12] by which altitude effects has been clearly identified from a pile of error-log data of patients, if global error-log data are available for the same systems/components, and some altitude dependency of the error rate is found for a particular error mode, the error mode is confirmed as being induced by terrestrial neutrons and its significance can be estimated directly.

Systematic classification standards for the system errors seem necessary to understand the major mechanisms and establish effective countermeasures. An example of such approaches is also seen in M. Brüggemann's work in which four types of single events are categorized based on the feature of outcome or correction methods required in the system [9.13].

9.5 Countermeasures in Multiple Hierarchies

Immunity of the device to radiation effects may be the ultimate goal of the research in the field, but any single countermeasure is not an absolute solution to make the devices immune. Combinations of countermeasures at the device/component/system level may be the only a realistic solution to the problem.

In this section, such possible countermeasures that are now available are reviewed.

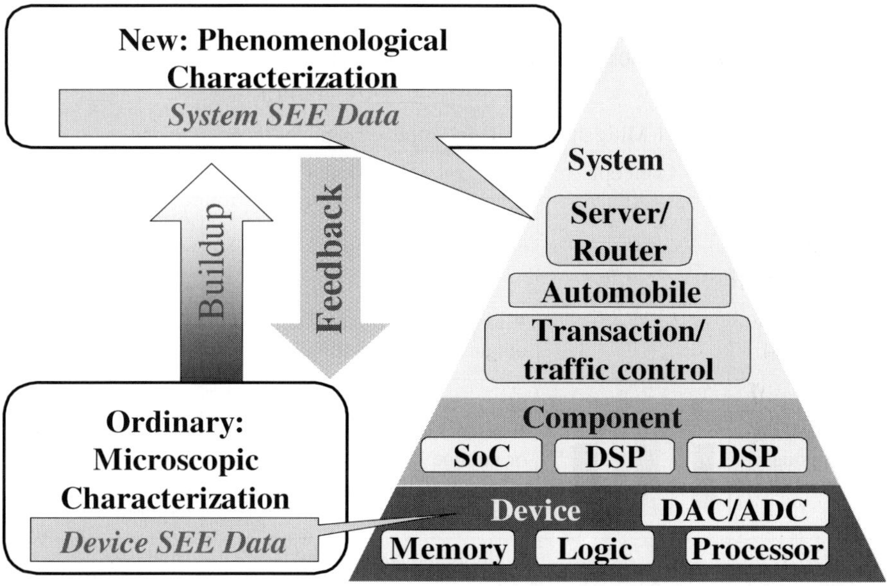

Fig. 9.3 A possible approach from the system side.

9.5.1 *Countermeasures at the process/device level*

Table 9.1 summarizes all the known and possible countermeasures to reduce terrestrial neutron soft-error at the process/device level. The countermeasures in the first two rows are mainly related to SRAM or logic devices. An area or speed penalty/overhead has to be considered carefully. Countermeasures in structure, layout or profile may be the most effective but obviously are the most difficult and costly. High performance and reliable 3D device simulation techniques are required for this purpose.

DICE is known as a soft-error-free latch [9.14]. The stable state of DICE is the state when the two node has the same data. The data can be re-written only when the states of two nodes are re-written. A soft-error may affect only one node, and as a result, the data of the DICE cannot be changed by ion bombardment. The double gate CMOSFET inverter [9.15] is a device in which two pMOSFETs and two nMOSFETs are sequentially connected to both sides of the storage node. Even if an ion hits one of the two nMOSFET/pMOSFET gates, the other gate stays intact, and the noise cannot propagate further.

Table 9.1 Mitigatin of SEU by improvements in process/device.

Method	Concept	Function	Note
Capacitance addition	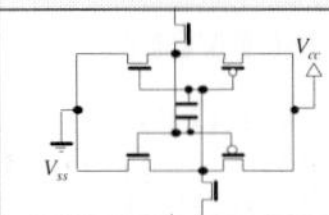	■ Q_{crit} increase	Area penalty
Resistance addition	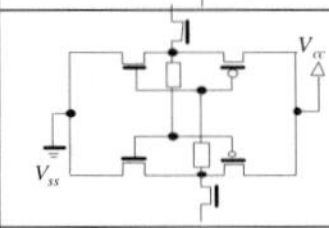	■ lengthen the time constant to obtain a margin to flip data	Area penalty
Structure, layout, profile	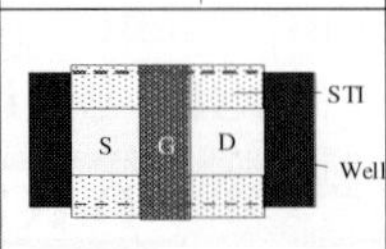	■ restrict charge collection path or volume ■ Eliminate seed point of bipolar action	High performance device simulation techniques are required.
DICE	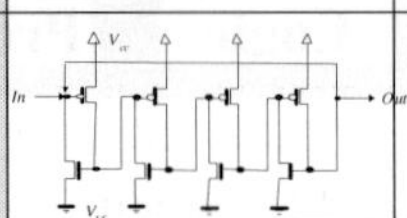	■ Hardened latch circuit	High cost
Double gate inverter	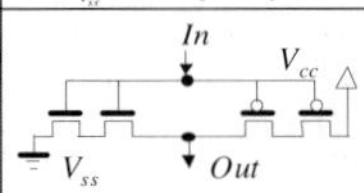	■ double sequential p- and n-channels for CMOSFET	Area penalty

9.5.2 *Countermeasures at the component level*

Figure 9.4 shows countermeasures at the component level. It is not assumed that no hardening techniques are applied at the device level. The three techniques of (1) parity, (2) ECC and (3) interleave technique are introduced.

(1) Parity

The state of a parity bit is set so as to make the remainder of the sum of bit data in a word divided by 2 (mod 2) zero when the data is written. If one of the bit states flips, the sum changes to and alerts the user that a bit in the word has flipped.

(2) ECC

Common ECC (Error Correction Code) is a method to correct one bit error and find double bit errors in a word by using redundant bits. A typical method is the use of the Hamming code [9.16]. For higher order detection and correction, the Reed Solomon code [9.17] can be applied, but it creates heavy area, speed, and power penalties.

(3) Interleave technique/scrubbing

The interleave technique (or scrubbing) [9.18] is applied to prevent MBUs by separating bits in a word with a distance that is large enough compared to the range of secondary ions produced by neutron spallation reactions.

Method	Function	Note
Parity	■ Detection of SBU	No correction
ECC	■ Correct a SBU ■ Detect double bit error in a word	Area, power dissipation penalty
Inter-leave	■ Keep interval between two bits in a word wide enough compared to ion range ■ The interval can be shorten by layout, structure adjustment	Area, power dissipation, speed penalty

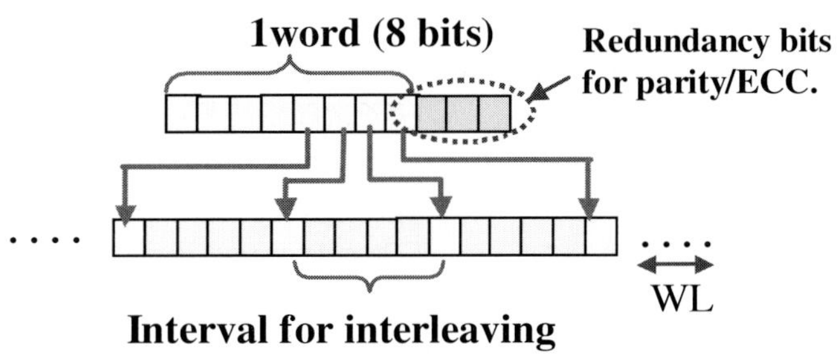

Fig. 9.4 Mitigation of SEUs by improvement in the components.

9.5.3 *Countermeasures at the system level*

Countermeasures at the system level are associated with high overhead and cost, in general. Such countermeasures are, however, the most enthusiastically investigated because they can be applied instantly without waiting for cost-effective, lower-level countermeasures. Such typical system level countermeasures are summarized in **Table 9.2**.

In the DMR (Double Module Redundancy) [9.19] technique, the same procedure is carried out in two parallel modules. The procedure is transferred to next module, only when the two outputs from the modules are consistent. This functionality may be realized by using the C-element [9.20].

In the TMR (Triple Module Redundancy) [9.21] technique, the same procedures are carried out in three parallel modules. If two or more outputs are the same, the majority is transferred to the next module through a voting module.

In the replication technique [9.22], a procedure is repeated twice. The result is transferred to the next module, only if the two results are consistent.

Table 9.2 Mitigation of SEU by improvement in systems.

Name	Concept	procedures	Merit/ demerit	Example
DMR	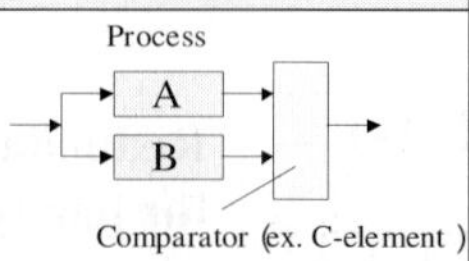	Execute the same process in parallel in A and B. If two results are the same, the process proceeds to the next step.	Speed, area, power penalty	
TMR	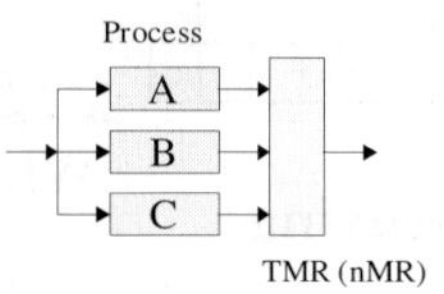	Execute the same process in A,B, and C. The majority result is passed to the next process via a voting module.	Area, power penalty	·NonStop Server (HP) ·Back-end cluster (Intel)
Repeat	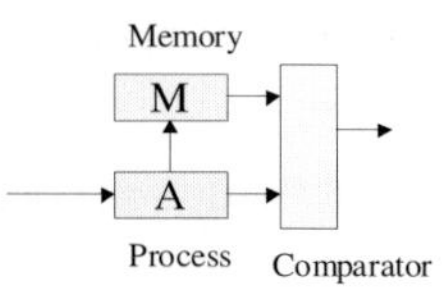	Repeat the same procedure twice. If two results are the same, the process proceeds to the next step.	Speed. Area penalty	

9.6 Interdisciplinary Co-operation Necessary for the Next Step of Challenges

Figure 9.5 describes the scales of time and space of research relevant to terrestrial neutron soft-error.

Time scale ranges from far below ata-second (INC) to more than ten million years (explosion of super Nova at the galactic nucleus). Space scales range from below fem-to meter (nucleon) up to 10^9 meter (galaxy).

Fields of research are covered from elementary particle physics to astronomy. To have a better understanding of neutron-induced failures in semiconductor devices, and to explore a new, vivid and attractive horizon in the semiconductor industry, interdisciplinary co-operation over a wide variety of many scientific fields, such as astronomy, cosmic ray physics, nuclear physics, accelerator physics, semiconductor physics, numerical simulation, and so on, is necessary. Such efforts are now being pursued and making vigorous connections among researchers/engineers worldwide.

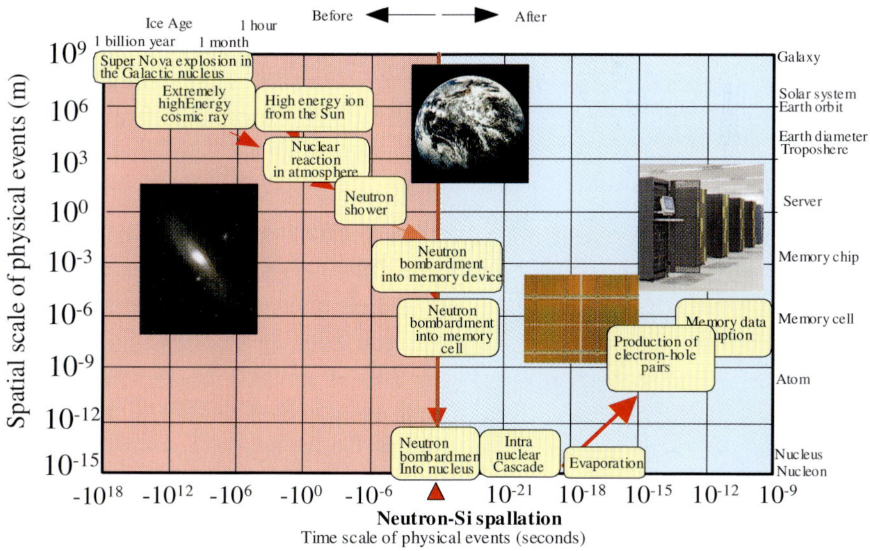

Fig. 9.5 Time and spatial scales of physical events related to terrestrial neutron-induced errors.

APPENDICES

A.1 Radiological Protection Quantities [2.6, 2.40]

Absorbed dose

The absorbed dose, D, is the quotient of $d\bar{\varepsilon}$ by dm, where $d\bar{\varepsilon}$ is the mean energy imparted by ionising radiation to matter of mass dm, thus

$$D = \frac{d\bar{\varepsilon}}{dm}.$$

The unit of absorbed dose is joule per kilogram [J kg^{-1}] and its special name is gray [Gy].

Dose equivalent

The dose equivalent, H, at a point is given by

$$H = QD = \int_L Q(L)\frac{dD}{dL}dL,$$

where $Q(L)$ is the quality factor for particles with linear energy transfer, L and $dD/dL \cdot dL$ is the absorbed dose between linear energy transfer, L and $L + dL$ at the point.

The unit of dose equivalent is joule per kilogram [J kg^{-1}] and its special name is sievert [Sv]. The quality factor, Q, is given as a function of linear energy transfer, L, in water with units of keV/μm.

Organ absorbed dose

The mean absorbed dose, D_T, in a specified tissue or organ of the human body, T, is given by

$$D_T = 1/m_T \int_{m_T} Ddm,$$

where m_T is the mass of the tissue or organ, and D is the absorbed dose in the mass element dm.

Equivalent dose

The equivalent dose, $H_{T,R}$, is the absorbed dose in an organ or tissue, T, multiplied by the relevant radiation weighting factor, thus:

$$H_{T,R} = w_R \cdot D_{T,R},$$

where $D_{T,R}$ is the absorbed dose averaged over the tissue or organ, T, due to radiation R, and w_R is the radiation weighting factor for radiation, R.

To determine the total equivalent dose, H_T, irradiated in a field consisting of several radiations with different values of w_R, the absorbed dose must be subdivided into blocks, multiplied by its own value of w_R, and summed over the different radiations, R, i.e.,

$$H_T = \sum_R w_R \cdot D_{T,R}.$$

The radiation weighting factor w_R is calculated from Q.

Effective dose

The effective dose, E, is the sum of the weighted equivalent doses in all the tissues and organs of the body. It is given by the expression:

$$E = \sum_T w_T \cdot H_T,$$

where H_T is the equivalent dose in tissue or organ T and w_T is the weighting factor for tissue T.

Ambient dose equivalent

The ambient dose equivalent, $H^*(d)$, at a point in a radiation field is the dose equivalent that would be produced by the corresponding expanded and aligned field in the ICRU sphere at a depth, d, on the radius opposing the direction of the aligned field. The recommended value of d is 10 mm for penetrating radiation and 0.07 mm for low-penetrating radiation.

The unit of ambient dose equivalent is joule per kilogram [J kg^{-1}] and its special name is sievert [Sv].

Personal dose equivalent

The personal dose equivalent, $H_p(d)$, is the dose equivalent in soft tissue at an appropriate depth, d, below specified point in the body. The unit of personal dose equivalent is joule per kilogram [J kg^{-1}] and its special name is sievert [Sv].

Irradiation geometries (Fig. A.1)

Antero-posterior geometry (AP): The irradiation geometry in which the ionizing radiation is incident on the front of the body in a direction orthogonal to the long axis of the body.

Postero-anterior geometry (PA): The irradiation geometry in which the ionizing radiation is incident on the back of the body in a direction orthogonal to the long axis of the body.

Lateral geometry (LAT): The irradiation geometry in which the ionizing radiation is incident from either side of the body in a direction orthogonal to the long axis of the body.

When it is necessary to be more specific, the following symbols are used: **RLAT** (from the right side to the left side of the body) or **LLAT** (from the left side to the right side of the body).

Rotational geometry (ROT): The geometry in which the body is irradiated by a parallel beam of ionizing radiation, from a direction orthogonal to the long axis of the body, which rotates at a uniform rate around the long axis. Alternatively, this geometry may be defined by rotating the body at a uniform rate about its long axis, while irradiating the body by a broad beam of ionizing radiation from a stationary source located on an axis at right angles to the long axis of the body.

Isotropic geometry (ISO): Defined by a radiation field in which the particle fluence per unit of solid angle is independent of direction.

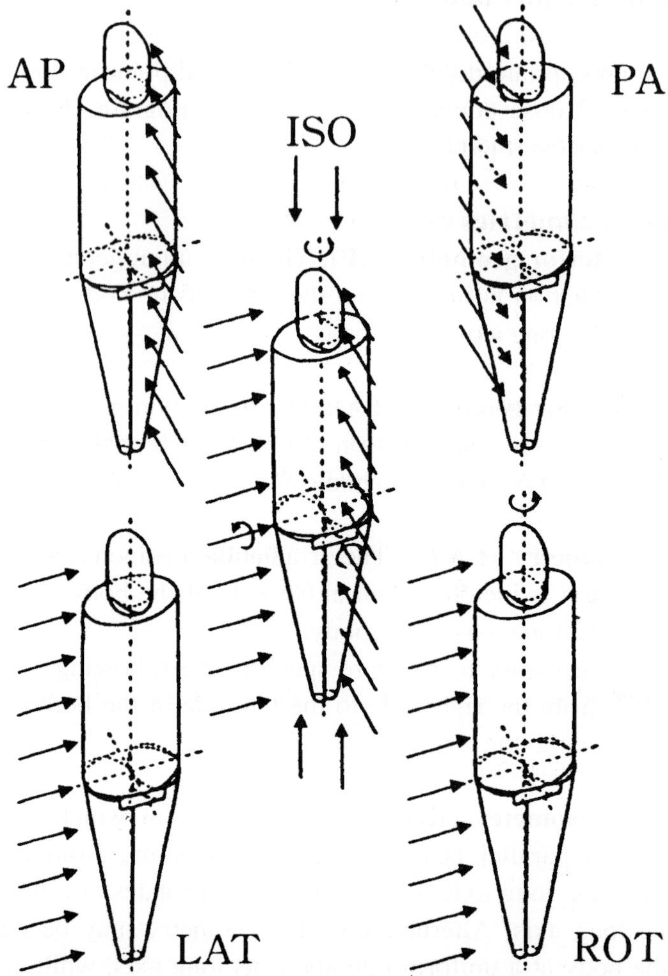

Fig. A.1 Definitions of irradiation geometries. [2.40]

A.2 Approximation Functions of Total Cross Section

In the Monte-Carlo simulator CORIMS [A.1]/SEALER [A.2], rough approximation functions are made for the total cross section of elements used in silicon semiconductor devices and implemented into the program as a lookup table. The original data are taken from JANIS [A.3] and LA150 [A.4]. The actual expressions are shown below. Many resonant peaks are neglected for simplicity. This may be unacceptable for the specialists in nuclear physics, but it was absolutely necessary to calculate failure rates within a reasonable CPU time. The authors expect a simple but much more accurate calculation formula will be available in future.

In many cases, the cross section is set to be constant beyond the maximum energy to which the evaluated data are available. The data base beyond the current maximum neutron energies is also expected to improved in the future.

The comparison between the original data and approximated values are made in **Figs. A.2–A.9**.

<Natural Si>

$$\sigma_{total} = \begin{cases} 0.00077E_n^4 - 0.0221E_n^3 + 0.2453E_n^2 - 1.2986E_n + 4.6868 & (1 \le E_n \le 10) \\ -3\times10^{-4} E_n^2 + 0.02E_n + 1.5938 & (10 < E_n \le 48) \\ 0.00008E_n^2 - 0.0262E_n + 2.8648 & (48 < E_n \le 150) \\ 0.728 & (150 < E_n) \end{cases}$$

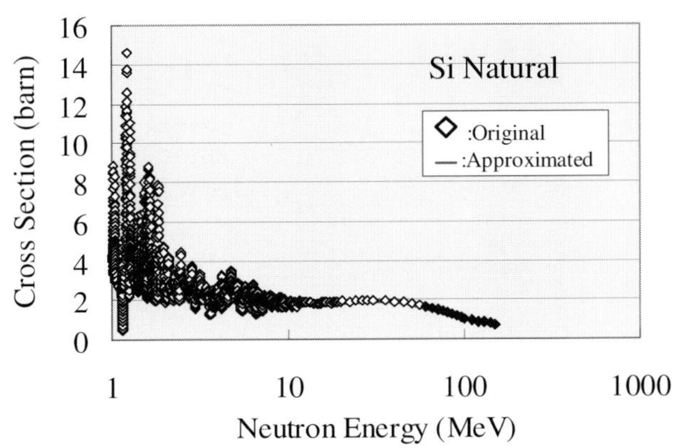

Fig. A.2 Total cross section of natural Si: comparison between the original data [A.3, A.4] and the approximated values.

<^{14}N>

$$\sigma_{total} = \begin{cases} -0.0049E_n^3 + 0.0898E_n^2 - 0.5798E_n + 2.7883 & (1 \le E_n < 8.8) \\ 0.00032E_n^3 - 0.0198E_n^2 + 0.3921E_n - 0.9013 & (8.8 \le E_n \le 23) \\ 1.7\times10^{-9}E_n^4 - 9.3\times10^{-7}E_n^3 + 0.00023E_n^2 - 0.0309E_n + 2.1034 & (23 \le E_n \le 150) \\ 0.377 & (150 < E_n) \end{cases}$$

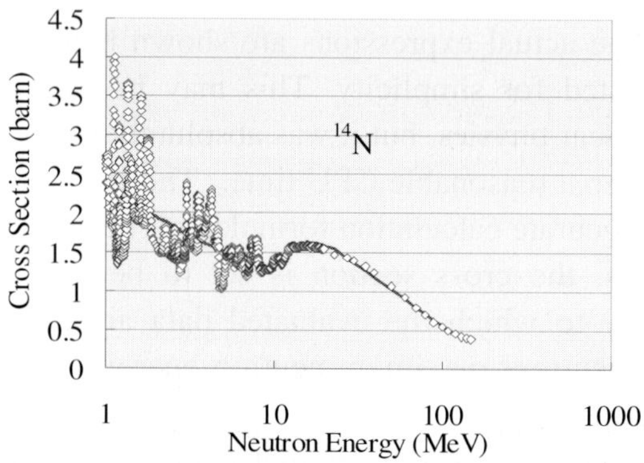

Fig. A.3 Total cross section of ^{14}N: comparison between the original data [A.3, A.4] and the approximated values.

<^{16}O>

$$\sigma_{total} = \begin{cases} -1.4827E_n + 4.6665 & (1 \le E_n < 2.49) \\ 4.772E_n^2 - 25.693E_n + 35.466 & (2.49 \le E_n < 3.44) \\ -0.0762E_n^3 + 1.5406E_n^2 - 10.333E_n + 24.248 & (3.44 \le E_n < 8.5) \\ 0.000352E_n^4 - 0.019E_n^3 + 0.366E_n^2 - 2.9128E_n + 9.3346 & (8.5 \le E_n < 19.4) \\ 0.0000002E_n^3 + 0.000008E_n^2 - 0.017E_n + 2.017 & (19.4 \le E_n < 150) \\ 0.322 & (E_n \ge 150) \end{cases}$$

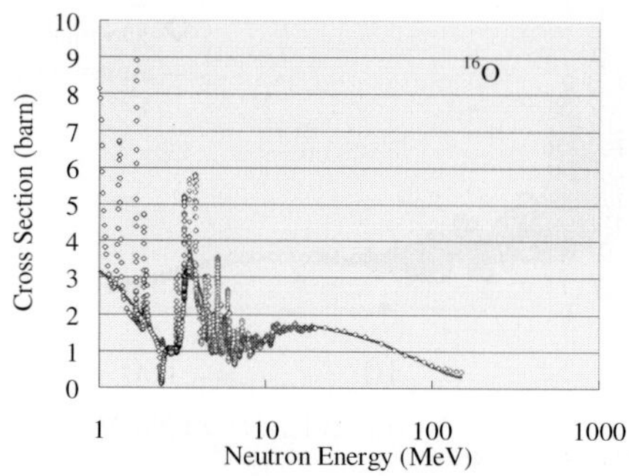

Fig. A.4 Total cross section of ^{16}O: comparison between the original data [A.3, A.4] and the approximated values.

<^{27}Al>

$$\sigma_{total} = \begin{cases} 0.0002E_n^4 - 0.0068E_n^3 + 0.0966E_n^2 - 0.7234E_n + 4.084 & (1 \le E_n < 10.1) \\ -1.3\times10^{-8}E_n^4 + 5.4\times10^{-6}E_n^3 - 3\times10^{-4}E_n^2 + 0.03E_n + 1.4707 & (10.1 \le E_n \le 150) \\ 0.739 & (150 < E_n) \end{cases}$$

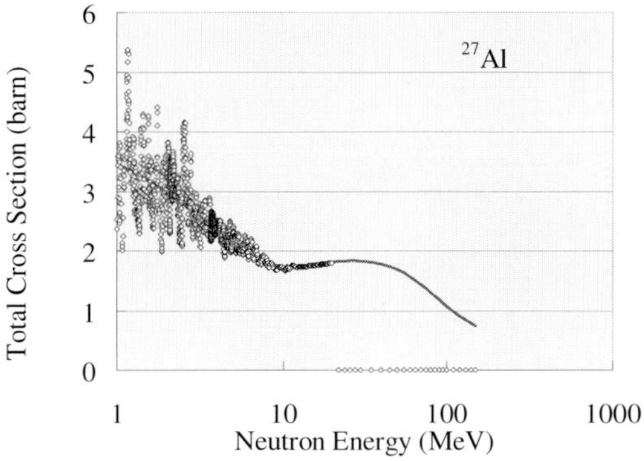

Fig. A.5 Total cross section of ^{27}Al: comparison between the original data [A.3, A.4] and the approximated values.

<Natuarl Cu>

$$\sigma_{total} = \begin{cases} -0.5683E_n + 3.9455 & (1 \le E_n < 1.64) \\ 0.0398E_n^4 - 0.4751E_n^3 + 2.1553E_n^2 - 4.1112E_n + 5.8066 & (1.64 \le E_n < 3.75) \\ -0.00081E_n^4 + 0.0289E_n^3 - 0.388E_n^2 + 2.2212E_n - 0.712 & (3.75 \le E_n < 12) \\ 8\times10^{-7}E_n^4 - 0.00026E_n^3 + 0.0182E_n^2 - 0.4808E_n + 6.8238 & (12 \le E_n < 31) \\ -3\times10^{-8}E_n^4 + 1.33\times10^{-5}E_n^3 - 0.0021E_n^2 + 0.125E_n + 0.1687 & (31 \le E_n < 150) \\ 1.4029 & (150 \le E_n) \end{cases}$$

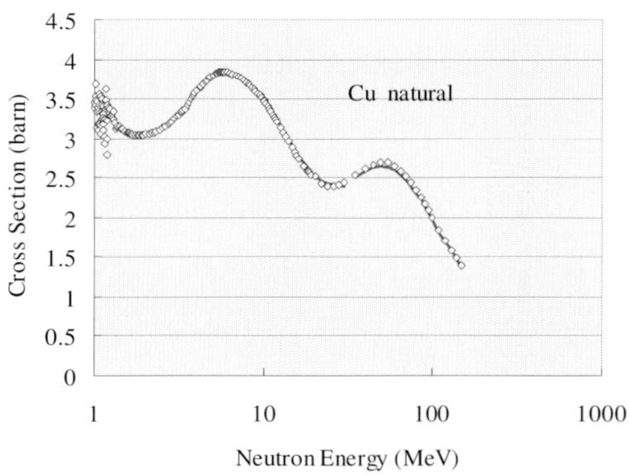

Fig. A.6 Total cross section of natural Cu: comparison between the original data [A.3, A.4] and the approximated values.

<Natural Ti>

$$\sigma_{total} = \begin{cases} -0.0031E_n^4 + 0.0778E_n^3 - 0.6936E_n^2 + 2.447E_n + 0.8364 & (1 \le E_n < 7) \\ -0.0049E_n^2 - 0.0731E_n + 3.9756 & (7 \le E_n < 11) \\ 0.0074E_n^2 - 0.2701E_n + 4.6786 & (11 \le E_n < 20) \\ 2.18 & (20 \le E_n) \end{cases}$$

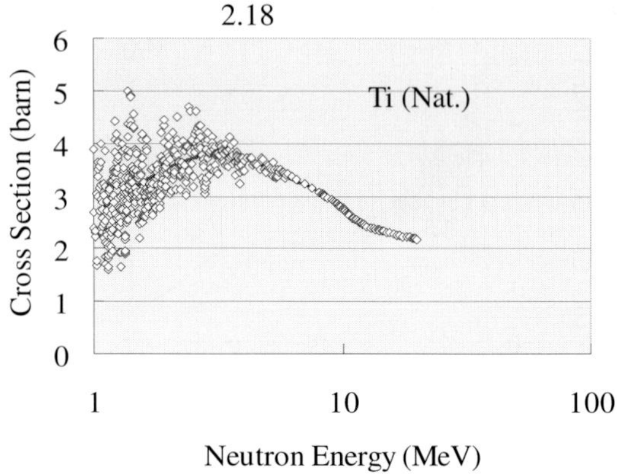

Fig. A.7 Total cross section of natural Ti: comparison between the original data [A.3, A.4] and the approximated values.

<^{181}Ta>

$$\sigma_{total} = \begin{cases} -0.0184E_n^4 + 0.3675E_n^3 - 2.3E_n^2 + 5.314E_n + 3.1325 & (1 \le E_n < 3.25) \\ 0.0053E_n^3 - 0.0579E_n^2 - 0.2282E_n + 7.7697 & (3.25 \le E_n < 8) \\ -0.0003E_n^3 + 0.007E_n^2 + 0.0135E_n + 4.6037 & (8 \le E_n < 20) \\ 5.3005 & (20 \le E_n) \end{cases}$$

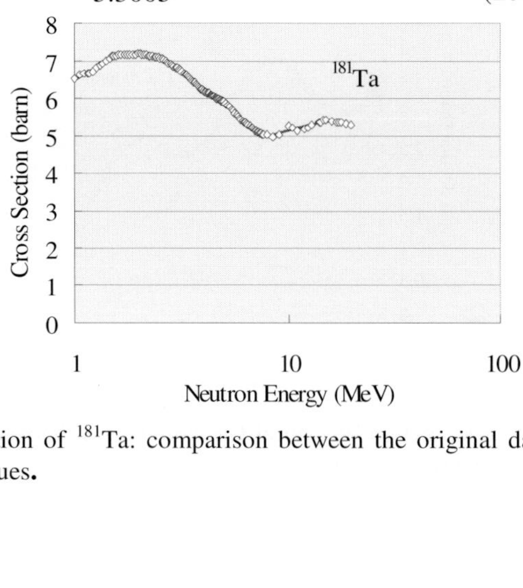

Fig. A.8 Total cross section of ^{181}Ta: comparison between the original data [A.3, A.4] and the approximated values.

<^{184}W>

$$\sigma_{total} = \begin{cases} 0.01E_n^3 - 0.157E_n^2 + 0.4E_n + 6.9369 & (1 \le E_n < 9.3) \\ -0.000235E_n^3 + 0.0057E_n^2 + 0.0184E_n + 4.6173 & (9.3 \le E_n < 22.5) \\ -5\times10^{-5}E_n^3 + 0.0088E_n^2 - 0.48E_n + 12.228 & (22.5 \le E_n < 82) \\ 5.49\times10^{-6}E_n^3 - 0.00184E_n^2 + 0.181E_n - 0.9537 & (82 \le E_n < 150) \\ 3.27 & (150 \le E_n) \end{cases}$$

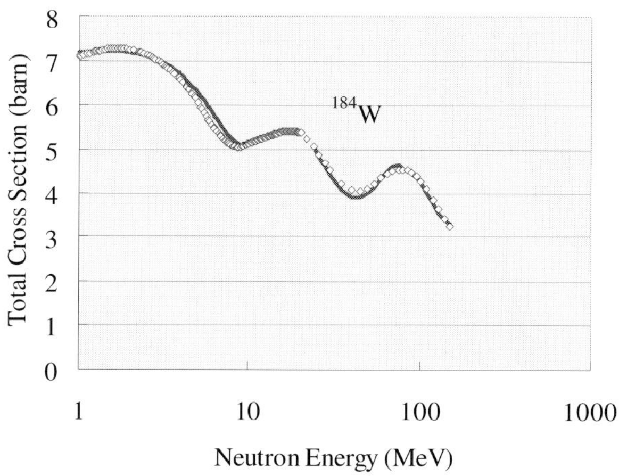

Fig. A.9 Total cross section of ^{184}Wi: comparison between the original data [A.3, A.4] and the approximated values.

A.3 Approximation Functions for Non-elastic Cross Section

The approximation functions for non-elastic reaction cross sections are constructed in similar manner to the total cross sections, so that the same limitations in accuracy should be noted.

The actual approximation functions implemented into the Monte-Carlo simulator CORIMS/SEALER are summarized below with figures that show the comparison between the data from the literature and approximated results.

<Natural Si>

$$\sigma_{nonela} = \begin{cases} 0.0006E_n^3 - 0.0266E_n^2 + 0.3594E_n - 0.4214 & (1 \le E_n < 10.6) \\ 0.000195E_n^3 - 0.011E_n^2 + 0.1901E_n + 0.1316 & (10.6 \le E_n < 20) \\ -0.0000014E_n^3 + 0.000354E_n^2 - 0.0312E_n + 1.4211 & (20 \le E_n < 110) \\ 0.42 & (110 \le E_n) \end{cases}$$

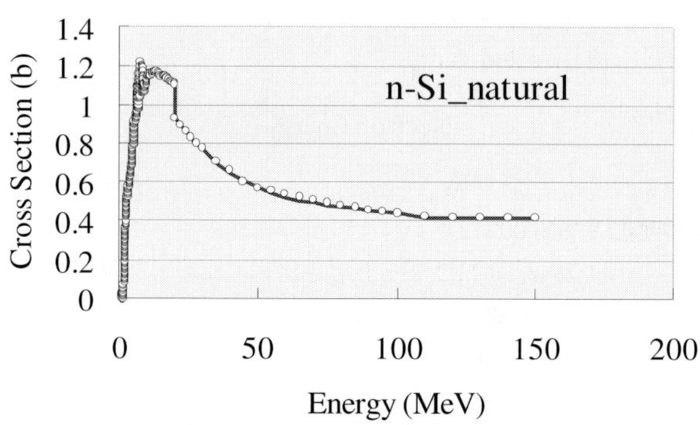

Fig. A.10 Non-elastic cross section of natural Si: comparison between the original data [A.3, A.4] and the approximated values.

<^{14}N>

$$\sigma_{nonela} = \begin{cases} -0.0202E_n^3 + 0.15144E_n^2 - 0.1999E_n + 0.0095 & (0 \le E_n < 5.53) \\ -0.0092E_n^3 + 0.1799E_n^2 - 1.0975E_n + 2.369 & (5.53 \le E_n < 9.352) \\ 0.001133E_n^3 - 0.0572E_n^2 + 0.9343E_n - 4.2959 & (9.352 \le E_n < 20) \\ -2.13\times10^{-7}E_n^3 + 8\times10^{-5}E_n^2 - 0.0102E_n + 0.7052 & (20 \le E_n < 150) \\ 0.256 & (E_n \ge 150) \end{cases}$$

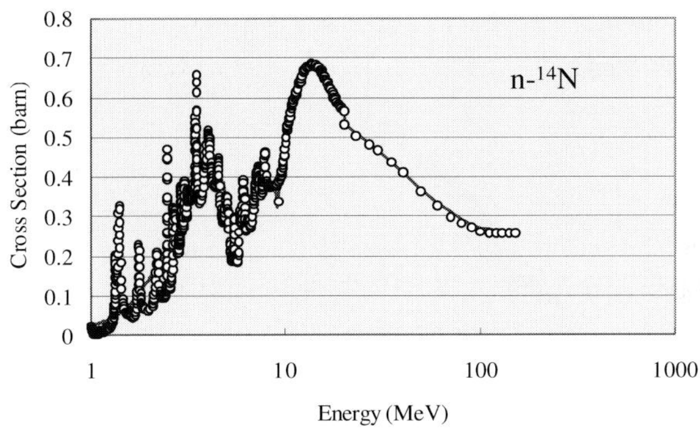

Fig. A.11 Non-elastic cross section of natural ^{14}N: comparison between the original data [A.3, A.4] and the approximated values.

<^{16}O>

$$\sigma_{nonela} = \begin{cases} 0 & (1 \le E_n < 3.6) \\ -0.0004E_n^3 + 0.0077E_n^2 + 0.0323E_n - 0.1886 & (3.6 \le E_n < 18.3) \\ -0.06E_n^2 + 2.3391E_n - 22.18 & (18.3 \le E_n < 20) \\ -3.96\times10^{-7}E_n^3 + 0.000128E_n^2 - 0.0142E_n + 0.8403 & (20 \le E_n < 120) \\ 0.295 & (E_n \ge 120) \end{cases}$$

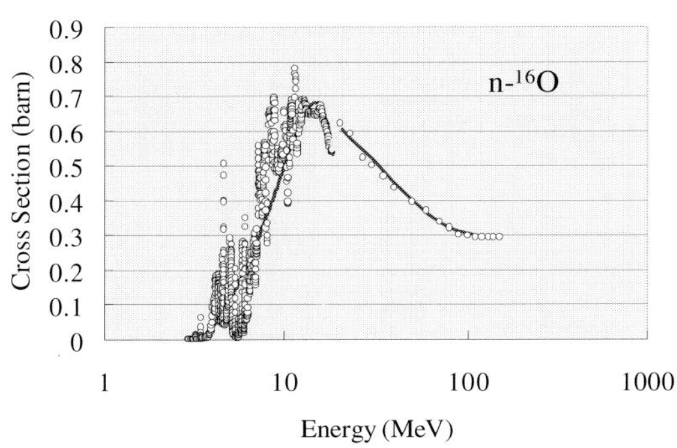

Fig. A.12 Non-elastic cross section of ^{16}O: comparison between the original data [A.3, A.4] and the approximated values.

<^{27}Al>

$$\sigma_{nonela} = \begin{cases} -0.286E_n^2 + 1.1322E_n - 0.8106 & (1 \le E_n < 2.328) \\ 0.0141E_n^3 - 0.23E_n^2 + 1.2822E_n - 1.6308 & (2.328 \le E_n < 7.3) \\ -0.0018E_n^3 + 0.0534E_n^2 - 0.4691E_n + 2.2278 & (7.3 \le E_n < 11.35) \\ -0.0048E_n^2 + 0.0849E_n + 0.7981 & (11.35 \le E_n < 16.4) \\ -5.4\times10^{-7}E_n^3 + 1.77\times10^{-4}E_n^2 - 0.01981E_n + 1.187 & (16.4 \le E_n < 130) \\ 0.42 & (130 \le E_n) \end{cases}$$

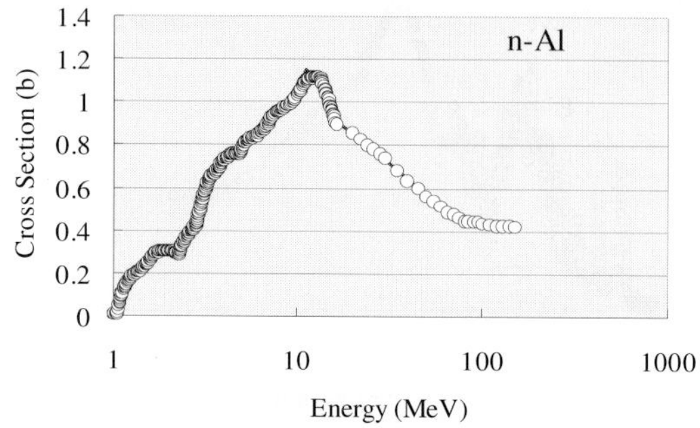

Fig. A.13 Non-elastic cross section of ^{27}Al comparison between the original data [A.3, A.4] and the approximated values.

<Natural Cu>

$$\sigma_{nonela} = \begin{cases} -0.0116E_n^4 + 0.1783E_n^3 - 1.04E_n^2 + 2.8931E_n - 1.8319 & (1 \le E_n < 5.6) \\ -0.0009E_n^3 + 0.0205E_n^2 - 0.1573E_n + 2.0616 & (5.6 \le E_n < 11.25) \\ -0.0001E_n^3 + 0.0072E_n^2 - 0.1694E_n + 2.7472 & (11.25 \le E_n < 20) \\ -0.00000021E_n^3 + 0.00009E_n^2 - 0.0147E_n + 1.64 & (20 \le E_n < 150) \\ 0.75 & (150 \le E_n) \end{cases}$$

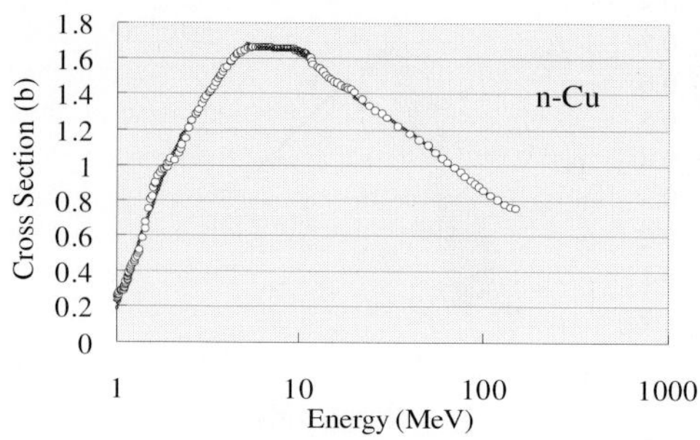

Fig. A.14 Non-elastic cross section of natural Cu: comparison between the original data [A.3, A.4] and the approximated values.

<Natural Ti>

$$\sigma_{nonela} = \begin{cases} 0.0568E_n^3 - 0.7915E_n^2 + 2.7178E_n - 1.8945 & (1 \le E_n < 2.45) \\ -0.1393E_n^3 + 1.1437E_n^2 - 2.6283E_n + 2.5491 & (2.45 \le E_n < 4.345) \\ -0.0478E_n^3 + 0.8019E_n^2 - 4.228E_n + 9.3117 & (4.345 \le E_n < 6.228) \\ 0.0001E_n^3 - 0.0036E_n^2 + 0.0108E_n + 1.3717 & (6.228 \le E_n < 12.91) \\ -0.0003E_n^2 + 0.0128E_n + 1.0591 & (12.91 \le E_n < 20) \\ 1.2 & (20 \le E_n) \end{cases}$$

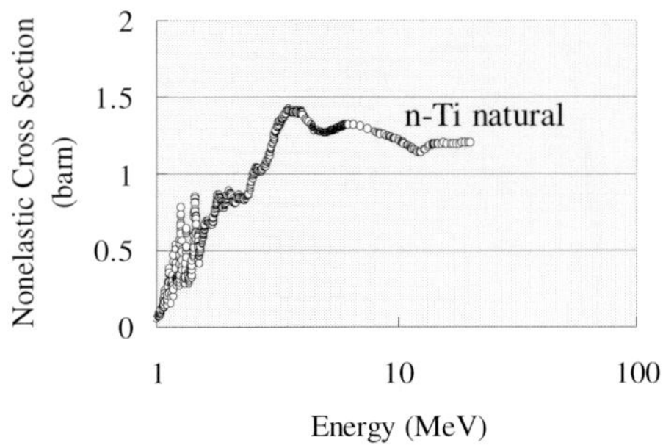

Fig. A.15 Non-elastic cross section of natural Ti: comparison between the original data [A.3, A.4] and the approximated values.

<^{181}Ta>

$$\sigma_{nonela} = \begin{cases} -1.6003E_n^2 + 4.6545E_n + 0.0046 & (1 \le E_n < 1.5) \\ 0.0743E_n^2 - 0.6374E_n + 4.0339 & (1.5 \le E_n < 5) \\ 0.0014E_n^2 - 0.0491E_n + 2.8936 & (5 \le E_n < 20) \\ 2.5 & (20 \le E_n) \end{cases}$$

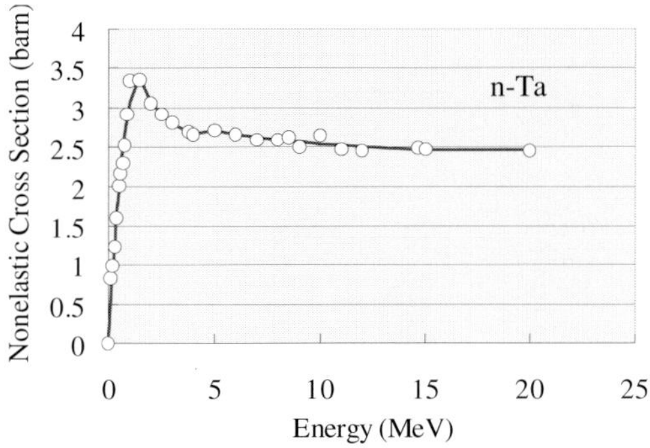

Fig. A.16 Non-elastic cross section of ^{181}Ta: comparison between the original data [A.3, A.4] and the approximated values.

<^{184}W>

$$\sigma_{nonela} = \begin{cases} -0.8447E_n^2 + 3.3126E_n - 0.0365 & (1 \le E_n < 2.0) \\ -0.0015E_n^2 - 0.0906E_n + 3.4482 & (2.0 \le E_n < 8.5) \\ 0.0014E_n^2 - 0.0452E_n + 2.8429 & (8.5 \le E_n < 20) \\ -8\times10^{-11}E_n^3 + 2.85\times10^{-5}E_n^2 - 0.0114E_n + 2.6906 & (20 \le E_n < 150) \\ 1.62 & (150 \le E_n) \end{cases}$$

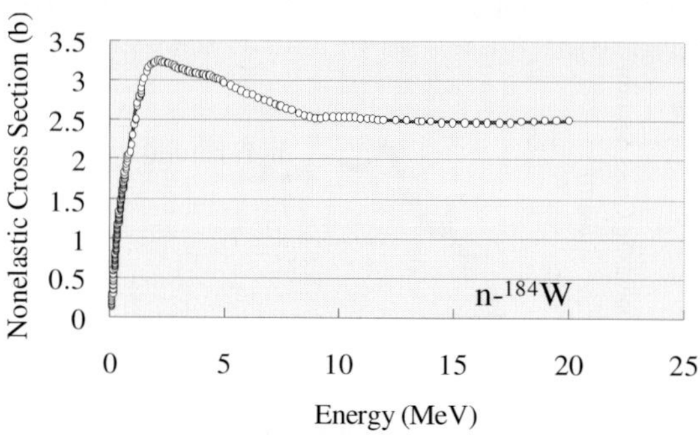

Fig. A.17 Non-elastic cross section of ^{184}W: comparison between the original data [A.3, A.4] and the approximated values.

A.4 Comparison of GEM Calculation Results for Inverse Reaction Cross Section with Literature Data

Figures A.18–A.23 show comparisons between literature data [A.5] and GEM [A.6] calculation results on inverse binary collision cross sections. The number of literature data does not seem to be statistically enough, so that it is difficult to make a conclusive remark but generally the calculated results from GEM seem to be reasonably consistent as a preliminary step.

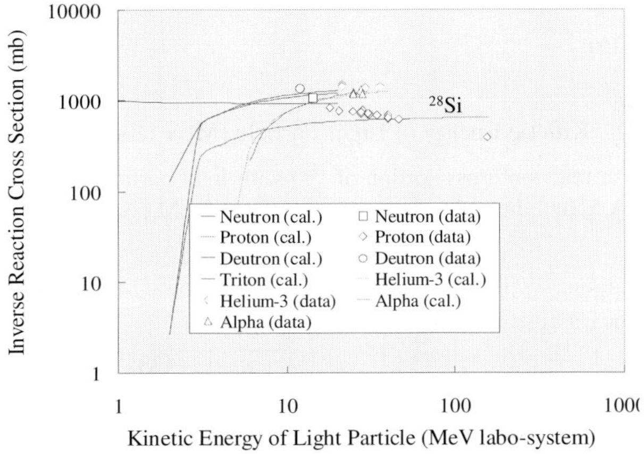

Fig. A.18 Inverse reaction cross section of ^{28}Si with light particles: comparison between literature data [A.5] and the approximated values from GEM [A.6].

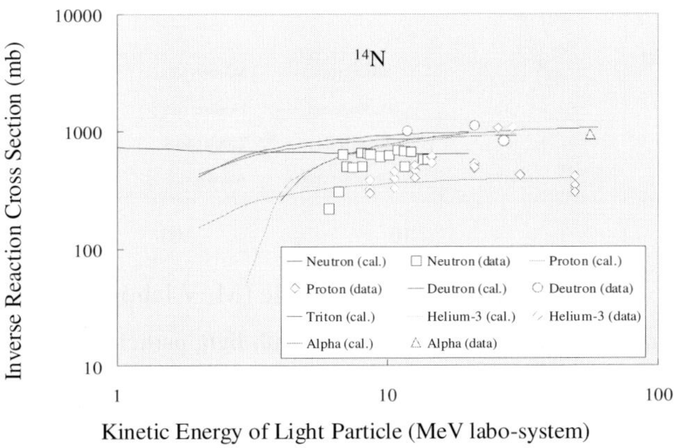

Fig. A.19 Inverse reaction cross section of ^{14}N with light particles: comparison between literature data [A.5] and the approximated values from GEM [A.6].

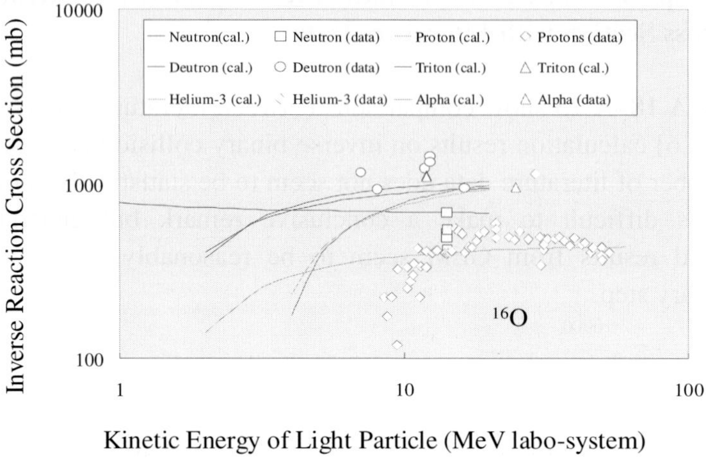

Fig. A.20 Inverse reaction cross section of ^{16}O with light particles: comparison between literature data [A.5] and the approximated values from GEM [A.6].

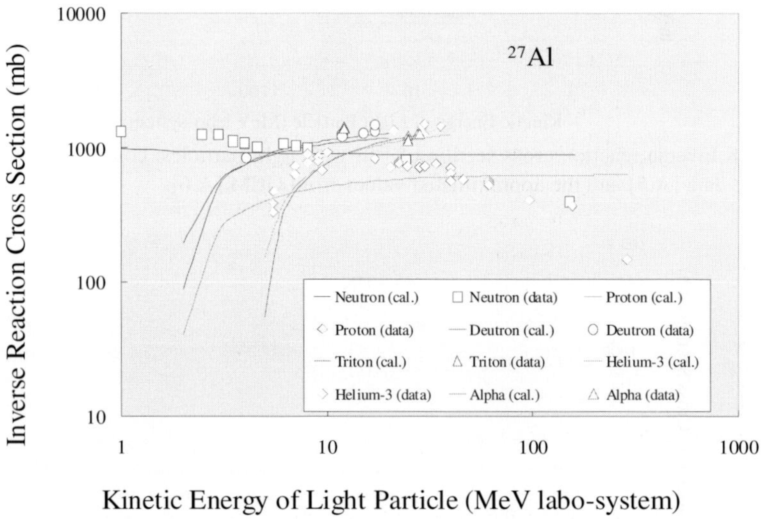

Fig. A.21 Inverse reaction cross section of ^{27}Al with light particles: comparison between literature data [A.5] and the approximated values from GEM [A.6].

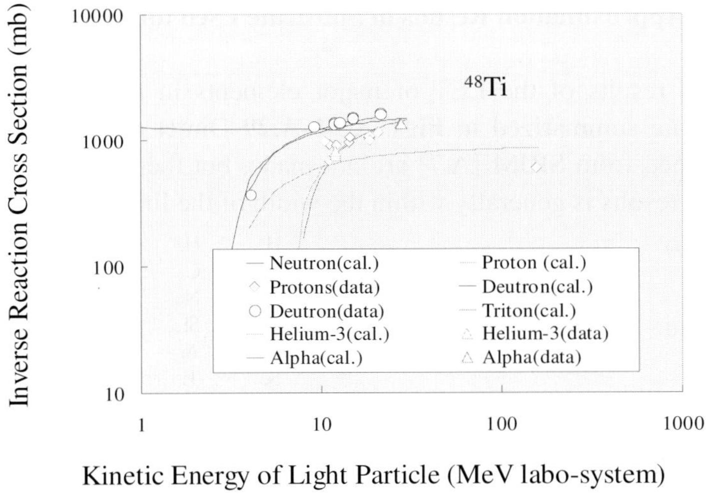

Fig. A.22 Inverse reaction cross section of ^{48}Ti with light particles: comparison between literature data [A.5] and the approximated values from GEM [A.6].

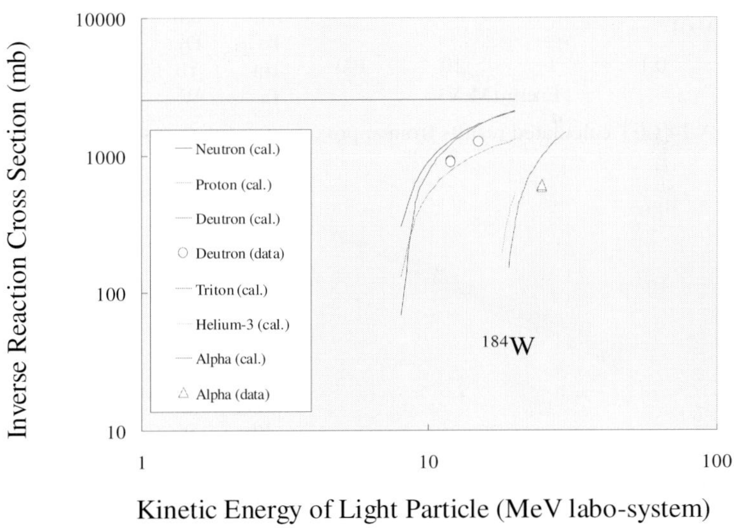

Fig. A.23 Inverse reaction cross section of ^{184}W with light particles: comparison between literature data [A.5] and the approximated values from GEM [A.6].

A.5 LET Approximation Results in Substrate Used for Silicon Devices

Calculated results of the LET of major elements in major component materials are summarized in **Figs. A.24–A.29** Direct comparisons with data obtained from SRIM [A.7] are not made, but the accuracy of each calculated results is generally within the width of the line.

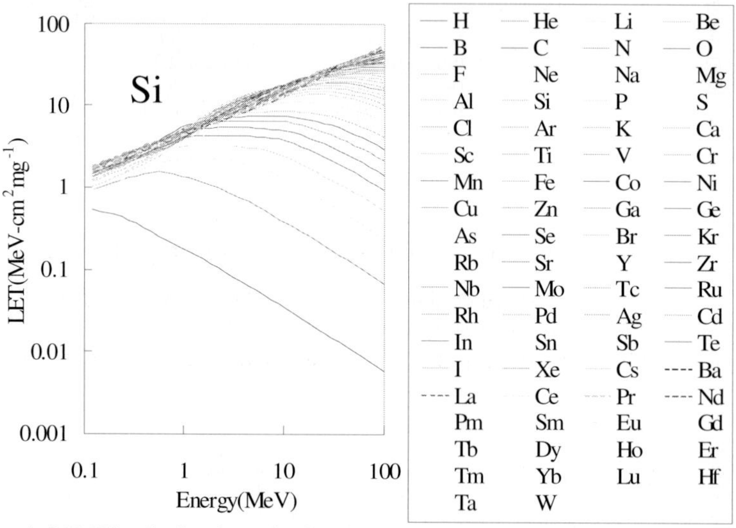

Fig. A.24 LET calculated results from approximation functions in a Si substrate.

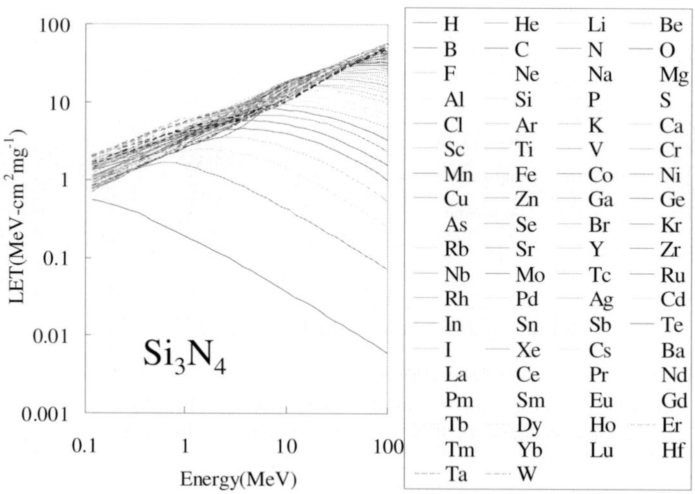

Fig. A.25 LET calculated results from approximation functions in a Si_3N_4 substrate.

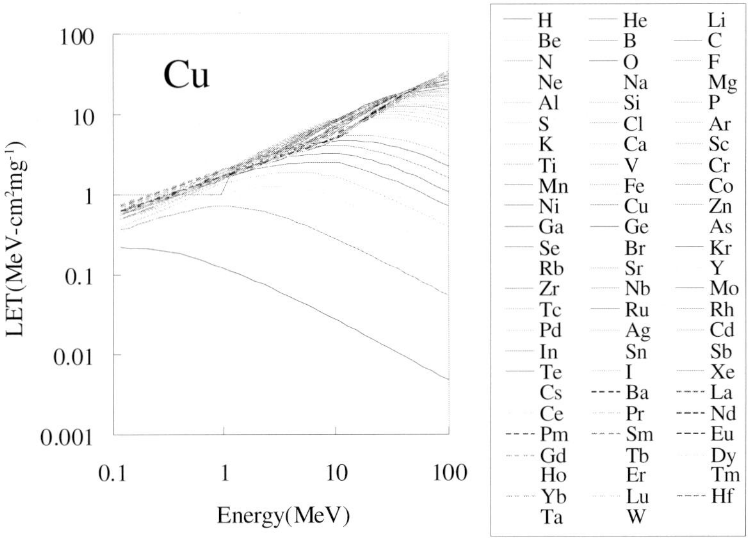

Fig. A.26 LET calculated results from approximation functions in a Cu substrate.

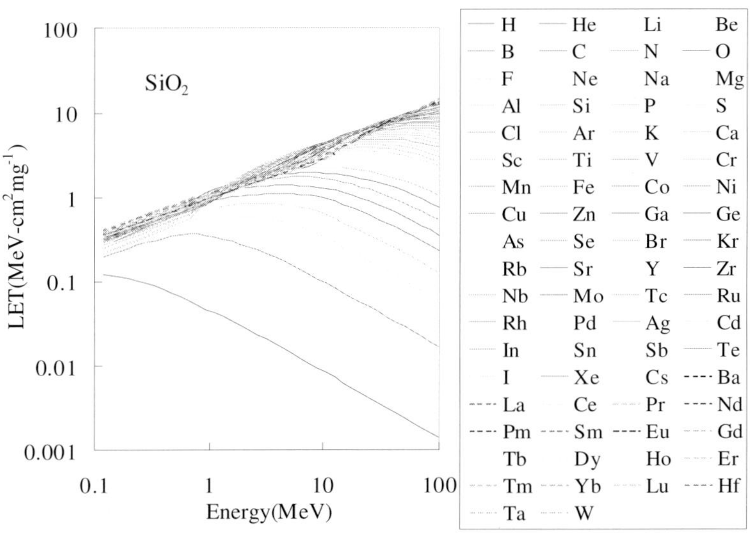

Fig. A.27 LET calculated results from approximation functions in a SiO_2 substrate.

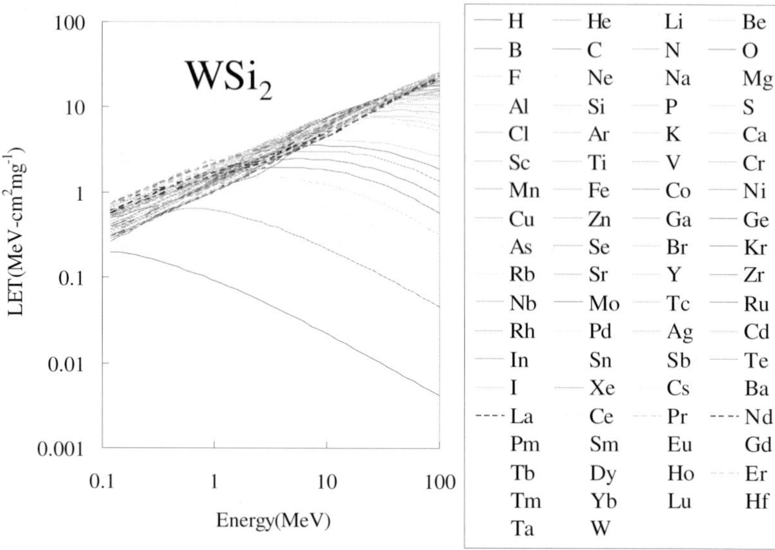

Fig. A.28 LET calculated results from approximation functions in a WSi_2 substrate.

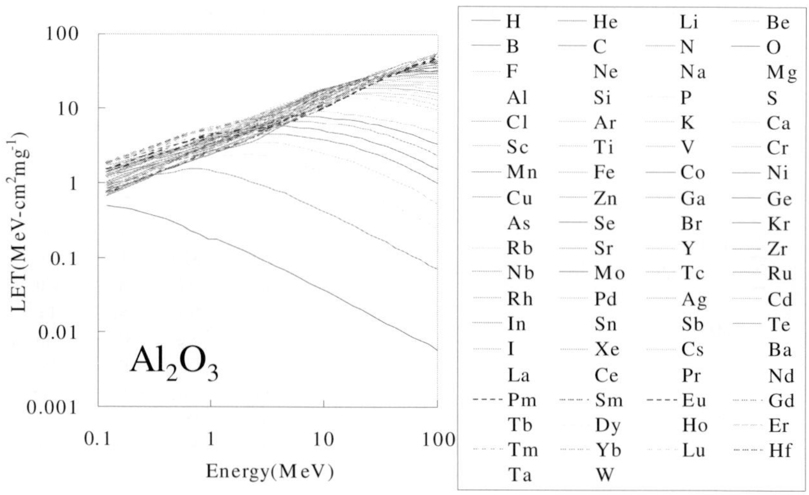

Fig. A.29 LET calculated results from approximation functions in an Al_2O_3 substrate.

A.6 Coefficients for the LET Calculation for Substrate Used in Silicon Devices

Parameters for LET calculation are given in Table A.30 for all combinations of a specific ion and substrate. In the table, each column stands for:

(1) First column: combination of ion and substrate. A stands for the ion and B stands for the substrate in A_B.
(2) Second column: I,II, and III stand for the electromagnetic loss in three energy regions of the incident ion. nI, nII, and nIII stand for the nuclear loss in three energy regions.
(3) Third column: The lower energy boundary in each energy region ($\log_{10}(E_{min})$).
(4) Fourth column: The upper energy boundary in each energy region ($\log_{10}(E_{max})$).
(5) Fifth column: coefficient for x^2 term.
(6) Sixth column: coefficient for x term.
(7) Seventh column: constant.

Coefficients in Eq. (6.6) for LET calculation are tabulated in the CD (Table A.30).

REFERENCES

Chapter 1

[1.1] E.L. Petersen, "The SEU Figure of Merit and Proton Upset Rate Calculations," *IEEE Trans. Nuc. Sci.*, Vol. 45, pp.2550-2562 (1998).

[1.2] S. Kuboyama, K. Sugimoto, S. Shugyo, S. Matsuda, and T. Hirao, "Single-Event Burnout of Epitaxial Bipolar Transistors," *IEEE Trans. Nuc. Sci.*, Vol. 45, pp.2527-2533 (1998).

[1.3] P.E. Dodd, *et al.*, Sexton, "Impact of Ion Energy on Single-Event Upset," *IEEE Trans. Nuc. Sci.*, Vol. 45, pp.2483-2491 (1998).

[1.4] T. Goka, H. Matsumoto, N. Nemoto, "SEE Flight Data from Japanese Satellites," *IEEE Trans. Nuc. Sci.*, Vol. 45, pp.2771-2778 (1998).

[1.5] R. Koga, "Single Event Sensitivities of 128- and 256-megabit Synchronous Dynamic Random Access Memories (SDRAMs)," *Proc. the 4th Int'l Workshop on Radiation Effects on Semiconductor Devices for Space Application, Tsukuba, 2000/5/15-18*, pp.81-88 (2000).

[1.6] E. Normand, "Extensions of Burst Generation Rate Method for Wider Application to Proton/Neutron-Induced Single Event Effects," *IEEE Trans. Nuc. Sci.*, Vol. 45, pp.2904-2914 (1998).

[1.7] E. Normand, and T.J. Baker, "Altitude and Latitude Variations in Avionics SEU and Atmospheric Neutron Flux," *IEEE Trans. Nuc. Sci.*, Vol. 40, pp.1484-1490 (1993).

[1.8] T.C. May, and Murray H. Woods, "Alpha-Particle-Induced Soft Errors in Dynamic Memories," *IEEE Trans. Electron Device*, Vol. ED-26, pp.2-9 (1979).

[1.9] J.F. Ziegler, and W.A. Lanford, "Effect of Cosmic Rays on Computer Memories," *Science*, Vol. 206, pp.776-788 (1979).

[1.10] E. Takeda, E. Murakami, K. Torii, Y. Okuyama, E. Ibe, K. Hinode, S. Kimura, "Reliability issues of silicon LSIs facing 100-nm technology node," *Microelectronics Reliability*, Vol. 42, No. 4-5, pp.493-506 (2002).

[1.11] A. Eto, M. Hidaka, Y. Okuyama, K. Kimura, M. Hosono, "Impact of Neutron Flux on Soft Errors in MOS Memories," *International Electron Devices Meeting, San Francisco, CA., December 6-9, 1998,* pp.367-370 (1998).

[1.12] D.C. Bossen, "CMOS Soft Errors and Server Design," *Workshop on Radiation Induced Soft Errors in Silicon Components and Computer Systems, IRPS, Dallas, Texas, April 7* (2002).

[1.13] C. Slayman, "Eliminating the Threat of Soft Errors - A System Vendor Perspective," *IRPS Panel Discussion on Eliminating the Threat of Soft Errors, Dallas, Texas, April 2, 2003,* No. 6 (2003).

[1.14] E. Takeda, *et al.*, "Reliability issues of silicon LSIs facing 100-nm technology node," *Microelectronics Reliability*, Vol. 42, pp.493-506 (2002).

[1.15] E. Ibe, "Current and Future Trend on Cosmic-Ray-Neutron Induced Single Event Upset at the Ground down to 0.1-Micron-Device," *TSL Workshop on Applied Physics, Uppsala, May, 3*, No. 1 (2001).

[1.16] E. Ibe, Y. Yahagi, F. Kataoka, Y. Saito, A. Eto, M. Sato, H. Kameyama, and M. Hidaka, "A Self-Consistent Integrated System for Terrestrial-Neutron Induced Single Event Upset of Semiconductor Devices at the Ground," *2002 ICITA, Buthurst, Australia,* No. 273-21 (2002).

[1.17] Actel, "Understanding Soft and Firm Errors in Semiconductor Devices," www.actel.com (2003).

[1.18] J.J. Fabula, A. Lesea, C. Carmichael, and S. Srimer, "The NSEU Response of Static Latch Based FPGAs," http://www.xilinx.com/xlnx/xil_prodcat_product.jsp?title=aero_radiation.

[1.19] O. Musseau, A. Torres, A.B. Campbell, A.R. Knudson, and S. Buchner, "Medium-Energy Heavy-Ion Single-Event-Burnout Imaging of Power MOSFETs," *IEEE Trans. Nucl. Sci.*, Vol. 4, pp.1415-1420 (1999).

[1.20] S. Huang, and G.A.J. Amaratunga, "Analysis of SEB and SEGR in Super-Junction MOSFETs," *IEEE Trans. Nuc. Sci.*, Vol. 47, pp.2640-2647 (2000).

[1.21] SIA, International Technology Roadmap for Semiconductors, 1999 Edition (1999).

[1.22] E. Ibe, Y. Yahagi, Y. Saito, H. Kameyama, H. Yukawa, A. Eto, and M. Hidaka, "Establishment of accelerated soft-error testing of sub-quarter micron silicon devices by using quasi-mono energy neutron beams," *TSL Progress Report 2000-2002*, pp.35-37 (2002).

[1.23] E. Ibe, "Self-Consistent Integrated System for Terrestrial Neutron Soft-error of Sub- Quarter Micron Devices," *Eliminating the threat of soft errors IRPS SER Panel Discussion, Dallas, Texas, April 2, 2003*, No. 5 (2003).

[1.24] Y. Yahagi, E. Ibe, Y. Saito, A. Eto, M. Sato, "Self-Consistent Integrated System for Susceptibility to Terrestrial-Neutron Induced Soft-error of Sub-quater Micron Memory Devices," *2002 International Integrated Reliability Workshop, Stanford Sierra Camp, S. Lake Tahoe*, pp.143-146 (2002).

[1.25] Y. Yahagi, E. Ibe, Y. Takahashi, Y. Saito, A. Eto, M. Sato, H. Kameyama, M. Hidaka, K. Terunuma, T. Nunomiya, and T. Nakamura, "Threshold Energy of Neutron-Induced Single Event Upset as a Critical Factor," *2004 IRPS, April 25-29, Phoenix, Arizona,* pp.669-670 (2004).

[1.26] J.F. Ziegler, H.P. Muhlfield, C.J. Montrose, H.W. Curtis, T.J. O. Gorman, and J.M. Ross, "Accelerated testing for Cosmic Soft-Error Rate," *IBM J. Res. Develop.*, Vol. 40, pp.51-72 (1996).

[1.27] E. Takeda, K. Takeuchi, E. Yamazaki., T. Toyabe, K. Ohshima, and K. Itoh, "Effective Funneling Lengh in Alpha-Particle Induced Soft Errors," *Extended*

Abstracts of the 18th Conference on Solid State Devices and Materials, Tokyo, No. B-7-1, pp.311-314 (1986).

[1.28] R. Jones, "Comparison Between SRAM SEE Cross-Sections From Ion Beam Testing With Thoese Obtained Using a New Picosecond Pulsed Laser Facility," *IEEE Trans. Nuc. Sci.*, Vol. 47, No. 3, pp.539-544 (2000).

[1.29] J. Barak, "Microbeam Mapping of Single Event Latchups and Single Event Upsets in CMOS SRAMs," *IEEE Trans. Nuc. Sci.*, Vol. 45, No. 3, pp.1595-1602 (1998).

[1.30] C. Inguimbert, C. Duzellier, R. Ecoffet, L. Guibert, J. Barak, M. Chabot, "Using a Carbon Beam as a Probe extract the Thickness of Sensitive Volumes," *IEEE Trans. Nuc. Sci.*, Vol. 47, No. 3, pp.551-558 (2000).

[1.31] R. Koga, P. Yu, S. Crain, K. Crawford, "Comparison of Single Event Effects (SEE) Sensitivities Induced by Protons and Heavy Ions," *Proc. the 5th Int'l Workshop on Radiation Effects on Semiconductor Devices for Space Application, October, 9-11 Takasaki City Gallery*, pp.149-154 (2002).

[1.32] E. Normand, "Extensions of Burst Generation Rate Method for Wider Application to Proton/Neutron-Induced Single Event Effects," *IEEE Trans. Nuc. Sci.*, Vol. 45, No. 6, pp.2904-2914 (1998).

[1.33] K. Johansson, P. Dyreklev, P. Granbom, N. Olsson, J. Blomgren, P-U. Renberg, "Energy-Resolved Neutron SEU Measurements from 22 to 160 MeV," *IEEE Trans. Nuc. Sci.*, Vol. 45, No. 6, pp.2519-2526 (1998).

[1.34] K. Chiba, I. Nashiyama, H. Asai, K. Sugimoto, Y. Iide, H. Shindo, S. Kuboyama, S. Matsuda, "Proton Irradiation Testing of Commercial Semiconductor Memories for the MDS-1 Project," *Proc. the 5th Int'l Workshop on Radiation Effects on Semiconductor Devices for Space Application, October, 9-11 Takasaki City Gallery*, pp.31-34 (2002).

[1.35] M. Olmos, "Investigation of thermal neutron induced soft error rates in commercial SRAMs with 0.35 μ m to 90 nm technologies," *2006 IRPS, March 26-30, 2006, San Jose, CA.*, No. 3B.2 (2006).

[1.36] T. Granlund, N. Olsson, "A Comparative Study between Proton and Neutron Induced SEUs in SRAMs," *2005 Radiation and Its Effects on Components and Systems, Sep.19-23, Palais des Congres, Cap d'Agde, France*, No. PE-2 (2005).

Chapter 2

[2.1] P. Goldhagen, "Overview of aircraft radiation exposure and recent ER-2 measurements", *Health Phys.* 79, 526-544 (2000).

[2.2] P. Goldhagen, M. Reginatto, T. Kniss, J.W. Wilson, R.C. Singleterry, I.W. Jones, W. Van Steveninck, "Measurement of the energy spectrum of cosmic-ray induced neutrons aboard an ER-2 high-altitude airplane", *Nucl. Instr. and Meth.*, A476, 42-51 (2002).

[2.3] S. Roesler, W. Heinrich, H. Schraube, "Calculation of radiation fields in the atmosphere and comparison to experimental data", *Radiat. Res.*, 149, 87-97 (1998).

[2.4] A. Kurochkin, B. Wiegel, B.R.L. Siebert, "Study of the radiation environment by galactic cosmic rays at flight altitudes at the summit of the Zugspitze and at PTB Braunschweig", *Radiat. Protec. Dosim.*, 83, 281-291 (1999).

[2.5] D. O'Sullivan, D. Bartlett, R. Grillmaier, W. Heinrich, L. Lindborg, H. Schraube, M. Silari, L. Tommasino, D. Zhou, "Investigation of radiation fields at aircraft altitudes", *Radiat. Protec. Dosim.*, 92, 195-198 (2000).

[2.6] International Commission on Radiological Protection, "1990 Recommendations of the International Commission on Radiological Protection", ICRP Publication 60, *Ann. ICRP* 21 (1991).

[2.7] M. Florek, J. Masarik, I. Szarka, D. Nikodemova, A. Hrabovcoba, "Natural neutron fluence rate and the equivalent dose in locations with different elevation and latitude", *Radiat. Protec. Dosim.*, 67, 187-192 (1996).

[2.8] H. Schraube, J. Jakes, A. Sannikov, E. Weitzenegger, S. Roesler, W. Heinrich, "The cosmic ray induced neutron spectrum at the summit of the Zugspitze (2963m)", *Radiat. Protec. Dosim.*, 70, 405-408 (1997).

[2.9] M. Hajek, T. Berger, W. Schoener, N. Vana, "Analysis of the neutron component at high altitude mountains using active and passive measurement devices", *Nucl. Instr. and Meth.*, A476, 69-73 (2002).

[2.10] B. Wiegel, A.V. Alevra, M. Matzke, U.J. Schrewe, J. Wittstock, "Spectrometry using the PTB neutron multisphere spectrometer (NEMUS) at flight altitudes and at ground level", *Nucl. Instr. and Meth.*, A476, 52-57 (2002).

[2.11] D.T. Bartlett, L.G. Hager, R.J. Tanner, J.D. Steele, "Measurements of the high energy neutron component of cosmic radiation fields in aircraft using etched track dosimeters", *Radiat. Meas.* 33, 243-253 (2001).

[2.12] U. Tveten, T. Haldorsen, J. Reitan, "Cosmic radiation and airline pilots: exposure pattern as a function of aircraft type", *Radiat. Protec. Dosim.*, 8**7**, 157-165 (2000).

[2.13] D.T. Bartlett, L.G. Hager, D. Irvine, M. Bagshaw, "Measurements on concorde of the cosmic radiation field at aviation altitudes", *Radiat. Protec. Dosim.*, 91, 365-376 (2000).

[2.14] U.J. Schrewe, "Global measurements of the radiation exposure of civil air crew from 1997 to 1999", *Radiat. Protec. Dosim.*, 91, 347-364 (2000).

[2.15] J.E. Kylloenen, L. Lindborg, G. Samuelson, "Cosmic radiation measurements on-board aircraft with the variance method", *Radiat. Protec. Dosim.*, 93, 197-205 (2001).

[2.16] B.J. Lewis, M.J. McCall, A.R. Green, L.G.I. Bennett, M. Pierre, U.J. Schrewe, K. O'Brien, E. Felsberger, "Aircraft exposure from cosmic radiation on commercial airline routes", *Radiat. Protec. Dosim.*, 93, 293-314 (2001).

[2.17] T. Nakamura, Y. Uwamino, T. Ohkubo, A. Hara, "Altitude variation of cosmic-ray neutrons", *Health Phys.*, 53, 509-517 (1987).

[2.18] J.F. Ziegler, "Terrestrial cosmic rays", *IBM J. Res. Develop.*, 40, 19-39 (1996).

[2.19] R. Gold, "Energy spectrum of fast cosmic-ray neutrons near sea level", *Phys. Rev.*, 165, 1406-1411 (1968).

[2.20] Y. Uwamino, T. Nakamura, A. Hara, "Two types of multi-moderator neutron spectrometers: gamma-ray insensitive type and high-efficiency type", *Nucl. Instr. Meth.*, A239, 299-309 (1985).

[2.21] T. Nunomiya, "Study on the deep penetration of neutrons produced by high-energy accelerator and cosmic ray", *Doctor Thesis of Tohoku University*, (2003), [in Japanese].

[2.22] L. S. Waters, (Ed.), "MCNPX User's Manual verion 2.4.0", *LA-CP-02-408*, Los Alamos National Laboratory, Los Alamos, New Mexico (2002).

[2.23] T. Nakamura, T. Nunomiya, M. Sasaki, "Development of active environmental and personal neutron dosemeters", Invited Paper of 9th Neutron Dosimetry Conference, *Radiat. Protec. Dosim.*, 110, 169-181 (2004).

[2.24] W.N. McElroy, S. Berg, T. Crockett, R.G. Hawkins, "A Computer-Automated Iterative Method for Neutron Flux Spectra Determination by Foil Activation", *AFWL-TR-67-41*, Air Force Weapons Laboratory (1967).

[2.25] J.F.Briesmeister Ed., "MCNP-A General Monte Carlo N-Particle Transport Code Version 4B", *LA-12625-M*, Los Alamos National Laboratory (1997).

[2.26] M. B. Chadwick, P. G. Young, S. Chiba, S. C. Frankle, G. M. Hale, H. G. Hughes, A. J. Koning, R. C. Little, R. E. MacFarlane, R. E. Prael, L. S. Waters, "Cross-Section Evaluations to 150 MeV for Accelerator-Driven Systems and Implementation in MCNPX", *Nucl. Sci. Eng.*, 131 293-328 (1999).

[2.27] M.Reginatto, P.Goldhagen, USDOE Report EML-595, New York, NY, USA (1998), available at http://www.eml.doe.gov/publications/reports/; "MAXED, A Computer Code For Maximum Entropy Deconvolution of Multisphere Neutron Spectrometer Data", *Health Phys.*, 77, 579-583, (1999).

[2.28] M.Reginatto, P.Goldhagen, S. Neumann, "Spectrum unfolding, sensitivity analysis and propagation of uncertainties with the maximum entropy deconvolution code MAXED", *Nucl. Instr. and Meth.*, A476, 242–246 (2002).

[2.29] T. Nakamura, "Recent development of advanced neutron detection technology", *J. Nucl. Radiochem. Sci.*, 4, R15-R24 (2003).

[2.30] N. Nakao, T. Nakamura, M. Baba, Y. Uwamino, N. Nakanishi, H. Nakashima, Sh. Tanaka, "Measurements of response function of organic liquid scintillator for neutron energy range up to 135 MeV", *Nucl. Instr. and Meth.*, A362, 454-465 (1995).

[2.31] N. Nakao, T. Kurosawa, T. Nakamura, Y. Uwamino, "Absolute measurements of the response function of an NE213 organic liquid scintillator for the neutron energy up to 206 MeV", *Nucl. Instr. and Meth.*, A463, 275-287 (2001).

[2.32] N. Nakao, T. Kurosawa, T. Nakamura, Y. Uwamino, "Development of a quasi-monoenergetic neutron field and measurements of the response function of an organic liquid scintillator for the neutron energy range from 66 MeV to 206 MeV", *Nucl. Instr. and Meth.*, A476, 176-180 (2002).

[2.33] M. Sasaki, N. Nakao, T. Nakamura, T. Shibata, A. Fukumura, "Measurements of the response functions of an NE213 organic liquid scintillator to neutrons up to 800 MeV", *Nucl. Instrum. Methods*, A480, 440-447 (2002).

[2.34] D. Satoh, T. Sato, A. Endo, Y. Yamaguchi, M. Takada, K. Ishibashi, "Measurement of response functions of a liquid organic scintillator for neutrons up to 800 MeV", *J. Nucl. Sci. Technol.*, 43, 714-719 (2006).

[2.35] R.A. Cecil, B.D. Anderson, R. Madey, "Improved predictions of neutron detection efficiency for hydrocarbon scintillators from 1 MeV to about 300 MeV", *Nucl. Instr. and Meth.*, 161, 439 (1979).

[2.36] D. Satoh, S. Kunieda, Y. Iwamoto, N. Shigyo, K. Ishibashi, "Development of SCINFUL-QMD code to calculate the neutron detection efficiencies for liquid organic scintillator up to 3GeV", *J. Nucl. Sci. Technol., Suppl.* 2, 657-660 (2002).

[2.37] T. Nakamura, A. Hara, T. Suzuki, "Realization of a high sensitivity neutron rem counter", *Nucl. Instr. Methods*, A241, 554-560 (1985).

[2.38] Engle, Jr., W. A. "A user's manual for ANISN, a one-dimensional discrete ordinates transport code with anisotropic scattering", *USAEC Report K-1693* (1967).

[2.39] J. Saegusa, M. Yoshizawa, Y. Tanimura, M. Yoshida, T. Yamano, H. Nakaoka, "Evaluation of energy responses for neutron dose-equivalent meters made in Japan", *Nucl. Instr. and Meth.*, A516, 193-202 (2004).

[2.40] International Commission on Radiological Protection, "Conversion Coefficients for use in Radiological Protection against External Radiation" ICRP Publication 74, *Annals of the ICRP*, 26, No.3-4, Elsevier Science, Oxford (1995).

[2.41] C. Birattari, A. Ferrari, C. Nuccetelli, M. Pellicioni, M. Silari, "An extended range neutron rem counter", *Nucl. Instr. and Meth.*, A297, 250-257 (1990).

[2.42] C. Birattari, A. Esposito, A. Ferrari, M. Pellicioni, T. Rancati, M. Silari, "The extended range neutron rem counter 'LINUS': Overview and latest developments", *Radiat. Protec. Dosim.*, 76, 135-148 (1998).

[2.43] A. Fasso, A. Ferrari, J. Ranft, P.R. Sala, "FLUKA: Present status and future developments", *Proc. 4th Int. Conf. on Calorimetry in High Energy Physics*, La Biodola, Italy, September 1993, Eds. A. Menzione, A. Scribano, (World Scientific) pp.493-502 (1994).

[2.44] M. Takada, S. Taniguchi, T. Nakamura, K. Fujitaka, "Development of a Phoswich detector to detect neutrons and charged particles for space application", *IEEE Trans. on Nucl. Sci.*, 45, 888-893 (1998).

[2.45] M. Takada, S. Taniguchi, T. Nakamura, N. Nakao, Y. Uwamino, T. Shibata, K. Fujitaka, "Neutron spectrometry in a mixed field of neutrons and protons with a Phoswich neutron detector", *Nucl. Instr. and Meth.*, A465, 498-524 (2001).

[2.46] T. Sato, D. Satoh, A. Endo, Y. Yamaguchi, "Development of Dose Monitoring System Applicable to Various Radiations with Wide Energy Ranges", *J. Nucl. Sci. Technol.*, 42, 768-778 (2005).

[2.47] A. Bertin, M. Bruschi, V. M. Bystritsky, et al., "Performance of a coincidence neutron spectrometer with double pulse-shape discrimination", *Nucl. Instr. and Meth.,* A337, 445 (1994).

[2.48] H. Iwase, K. Niita, T. Nakamura, "Development of a general-purpose particle and heavy ion transport Monte Carlo code", *J. Nuc. Sci. Technol.*, 39(11), 1142-1151 (2002).

[2.49] M. Tsutsumi, K. Saito, S. Moriuchi, "Spectrum-dose conversion operators, G(E) functions of NaI(Tl) scintillators adapted for effective dose equivalent quantities", *JAERI-M 91-204*, Japan Atomic Energy Research Institute (1991), [in Japanese].

[2.50] E. Kim, A. Endo, Y. Yamaguchi, et al., "Measurement of neutron dose with an organic liquid scintillator coupled with a spectrum weight function", *Radiat. Protec. Dosim.*, 102[1], 31 (2002).

[2.51] T. Nunomiya, S. Abe, N. Hirabayashi, T. Nakamura, "Sequential measurements of cosmic-ray neutron energy spectrum and ambient dose equivalent on the ground", *J. Nucl. Sci. Technol., Suppl.* 4, 466-469 (March 2004), *Proc. 2nd iTRS Intern. Symp. Radiation Safety and Detection Technology*, July 24-25, 2003, Tohoku Univ., Sendai, Japan.

[2.52] T. Nakamura, T. Nunomiya, S. Abe, K. Terunuma, H. Suzuki, "Sequential measurements of cosmic-ray neutron energy spectrum and dose rate at sea level in Sendai, Japan", *J. Nucl. Sci. Technol.*, 42, 843-853 (2005).

[2.53] R.H. Johnson, B.W. Wehring, "FORIST: Neutron Spectrum Unfolding Code - Iterative Smoothing Technique", *ORNL/RSIC-40*, Oak Ridge National Laboratory (1976).

[2.54] K. Nagaoka, Private communications, August 2006: M. Kowatari, Y. Ohta, S. Satoh, K. Nagaoka, J. Abukawa, S. Tachimori, T. Nakamura, "Evaluation of geomagnetic dependence of the cosmic-ray induced environmental neutrons in Japan", *J. Nucl. Sci. Technol.*, 44, 114-120 (2007).

[2.55] D.J. Thomas, A.G. Bardell, E.M. Macaulay, "Characterisation of a gold foil-based Bonner sphere set and measurements of neutron spectra at a medical accelerator", *Nucl. Instr. and Meth.,* A476, 31-35 (2002); B. Wiegel, A.V. Alevra, "NEMUS—the PTB Neutron Multisphere Spectrometer: Bonner spheres and more", *Nucl. Instr. and Meth.,* A476, 36-41 (2002).

[2.56] M. Matzke, "Propagation of uncertainties in unfolding procedures", *Nucl. Instr. and Meth.*, A476, 230-241 (2002).

[2.57] M.S. Gordon, P. Goldhagen, K.P. Rodbell, T.H. Zabel, H.H.K. Tang, J.M. Clem, P. Bailey, "Measurement of the flux and energy spectrum of cosmic-ray induced neutrons on the ground", *IEEE Trans. Nucl. Sci.*, 51, 3427-3434 (2004).

[2.58] S. Roesler, W. Heinrich, H. Schraube, "Monre Carlo calculation of the radiation field at aircraft altitudes", *Radiat. Protec. Dosim.*, 98, 367-388 (2002).

[2.59] M. Kowatari, K. Nagaoka, S. Satoh, Y. Ohta, J. Abukawa, S. Tachimori, T. Nakamura, "Evaluation of the altitude variation of the cosmic-ray induced environmental neutrons in the Mt.Fuji area", *J. Nucl. Sci. Technol.*, 42, 495-502 (2005).

[2.60] R.J. Sheu, S.H. Jiang, "Cosmic-ray induced neutron spectra and effective dose rates near air/ground and air /water interfaces in Taiwan," *Health. Phys.,* 84, 92-99 (2003).

[2.61] A.V. Sannikov, E.N. Savitskaya, "Ambient dose equivalent conversion factors for high energy neutrons based on the ICRP 60 recommendations", *Radiat. Protec. Dosim.*, 70, 383-386 (1997).

[2.62] V. Mares, Private Communications, March, 2005.

[2.63] K. Nagaoka, I. Hiraide, K. Sato, T. Yamagami, T. Nakamura, T. Yabutani, "Measurements of neutron dose rates with a balloon in Japan", Proceedings of 10th Neutron Dosimetry Conference, *Radiat. Protec. Dosim.*, Advance Access published May 24, 2007, doi:10.1093/rpd/ncm007.

[2.64] H. Matsumoto, T. Goka, K. Koga, S. Iwai, T. Uehara, O. Sato, S. Takagi, "Real-Time Measurement of the Low-Energy-Range Neutron Spectra inside the Space Shuttle STS-89 (S/MM-8)", *Radiat. Meas.*, 33, 321-333 (2001).

[2.65] H. Koshiishi, H. Matsumoto, K. Koga, T. Goka, "Evaluation of low-energy neutron environment inside the International Space Station" *Annual Report 2002*, Office of Research and Development, National Space Development Agency of Japan, 2002, 151-154 (2002); *Radiat. Meas.*, 42, 1510-1520 (2007).

[2.66] T.W. Armstrong, K.C. Chandler, J. Barish, "Calculations of Neutron Flux Spectra Induced in the Earth's Atmosphere by Galactic Cosmic Rays", *J. Geophysical Re.,* 78, 2715-2726 (1973).

[2.67] K. Shin, Y. Uwamino, T. Hyodo, "Propagation of errors from response functions to unfolded spectrum", *Nucl. Technol.*, 53, 78-85 (1981). Originally by W.R. Burrus, V.V. Verbinski, *Nucl. Instr. and Meth.*, 67, 181 (1969).

[2.68] K. O'Brien, H.A. Sandmeier, G.E. Hansen et al., "Cosmic-ray induced neutron background sources and fluxes for geometries of air over water, ground, iron and aluminum", *J. Geophys. Res.*, 83[A1], 114-120 (1978).

[2.69] ACE Science Center, California Institute of Technology, http://www.srl.caltech.edu/ACE/index.html.

[2.70] J.W. Bieber et al. University of Delaware Bartol Research Institute Neutron Monitor Program. [Online]. http://www.bartol.udel.edu/~neutronm.

[2.71] International Commission on Radiation Units and Measurements, "Conversion Coefficients for use in Radiological Protection Against External Radiation", *ICRU Report 57*, U.S.A. (1998).

[2.72] A. Ferrari, M. Pelliccioni, M. Pillon, "Fluence to Effective Dose Conversion Coefficients for Neutrons up to 10 TeV", *Rad. Protec. Dosim.,* 71, 165 – 173 (1997); A. Ferrari, M. Pelliccioni, "Fluence to Dose Equivalent Conversion Data and Effective Quality Factors for High Energy Neutrons", *Rad. Protec. Dosim.*, 76, 215-224 (1998).

[2.73] United Nations Scientific Committee on the Effects of Atomic Radiation, Sources and effects of ionizing radiation, UNSCEAR 2000 Report to the General Assembly, with Scientific Annexes, Vol. I, United Nations Publication, New York (2000).

[2.74] Air Density and Density Altitude Calculations, http://wahiduddin.net/calc/density_altitude.htm.

[2.75] M. Matsumoto, M. Furukawa, S. Tokonami, K. Fujitaka, T. Nakamura, "Distribution of neutron component of cosmic rays with altitude at Mt. Fuji", *Radioisotopes*, 44, 33-34 (1995), [in Japanese].

[2.76] H. Schraube, V. Mares, S. Roesler, W. Heinrich, "Experimental verification and calculation of aviation route doses", *Radiat. Protec. Dosim.*, 86, 309-315 (1999).

[2.77] W. Friedberg,, K. Copeland, F. E. Duke, K. O'Brien, E. B. Darden, "Guidelines and technical information provided by the U.S. Federal Aviation Administration to promote radiation safety for air carrier crew members", *Radiat. Protec. Dosim.*, 86, 323-327 (1999).

[2.78] K. O'Brien, W. Friedberg, H. H. Sauer, D. F. Smart, "Atmospheric cosmic rays and solar energetic particles at aircraft altitudes", *Environ. Int.*, 22 (Suppl.1), S9-44 (1996).

[2.79] A. Ferrari, M. Pelliccioni, T. Rancati, "Calculation of the radiation environment caused by galactic cosmic rays for determining air crew exposure", *Radiat. Protec. Dosim.*, 93, 101-114 (2001).

[2.80] J. M. Clem, G. De Angelis, P. Goldhagen, J. W. Wilson, "New calculations of the atmospheric cosmic radiation field – results for neutron spectra", *Radiat. Protec. Dosim.*, 110, 423-428 (2004).

[2.81] B.J. Lewis, C.M. Desormeaux, A.R. Green, L.G.I. Bennett, A. Butler, M. McCall, J.C. S. Vergara, "Assessment of aircrew radiation exposure by further measurements and model development", *Radiat. Protec. Dosim.*, 111, 151-171 (2004): B. J. Lewis, L. G. I. Bennett, A. R. Green, M. J. McCall, B. Ellaschuk, A. Butler, M. Pierre, "Galactic and solar radiation exposure to aircrew during a solar cycle", *Radiat. Protec. Dosim.*, 102, 207-227 (2002).

[2.82] M.E. Shea, D.F. Smart, "Tables of asymptotic directions and vertical cutoff rigidities for a five degree by fifteen degree world grid as calculated using the international geomagnetic reference field for epoch 1975.0", Air Force Geophysics Laboratory, Hanscom AFB, Mass., Rep. AFCRL-TR-75-0185 (1975).

[2.83] I. Usokin, Oulu Cosmic Ray Station, [Online]: http://cosmicrays.oulu.fi//.

[2.84] C. Lopate, Climax and Haleakala Neutron Monitor Datasets, [Online]: http://ulysses.sr.unh.edu/NeutronMonitor/neutron_mon.html.

[2.85] J. Clem, L. Dorman, "Neutron monitor response functions", *Space Sci. Rev.*, 93, 335-363 (2000).

[2.86] A. Belov, A. Struminsky, V. Yanke, "Neutron monitor response functions for galactic and solar cosmic rays", *Proc. 1999 ISSI Workshop on Cosmic Rays and Earth* (1999).

[2.87] V. Mares, S. Roesler, H. Schraube, "Averaged particle dose conversion coefficients in air crew dosimetry", *Radiat. Protec. Dosim.*, 110, 371-376 (2004).

[2.88] A. Ferrari, M. Pelliccioni, R. Villari, "Evaluation of the influence of aircraft shielding on the aircrew exposure through an aircraft mathematical model", *Radiat. Protec. Dosim.*, 108, 91-105 (2004).

[2.89] T. Sato, K. Niita, "Analytical function to predict cosmic-ray neutron spectra in the atmosphere", *Radiation Research*, 166, 544-555 (2006).

[2.90] T. Fukahori, Y. Watanabe, N. Yoshizawa, F. Maekawa, S. Meigo, C. Konno, N. Yamano, A.Yu. Konobeyev, S. Chiba, "JENDL High Energy File", *J. Nucl. Sci. Technol., Suppl.* 2, 25-30 (2002).

[2.91] A. J. Tylka, J. H. Adams, Jr., P. R. Boberg, B. Brownstein, W. F. Dietrich, E. O. Flueckiger, E. L. Petersen, M. A. Shea, D. F. Smart, E. C. Smith, "CREME96: A revision of the cosmic ray effects on micro-electronics code", *IEEE Trans. Nucl. Sci.*, 44, 2150-2160 (1997).

[2.92] P. Goldhagen, J. M. Clem, J. W. Wilson, "The energy spectrum of cosmic-ray induced neutrons measured on an airplane over a wide rage of altitude and latitude", *Radiat. Protec. Dosim.*, 110, 387-392 (2004).

[2.93] http://www.jaea.go.jp/04/nsed/ers/radiation/rpro/EXPACS/expacs-eng.html.

Chapter 3

[3.1] R. Baumann, "The Impact of Technology Scaling on Soft Error Rata Performance and Limits to the Efficacy of Error Correction", *IEEE International Electron Device Meeting, Dec. 8-11, 2002*, pp.329-332 (2002).

[3.2] JESD89A "Measurement and Reporting of Alpha Particles and Terrestrial Cosmic Ray-Induced Soft Errors in Semiconductor Devices," Revision of JEDEC Standard No. 89, October 2006.

[3.3] A. K. Sharma, "Introduction to Advanced Semiconductor Memories, "Advanced Semiconductor Memories: Architectures, Designs, and Applications," IEEE (2003).

[3.4] S. Meguro, S. Ikeda, K. Nagasawa, A, Koike, T. Yasui, "Hi-CMOS technology" IEEE International Electron Device Meeting, pp.59-62 (1984).

[3.5] S. Ikeda, Y. Yoshida, K. Kobayashi, Y. Mitsui, "Failure Analysis of 6T SRAM on Low-Voltage and High-Frequency Operation", *IEEE Transactions on Electron Device,* Vol. 50, No.5, May, 2003, pp.1270-1276.

[3.6] S. Ikeda, Y. Yoshida, S. Kamohara, K. Imato, K. Ishibashi, K. Takahashi, "Threshold Voltage-Related Soft error Degradation in a TFT SRAM Cell", *IEEE Transactions on Electron Devices* Vol. 50, No. 2, pp.391-396 (2003).

[3.7] A. Eto, M. Hidaka, Y. Okuyama, K. Kimura, M. Hosono, "Impact of Neutron flux on Soft Errors in MOS Memories", *Technical. Digest International Electron Device. Meeting, San Francisco, 1998,* p.367.

[3.8] D. Krawzsenek, P. Hsu, H. Anthony, C. Land, "Single Event Effects And Total Ionizing Dose Results of a Low Voltage EEPROM", *Radiation Effects Data Workshop, 2000.*

[3.9] Fred W. Sexton, Paul E. Dodd, Marty R. Shaneyfelt, and Jim R. Schwank, "Integrated Circuit Vulnerability to E-Beam Decontamination Process" by Sandia National Laboratories.

[3.10] Kenneth A. LaBel, Amy K. Moran, Donald K. Hawkins, Anthony B. Sanders, "Current Single Event Effect test Results For Candidate Spacecraft Electronics" *IEEE, Radiation Effects Data Workshop, 19, July 1996*, pp.19-27.

[3.11] http://klabs.org/richcontent/Misc_Content/meetings/eeprom_oct_2005/index.

[3.12] T. Terasawa, "Cosmic plasmas: their research frontiers," *Sci. Tech. Adv. Materials*, Vol. 2, pp.461-472, (2001).

[3.13] J. F. Ziegler, "Terrestrial cosmic ray intensities," *IBM J. Res. Develop.*, Vol. 42, pp.117-139 (1998).

[3.14] http://www.sel.noaa.gov/index.html.

[3.15] http://nssdc.gsfc.nasa.gov/space/cgm/cgm.html.

[3.16] http://www.cosmicrays.org/muon-cutoff-rigidity.php.

[3.17] http://wahiduddin.net/calc/density_altitude.htm.

[3.18] http://www.oulu.fi/~spaceweb/textbook/magnetosphere.html.

[3.19] T. Nakamura, T. Nunomiya, S. Abe, and K. Terunuma, "Sequential measurements of cosmic-ray neutron energy spectrum and ambient dose equivalent on the ground," in Proc. *11th Int'l Congress of the Int'l Radiation Protection Association,* Madrid, Spain, 23-28 May, 2004, Paper No. 3C6 (in CD-ROM).

[3.20] H. Kobayashi, H. Usuki, K. Shiraishi, H. Tsuchiya, N. Kawamoto, "Comparison Between Neutron-induced System-SER and Accelerated-SER in SRAMs", in Proc. 43th Int'l Reliability Physics Symposium, IEEE, Phoenix, Arizona, USA, 25-29, April, 2004.

[3.21] R. Baumann, "Radiation-Induced Soft errors in Advanced Semiconductor Technologies", IEEE Transactions on Device and Materials Reliability, Vol. 5, No. 3, September 2005, pp.305-316.

[3.22] P. Hazucha, and C. Svensson, "Impact of CMOS Technology Scaling on the Atmospheric Neutron Soft Error Rate", *IEEE, Transactions on Nuclear Science*, Vol. 47, No. 6, Dec. 2000, pp.2586-2594.

[3.23] http://mcnpx.lanl.gov/.

[3.24] http://pcfluka.mi.infn.it/.

[3.25] http://wwwasd.web.cern.ch/wwwasd/geant/.

[3.26] H. Kameyama, Y. Yahagi and E. Ibe, "A quantitative analysis of neutron-induced multi-cell upset in deep sub micron SRAMs and of the impact due to anomalous noise", *Proc. 45th Int'l Reliability Physics Symposium, IEEE, Phoenix, Arizona, USA, 15-19 April, 2007.*

[3.27] J. Maiz, S. Hareland, K. Zhang and P. Armstrong "Characterization of Multi-bit Error events in Advanced SRAMs" IEEE, International Electron Device Meeting 2003, pp.519-522.

[3.28] K. Osada, Y. Saitoh, E. Ibe, K. Ishibashi, "16.7A/cell Tunnel-Leakage-Suppressed 16Mb SRAM for Handling Cosmic-Ray-Induced Multi-Errors", *ISSCC 2003*, No. 17.2.

[3.29] N.Seifert, P. Slankard, M. Kirsch, B. Narasimham, V. Zia, C. Brookreson, A. Vo, S. Mitra, B. Gill, J. Maiz, "Radiation-Induced Soft Error Rates of Advanced CMOS Bulk Devices", *Proc. 44th Int'l Reliability Physics Symposium, IEEE, San Jose, USA, 26-30 March, 2006,* pp.217-225.

[3.30] http://www.oulu.fi/~spaceweb/textbook/ionosphere.html.

[3.31] http://helios.izmiran.troitsk.ru/cosray/main.htm.

[3.32] http://ulysses.sr.unh.edu/NeutronMonitor/Misc/ClimaxGifs/Climax2003.gif.

[3.33] http://neutronm.bartol.udel.edu/.

[3.34] J. L. Autran, P. Roche, J. Borel, C. Sudre, K. Castellani-Coulie, D. Munteanu, T. Parrassin, G. Gasiot and J. P. Schoellkopf, "Altitude SEE Test European Platform (ASTEP): Project Overview, First Results in CMOS 130 nm and Perspectives," presented at *the Radiation Effects on Components and Systems 2006 Workshop (RADECS 2006 Workshop)*, Athens, Greece, 27-29 September, 2006, Paper No. C-5.

[3.35] K. Terunuma, "Basic study on the measurement of neutron activation cross section of light nucleus for several tens and cosmic-ray-neutron soft errors of the semiconductor devices" Tohoku University, Master Thesis, 2003 March, (in Japanese).

Chapter 4

[4.1] James F. Ziegler & Helmer Puchner, "*SER-Histrory, Trends and Challenge*" Cypress Semiconductor Co. 2004.

[4.2] JEDS89 "*Measurement and Reporting Of Alpha Particles and Terrestrial Cosmic Ray- induced Soft Errors in Semiconductor Devices*," JEDEC Standard No. 89, August 2001, and *JESD89-Rev* (http://www.seutest.co) .

[4.3] C.A. Uttley, "*Neutron Sources for Basic Physics and Application*", ed. S. Cierjacks (Pergamon, 1983).

[4.4] H. Liskien and A. Paulsen; "Neutron Production Cross Sections for the T(d,n)Be and ^{7}Li(p,n)^{7}Be*", *Nuclear Data Tables 11, (7), 569-614, (1973).*

[4.5] H. Liskien and A. Paulsen; "Neutron Production Cross Sections and energies for the ^{7}Li(p,n)^{7}Be and ^{7}Li(p,n)^{7}Be*", *Atomic and Nuclear Data Tables* 15, 57-84, (1975).

[4.6] J. Morin, P. Zynomski, J.-C. Amoud, and J. David., "Moyes d'irradiation neutroniqueau CEA Valduc" *Proc. RADECS 1994*, pp.10-15.

[4.7] M. Baba, M. Takada, T. Iwasaki, S. Matsuyama, T. Nakamura, H. Ohguchi, T. Nakao, T. Sanami, N. Hirakawa, "Development of monoenergetic neutron calibration fields between 8 keV to 15 MeV." *Nucl. Instrum. Methods,* A376 (1996) 115.

[4.8] M. Baba., N. Hirakawa, T. Iwasaki, K. Kanda, "Measurement of Fast neutron Induced Fission Cross Section for Actinide Nuclides", *J. Nuc. Sci. Technol.* 24 (1987) 475.

[4.9] Y. Yahagi, Ph.D Dissertation (Tohoku University), 2005, in Japanese.

[4.10] http://www.npl.co.uk/jonrad/services/rn0201.html.

[4.11] J.A. Jungerman and F.P. Brady, *Nucl. Instrum. Methods,* 89 (1970) 167.

[4.12] A. Bol., P. Leleux, P. Lipnik, P. Macq and A. Ninane, "A novel design for a fast intense neutron beam" , *Nucl. Instrum. Methods,* 216 (1983) 169-173.

[4.13] S.S. Hanna, C.J. Martoff, D. Pocanic, K. Wang, R.C. Byrd, C.C. Foster, I.J. van Heerden, "A Monochromatic neutron facility for (n,p) reactions", *Nucl.Instrum.Methods,* A401 (1997) 345-354.

[4.14] H. Conde, S. Hultqvist, N. Olsson, T. Ronnqvist, R. Zorro, J. Blomgren, G. Tibell, Hakansson, O. Jonsson, A. Lindhom, L. Nilsson, P.-U. Renberg, A. Brocksteds, P. Ekstrom, M. Osterlund, F.P. Brady, and Z. Szeflinski, "A facility for studies of neutron-induced reactions in the 50–200 MeV range", *Nucl. Instrum. Methods,* A292 (1990) 121-128.

[4.15] S. Pomp, A.V. Prokofiev, J. Blomgren, O. Bystrom, C. Ekstrom, N. Haag, A. Hindebran, C. Johansson, O. Jonsson, P. Mermod, L. Nilsson, D. Reistad, N. Olsson, P.-U. Renberg, M. Österlund, U. Tippawan, D. Wessman, and V. Ziemann, "The New Uppsala Neutron Beam Facility", Proc. Int. Conf. on Nuclear Data for Sci. Technol., 2005, Santa Fe, (American Institute of Physics, 2005) 780-783.

[4.16] Indiana Univeristy Cyclotron Facility: http://www.iucf.indiana.edu/.

[4.17] Su. Tanaka, T. Nakamura, M. Baba, M. Imamura, H. Hirayama, K. Shin, Sh. Tanaka, R. Tanaka, *Proc. 2nd Int'l Symposium on Advanced Nuclear Energy Research -Evolution by Accelerators-*, (January 24-26, 1992, Mito), p.342.

[4.18] H. Sakai, H. Okamura, H. Otsu, T. Wakasa, and S. Ishida, "Facility for the (p,n) polarization transfer", *Nucl. Instrum. Methods*, A369, (1996) 120-134. http://www.rcnp.osaka-u.ac.jp/Divisions/np1-a/n0/index.html.

[4.19] A. Terakawa, H. Suzuki, K. Kumagai, Y. Kikuchi, T. Uemori, H. Fujisawa, N. Sugimoro, K. Itoh, M. Baba, H. Orihara, K. Maeda, "New fast-neutron time-of-flight facilities at CYRIC", *Nucl. Instrum. Methods*, A491 (2002) 419-425. http://www.cyric.tohoku.ac.jp.

[4.20] M. Baba, "QUASI-MONOENERGETIC NEUTRON SOURCES", *Proc. Int. Symp. On Fast Neutron Detection and its Application (FNDA)*, 2006, Proc. Science.

[4.21] N. Nakao, Y. Uwamino, T. Nakamura, T. Shibata, N. Nakanishi, M. Takada, E. Kim, T, Kurosawa, "Development of a quasi-monoenergetic neutron field using the ^{7}Li(p,n)^{7}Be reaction in the 70-210 MeV energy range at RIKEN", *Nucl. Instrum. Methods,* A420 (1999) 218-231.

[4.22] M. Baba, Y. Nauchi, T. Iwasaki, T. Kiyosumi, M. Yoshioka, S. Matsuyama, N. Hirakawa, T. Nakamura, Su. Tanaka, S. Meigo H. Nakashima, Sh. Tanaka, N. Nakao, "Characterization of a 40-90 MeV ^{7}Li(p,n) Neutron Source at TIARA using a Proton Recoil Telescope and a TOF Method", *Nucl. Instrum. Methods,* A428 (1999) 454-465.

[4.23] G.G. Ohlsen, "Kinematic Realations in Reactions of the Form A+B → C+D+E", *Nucl. Instrum. Methods,* 37 (1964) 240-248.

[4.24] S. Taniguchi, N. Nakao, T. Nakamura, H. Yashima, Y. Iwamoto, D. Satoh, Y. Nakane, H. Nakashima, T. Itoga, A. Tamii, K. Hatanaka, "Development of a Quasi-Monoenergetic Neutron Field Using the ^{7}Li(p,n)^{7}Be Reaction in the Energy Range from 250 to 390 MeV at RCNP", Proc. tenth Int. Symp. On Neutron Dosimetry, *Radiat. Protec. Dosim.*, Advance Access published May 13, 2007, doi:10.1093/rpd/ncm007.

[4.25] A.V. Prokofiev, M.B. Chadwick, S.G. Mashnik, N. Olsson and L. Waters, "Development and Validation of the Li(p,n) Nuclear Data Library and Its Application in Monitoring of Intermediate Energy Neutrons," *Nuclear Science and Technology*, Supplement 2, pp.112-115 (2002).

[4.26] T.N. Taddeucci, C.A. Goulding, T.A. Carey, R.C. Byrd, C.D. Goodman, C. Caarde, J. Larsen, D. Horen, J. Rapaport, E. Sugarbaker, *Nucl. Phys.,* A469 (1987) 125.

[4.27] S.G. Mashnik, M.B. Chadwick, P.G. Yoing, R.E. MacFarlen, and L.S. Waters, "^{7}Li(p,n) Neutron Data Library for Incident Proton Energies to 150 MeV", *LA-UR*-00-1067 (Los Alamos National Laboratory).

[4.28] A.V. Prokofiev, A.N. Smirnov, V.G. Khlopin, and P.-U. Renberg, "A monitor of intermediate-energy neutrons based on thin film breakdown counters," *TSL Progress Report 1998-1999*, pp.98-99 (2000).

[4.29] http://www.tsl.uu.se/tsl_electronics_testing.html.

[4.30] T. Hirao, I. Nashiyama, T. Kamiya and Nishijima, "Effects of micro-beam induced damage on single-event current measurements", *Nucl. Instrum. Methods,* B1.4, (1995) 508-514.

[4.31] R. Nolte, M.S. Allie, P.J. Binns, F. Brooks, A. Buffler, V. Dangendorf, J.P. Meulders, F. Roos, H. Schumacher, B. Wiegel, "High-energy neutron reference fields for the calibration of detectors used in neutron spectrometry", *Nucl. Instrum. Methods*, A476, (2002) 369-373.

[4.32] J.S. Sisterson, F.D. Brooks, A. Buffler, M.S. Allie, D.T.L. Jones, M.B. Chadwick, "Cross-section Measurements for neutron-induced reactions in copper at neutron energies of 70.7 and 110.8 MeV", *Nucl. Instrum. Methods,* B240 (2005) 617-624.

[4.33] M. Baba, H. Okamura, M. Hagiwara, T. Itoga, S. Kamada, Y. Yahagi, E. Ibe, "Installation and application of an intense ^{7}Li(p,n) neutron source for 20-90 MeV region", Proc. Tenth Int. Symp. On Neutron Dosimetry, to be published in Radiation Protection Dosimetry.

[4.34] S. Leray, F. Bone, S. Crespin, J. Frehaut, X. Ledoux, E. Martinez, Y. Patin, E. Petibon, P. Pras, A. Boudard, R. Legrain, Terrien, F. Brochard, D. Drake, J.C. Duchazeabeniex, J.M. Durand, S.I. Meigo. G. Millert, D.M. Whittal, W. Wlazlo, D. Durand, C. Le Brun, F.R. Lecolley, F. Lefevres, M. Louvel, C. Varignon, F. Hanappe, S. Menard, L. Stuttge, "Spallation neutron production by 0.8, 1.2, and 1.6 GeV protons on various targets" *Physical Rev.,* C65 (0446219) 65 (2002).

[4.35] Spallation Neutron Source, Oak Ridge National Laboratory, USA, http://www.sns.gov/.

[4.36] Japan Particle Accelerator Research Complex, JAEA and KEK joint project, http://j-parc.jp/.

[4.37] LANSCE Home Page: http://www.lanl.gov.

[4.38] S.A. Wender, S. Balestrini, A. Brown, R.C. Haight, C.M. Laymon, T.M. Lee, and N.W. Hill, "A fission ionization detector for neutron flux measurements at a spallation source," *Nucl. Instrum. Method in Phys. Res., Sec. A*, Vol. A336, pp.226-231 (1993).

[4.39] Y. Yahagi, E. Ibe, S. Yamamoto, Y. Yoshino, M. Sato, Y. Takahashi, H. Kameyama, A. Saito and M. Hidaka, "Versatility of SEU Function and Its Derivation from the Irradiation Tests with Well-defined White Neutron Beams," *IEEE Trans. Nucl. Sci.*, Vol. 52, No. 5, pp.1562-1567 (2005).

[4.40] E.W. Blackmore, P.E. Dodd, and M.R. Shaneyfelt, "Improved capabilities for proton and neutron irradiations at TRIUMF," *in Proc. 2003 IEEE Radiation Effects Data Workshop, 21-25, July, 2003*, pp.149-155 (2003).

[4.41] http://www.rcnp.osaka-u.ac.jp/Divisions/np1-a/n0/information/index.html.

[4.42] Kichiji Hatanaka (RCNP) private communication to E. Ebe (2006).

Chapter 5

[5.1] K. Johansson, P. Dyreklev, B. Granbom, N. Olsson, J. Blomgren and P.-U. Renberg, "Energy-Resolved Neutron SEU Measurements from 22 to 160 MeV," *IEEE Trans. Nucl. Sci.*, Vol. 45, No. 6, pp.2519-2526 (1998).

[5.2] M.B. Chadwick and E. Normand, "Use of New ENDF/B-VI Proton and Neutron Cross Sections for Single Event Upset Calculations," *IEEE Trans. Nucl. Sci.*, Vol. 46, No. 6, pp.1386-1394 (1999).

[5.3] C. Inguimbert, S. Duzellier, R. Ecoffet, L. Guibert, J. Barak and M. Chabot, "Using a Carbon Beam as a Probe extract the Thickness of Sensitive Volumes," *IEEE Trans. Nucl. Sci.*, Vol. 47, No. 3, pp.551-558 (2000).

[5.4] S. Duzellier, R. Ecoffet, D. Falguere, T. Nuns, L. Guibert, W. Hajdas and M.C. Calvet, "Low Energy Proton Induced SEE in Memories," *IEEE Trans. Nucl. Sci.*, Vol. 44, No. 6, pp.2306-2310 (1997).

[5.5] R. Jones, A.M. Chugg, C.M.S. Jones, P.H. Duncun, C.S. Dyer and C. Sanderson, "Comparison Between SRAM SEE Cross-Sections From Ion Beam Testing With Those Obtained Using a New Picosecond Pulsed Laser Facility," *IEEE Trans. Nucl. Sci.*, Vol. 47, No. 3, pp.539-544 (2000).

[5.6] J. Barak, "Empirical Modeling of Proton Induced SEU Rates," *IEEE Trans. Nucl. Sci.*, Vol. 47, No. 3, pp.545-550 (2000).

[5.7] A. Eto, M. Hidaka, Y. Okuyama, K. Kimura and M. Hosono, "Impact of Neutron Flux on Soft Errors in MOS Memories," in *Tech. Dig of International Electron Device Meeting (IEDM) 1998.* Dec. 1998, San Francisco, CA, Dec. 1998, pp.367-370 (1988).

[5.8] R. Koga, "Single Event Sensitivities of 128- and 256-megabit Synchronous Dynamic Random Access Memories (SDRAMs)," in *Proc. the 4th Int'l Workshop on Radiation Effects on Semiconductor Devices for Space Application (RASEDA)*, Tsukuba, May, 15-18, 2000, pp.81-88 (2000).

[5.9] E. Ibe, Y. Yahagi, F. Kataoka, Y. Saito, A. Eto, M. Sato, H. Kameyama and M. Hidaka, "A Self-Consistent Integrated System for Terrestrial-Neutron Induced Single Event Upset of Semiconductor Devices at the Ground," presented at the *IEEE 1st International Conference on Information Technology and Applications (ICITA)*, Bathurst, Australia, Nov. 25-28, 2002, Paper No. 273-21 (in CD-ROM) (2002).

[5.10] Y.Z. Xu, H. Puchner, A. Chatila, O. Pohland, B. Bruggeman, B. Jin, D. Radaelli, and S. Daniel, "Process Impact on SRAM alpha-Particle SEU Performance," in *Proc. IEEE 2004 International Reliability Physics Symposium (IRPS)*, Phoenix, AZ, 2004, pp. 294-299 (2004).

[5.11] P. Roche, F. Jacquet, C. Caillat, and J.-P. Schoellenkopf, "An Alpha Immune and Ultra Low Neutron SER High Density SRAM," in *Proc. IEEE 2004 International Reliability Physics Symposium (IRPS)*, Phoenix, AZ, 2004, pp.671-672 (2004).

[5.12] J.M. Sisterson, F.D. Brooks, A. Buffer, M.S. Allie, D.T.L. Jones, and M.B. Chadwick, "Cross-section measurements for neutron-induced reactions in copper at neutron energies of 70.7 and 110.5 MeV," *Nucl. Instrum. Methods in Phys. Res. B*, Vol. B240, pp.617-624 (2005).

[5.13] W.L. Bendel and E.L. Petersen, "Proton Upsets in Orbit," *IEEE Trans. Nucl. Sci.*, Vol. NS-30, No. 6, pp.4481-4485 (1983).

[5.14] Y. Shimano, T. Goka, S. Kuboyama, K. Kawachi, T. Kanai and Y. Takami, "The Measurement and Prediction of Proton Upset," *IEEE Trans. Nucl. Sci.*, Vol. 36, No. 6, pp.2344-2348 (1989).

[5.15] E.L. Petersen, J.C. Pickel, J.H. Adams, Jr., and E.C. Smith, "Rate Prediction for Single Event Effects," *IEEE Trans. Nucl. Sci.*, Vol. 39, No. 6, pp.1577-1599 (1992).

[5.16] M. Olmos, A. Prokofiev, and R. Gillard, "Unfolding Procedure for SER Measurements Using Quasi-Monoenergetic Neutrons," in *Proc. IEEE 2005 International Reliability Physics Symposium (IRPS)*, San Jose, CA, 2005, pp.696-697 (2005).

[5.17] E. Ibe, Y. Yahagi, Y. Saito, H. Kameyama, H. Yukawa, A. Eto, and M. Hidaka, "Establishment of accelerator soft-error testing of sub-quarter micron silicon devices by using quasi-mono energy neutron beams," *TSL Progress Report 2001-2002*, pp.35-37 (2002).

[5.18] W.J. Stapor, J.P. Meyers, J.B. Langworthy and E.L. Petersen, "Two Parameter GENDEL Model Calculations for Predicting Proton Induced Upsets," *IEEE Trans. Nucl. Sci.*, Vol. 37, No. 6, pp.1966-1973 (1990).

[5.19] J.F. Ziegler and W.A. Lanford, "Effect of Cosmic Rays on Computer Memories," *Science*, Vol. 206, pp.776-788 (1978).

[5.20] E. Normand, J.L. Wert, W.R. Doherty, D.L. Oberg, P.R. Measel, and T.L. Criswell, "Use of PuBe Source to Simulate Neutron-Induced SEU in Static RAMs," *IEEE Trans. Nucl. Sci.*, Vol. 35, No. 6, pp.1523-1528 (1988).

[5.21] S. Agosteo, L. Castellani, G.D'Angello, A. Favalli, I. Lippi, R. Martinelli and P. Zotto, "First Evaluation of Neutron Induced Single Event Effects on the CMS Barrel Muon Electronics," in *Proc. the Sixth Workshop on Electronics for LHC Experiments*, Krakow, Poland, 11-15 Sept. 2000, (2000). http://lebwshop.home.cern.ch/lebwshop/LEB00_Book/rad_00.htm.

[5.22] C. Dyer, S. Clucas, F. Lei, P. Truscott, R. Nartello, and C. Comber, "Comparative Simulations of Single Event Upsets Induced by Protons and Neutrons in Commercial SRAMs," presented at *the 7th European Conference on Radiation and Its Effects on Components and Systems (RADECS)*, Noordwijk, The Netherlands, 15-19, Sept. 2003, Session C-4.

[5.23] Nuclear Data Center, Japan Atomic Energy Agency, JENDL-3.3, http://wwwndc.tokai-sc.jaea.go.jp/jendl/j33/j33.html.

[5.24] E. Ibe, Y. Yahagi, and H. Yamaguchi, "Monte-Carlo Simulation for the Effects of Composite Materials on Single Event Effects of Sub-100 nm Semi-conductor Devices," in Proc. of *2004 Symposium on Nuclear Data*, Nov. 11-12, 2004, Tokai Research Establishment (2004).

[5.25] F. Kreupl, A.P. Graham, M. Liebau, G.S. Duesberg, R. Seidel and E. Unger, "Carbon Nanotubes for Interconnect Applications," *Tech. Digest of IEEE International Electron Device Meeting (IEDM) 2004*, San Francisco, CA, Dec. 13-15, 2004, pp.683-686 (2004).

[5.26] Y.-M. Lin, J. Appenzeller, and P. Avouris, "Novel Carbon Nanotube FET Design with Tunable Polarity," *Tech. Digest of IEEE International Electron Device Meeting (IEDM) 2004*, San Francisco, CA, Dec. 13-15, 2004, pp.687-690 (2004).

[5.27] J.F. Ziegler, "Terrestial Cosmic Rays," *IBM J. Res. Develop.*, Vol. 40, No. 1, pp.19-39, (1996).

[5.28] JESD89 "Measurement and Reporting of Alpha Particles and Terrestrial Cosmic Ray-Induced Soft Errors in Semiconductor Devices," JEDEC Standard No. 89, August 2001.

[5.29] W.N. Hess, H.W. Patterson, R. Wallace and E.L. Chupp, "Cosmic-Ray Neutron Energy Spectrum," *Phys. Rev.*, Vol. 116, No. 2, pp.445-457, (1959).

[5.30] M.S. Gordon, P. Goldhagen, K.P. Rodbell, T.H. Zabel, H.H.K. Tang, J.M. Clem, and P. Bailey, "Measurement of the Flux and Energy Spectrum of Cosmic-Ray Induced Neutrons on the Ground," *IEEE Trans. Nucl. Sci.*, Vol. 51, No. 6, pp.3427-3434 (2004).

[5.31] T. Nakamura, Y. Uwamino, T. Ohkubo and A. Hara, "Altitude Variation of Cosmic-ray Neutrons," *Health Physics*, Vol. 53, No. 5, pp.509-517, (1987).

[5.32] T. Nakamura, T. Nunomiya, N. Hirabayashi, H. Suzuki, Y. Sato, and D.A.H. Rasolonjatovo, "Sequential measurements of energy spectrum and intensity for cosmic ray neutrons," in Proc. *7th Int'l Symp. Natural Radiation Environment (NRE-VII)*, Rhodes, Greece, 20-24 May, 2002./ T. Nakamura, T. Nunomiya, S. Abe, and K. Terunuma, "Sequential measurements of cosmic-ray neutron energy spectrum and ambient dose equivalent on the ground," in Proc. *11th Int'l Congress of the Int'l Radiation Protection Association,* Madrid, Spain, 23-28 May, 2004, Paper No. 3C6 (in CD-ROM).

[5.33] T. Sato, K. Niita, "Analytical function to predict cosmic-ray neutron spectra in the atmosphere," Radiation Research, Vol. 166, pp.544-555 (2006).

[5.34] LANSCE Home Page, http://www.lanl.go.jp.

[5.35] H. Kobayashi, H. Usuki, K. Shiraishi, H. Tsuchiya, N. Kawamoto, G. Merchant and J. Kase, "Comparison between Neutron-Induced System-SER and Accelerated-SER in SRAMs," in *Proc. IEEE 2004 International Reliability Physics Symposium*, Phoenix, AZ, 2004, pp. 288-293 (2004).

[5.36] Y. Yahagi, E. Ibe, S. Yamamoto, Y. Yoshino, M. Sato, Y. Takahashi, H. Kameyama, A. Saito and M. Hidaka, "Versatility of SEU Function and Its Derivation from the Irradiation Tests with Well-defined White Neutron Beams," *IEEE Trans. Nucl. Sci.,* Vol. 52, No. 5, pp.1562-1567 (2005).

[5.37] S.A. Wender, S. Balestrini, A. Brown, R.C. Haight, C.M. Laymon, T.M. Lee, and N.W. Hill, "A fission ionization detector for neutron flux measurements at a spallation source," *Nucl. Instrum. Method in Phys. Res., Sec. A*, Vol. A336, pp.226-231 (1993).

[5.38] Y. Yahagi, E. Ibe, Y. Saito, A. Eto, M. Sato, H. Kameyama, M. Hidaka, K. Terunuma, T. Nunomiya, and T. Nakamura, "Self-consistent integrated system for susceptibility to terrestrial neutron induced soft-error of sub-quarter micron device," *in 2002 IEEE International Integrated Reliability Workshop (IIRW) Final Report, Stanford Sierra Camp, Lake Tahoe, CA, USA, Oct., 21-24 2002, pp.143-146* (2002).

[5.39] J. Maiz, S. Hareland, K. Zhang, and P. Armstrong, "Characterization of Multi-bit Soft Error events in advanced SRAMs," in *Tech. Digest of IEEE International Electron Device Meeting (IEDM) 2003, Washington, DC, December 7-10, 2003,* No. 21.4 (2003).

[5.40] Y. Tosaka, H. Ehara, M. Igeta, T. Uemura, H. Oka, Matsuoka, N., and Hatanaka, K., "Comprehensive Study of Soft Errors in Advanced CMOS Circuits with 90/130nm Technology," in *Tech. Digest of IEEE International Electron Device Meeting (IEDM) 2004*, San Francisco, CA, Dec. 13-15, 2004, pp.941-944 (2004).

[5.41] E. Ibe, H. Kameyama, Y. Yahagi, K. Nishimoto, Y. Takahashi, "Distinctive Asymmetry in Neutron-Induced Multiple Error Patterns of 0.13um process SRAM," *The 6th International Workshop on Radiation Effects on Semiconductor Devices for Space Application (RASEDA), Tsukuba, October 6-8, 2004,* pp.19-23 (2004).

[5.42] E. Ibe, S.S. Chung, S. Wen, H. Yamaguchi, Y. Yahagi, H. Kameyama, and S. Yamamoto, "Multi-Error Propagation Mechanisms Clarified in CMOSFET SRAM Devices under Quasi-Mono Energetic Neutron Irradiation," *presented at 2006 NSREC, Ponte Vedra Beach, FL., July 17-21*, (2006).

[5.43] A.K. Sharma, "Speciality SRAMs," in "Advanced Semiconductor Memories Archtechtures, Designs, and Applications," IEEE Press, pp.91-122 (2003).

[5.44] Y. Yahagi, E. Ibe, Y. Takahashi, Y. Saito, A. Eto, M. Sato, H. Kameyama, M. Hidaka, K. Terunuma, T. Nunomiya and T. Nakamura, "Threshold Energy of Neutron Induced SEU as a Critical Factor," in *Proc. IEEE 2004 International Reliability Physics Symposium (IRPS)*, Phoenix, AZ, 25-29 April, 2004, pp. 669-670 (2004).

[5.45] E. Ibe, S.S. Chung, S.J. Wen, H. Yamaguchi, Y. Yahagi, H. Kameyama, S. Yamamoto, and T. Akioka, "Spreading Diversity in Multi-cell Neutron-Induced Upsets with Device Scaling," *presented at IEEE Custom Integrated Circuits Conference 2006, San Jose, CA, Sept. 10-13 2006, 10.4 (2006), also published* in *Proc. IEEE Custom Integrated Circuits Conference 2006, San Jose, CA, Sept. 10-13 2006,* pp.437-444 (2006).

[5.46] Y. Yahagi, E. Ibe, H. Kameyama, M. Sato, T. Akioka, S. Yamamoto, and H. Yamaguchi, "A Novel Feature of Neutron-Induced Multi-Cell Upset in 130 and 180 nm SRAMs," *presented at 2006 Workshop on Radiation and Its Effects on Components and Systems (RADECS), Athens, Greece, Sept., 27-29 2006, PD-5* (2006). / Y. Yahagi, E. Ibe, H. Kameyama, M. Sato, T. Akioka, S. Yamamoto, and H. Yamaguchi, "A Novel Feature of Neutron-Induced Multi-Cell Upsets in 130 and 180 nm SRAMs," *IEEE Trans. Nucl. Sci.*, Vol. 54, No. 4, pp.1030-1036 (2007).

[5.47] H. Yamaguchi, E. Ibe, Y. Yahagi, S. Yamamoto, T. Akioka and H. Kameyama, "Novel Mechanism of Neutron-induced Multi-cell Error in CMOS Devices Tracked down from 3D Device Simulation," in Proc. *2006 International Conference on Simulaton of Semiconductor Process and Devices (SISPAD), Monterey, CA, USA, Sep. 6-8 2006,* pp.184-187 (2006).

Chapter 6

[6.1] E. Ibe, "Current and Future Trend on Cosmic-Ray-Neutron Induced Single Event Upset at the Ground down to 0.1-Micron-Device," *TSL Workshop on Applied Physics, Uppsala, May, 3,* No. 1 (2001).

[6.2] E. Ibe, Y. Yahagi, F. Kataoka, Y. Saito, A. Eto, M. Sato, H. Kameyama, and M. Hidaka, "A Self-Consistent Integrated System for Terrestrial-Neutron Induced Single Event Upset of Semiconductor Devices at the Ground," *2002 ICITA, Buthurst, Australia,* No. 273-21 (2002).

[6.3] E. Ibe, H. Kameyama, Y. Yahagi, K. Nishimoto, and Y. Takahashi, "Distinctive Asymmetry in Neutron-Induced Multiple Error Patterns of 0.13μm process SRAM," *The 6th International Workshop on Radiation Effects on Semiconductor Devices for Space Application, Tsukuba, October 6 - 8, 2004,* pp.19-23 (2004).

[6.4] E. Ibe, Y. Yahagi, H. Kameyama, and Y. Takahashi, "Single Event Effects of Semiconductor Devices at the Ground," *Ionizing Radiation*, Vol. 30, No. 7 (2004).

[6.5] E. Ibe, Y. Yahagi, H. Yamaguchi, "Monte-Carlo Simulation for the Effects of Composite Materials on Single Event Effects of Sub-100 nm Semiconductor Devices," *Proc. 2004 Symposium on Nuclear Data, November 11-12, 2004, JAERI, Tokai, Japan, Vol. JAERI-Conf 2005-003*, No. 2.14, pp.100-105 (2005).

[6.6] E. Ibe, Y. Yahagi, H. Yamaguchi, and H. Kameyama, "SEALER: Novel Monte-Carlo Simulator for Single Event Effects of Composite-Materials Semiconductor Devices," *2005 Radiation and Its Effects on Components and Systems, Sep.19-23, Palais des Congres, Cap d"Agde, France*, No. E-4 (2005).

[6.7] H.H.K. Tang, G.R. Srinivasan, and N. Azziz, "Cascade statistical model for nucleon-induced reactions on light nuclei in the energy range 50-MeV-1GeV," *Physical Review C*, Vol. 42, No. 4, pp.1598-1622 (1990).

[6.8] I. Dostrovsky, Z. Fraenkel, and G. Friedlander, "Monte Carlo Calculations of Nuclear Evaporation Process. III. Applications to Low-Energy Reactions," *Physical Review*, Vol. 116, No. 3, pp.683-702 (1959).

[6.9] S. Furihata, "Parameters Used in GEM," Development of a Generalized Evaporation Model and Study of Residual Nuclei Productions, Thesis for PhD, Tohoku Univ. pp.18-20 (2004).

[6.10] C. Hu, "Alpha-Particle-Induced Field and Enhanced Collection of Carriers," *IEEE Electron Device Letters*, Vol. EDL-3, No. 2, pp.31-34 (1982).

[6.11] http://www.nea.fr/janis/welcome.html.

[6.12] M.B. Chadwick, and E. Normand, "Use of New ENDF/B-VI Proton and Neutron Cross Sections for Single Event Upset Calculations," *IEEE Trans. Nucl. Sci.*, Vol. 46, No. 6, pp.1386-1394 (1999).

[6.13] C.M. Perey, "Compilation of Phenomenological Optical-Model Parameters 1954-1975," *Atomic Data and Nuclear Data Tables,* Vol. 17, No. 7, pp.1-101 (1976).

[6.14] J.F. Ziegler, "The Stopping of Energetic Light Ions in Elemental Matter," *J. Appl. Phys.*, Vol. 85, pp.1249-1272 (1999).

Chapter 7

[7.1] H.H.K. Tang, G.R. Srinivasan, and N. Azziz, "Cascade statistical model for nucleon-induced reactions on light nuclei in the energy range 50-MeV-1GeV," *Physical Review C,* Vol. 42, No. 4, pp.1598-1622 (1990).

[7.2] F. Bertland, R. Peele, *Phys. Rev.* C 8, p.1045 (1973).

[7.3] E. Ibe, S. Chung, S. Wen, H. Yamaguchi, Y. Yahagi, H. Kameyama, S. Yamamoto, and T. Akioka, "Spreading Diversity in Multi-cell Neutron-Induced Upsets with Device Scaling," 2006 CICC, San Jose, CA., September 10-13, 2006, pp.437-444 (2006).

[7.4] E. Ibe, Y. Yahagi, H. Kameyama, and Y. Takahashi, "Single Event Effects of Semiconductor Devices at the Ground," *Ionizing Radiation*, Vol. 30, No. 7 (2004).

[7.5] R.C. Baumann, and E.B. Smith, "Neutron-Induced Boron Fission as a Major Source of Soft Errors in Deep Submicron SRAM Devices," *2000 IEEE Int'l Reliability Physics Symposium, San Jose, CA, April 10-13*, pp.152-157 (2000).

[7.6] E. Ibe, Y. Yahagi, F. Kataoka, Y. Saito, A. Eto, M. Sato, H. Kameyama, and M. Hidaka, "A Self-Consistent Integrated System for Terrestrial-Neutron Induced Sungle Event Upset of Semiconductor Devices at the Ground," *2002 ICITA, Bathurst, Australia*, No. 273-21 (2002).

[7.7] A. Eto, M. Hidaka, Y. Okuyama, K. Kimura, M. Hosono, "Impact of Neutron Flux on Soft Errors in MOS Memories," *IEDM, San Francisco, CA., December 6-9, 1998*, pp.367-370 (1998).

[7.8] S. Satoh, Y. Tosaka, T. Itakura, "Scaling Law for Secondary Cosmic-Ray Neutron-Induced Softr Errors in DRAMs," *Extended Abstracts of the 1998 Int'l Conf. on Solid State Devices and Materials, Hiroshima*, pp.40-41 (1998).

[7.9] E. Ibe, "Self-Consistent Integrated System for Terrestrial Neutron Soft-error of Sub- Quarter Micron Devices," *Eliminating the threat of soft errors IRPS SER Panel Discussion, Dallas, April 2, 2003*, No. 5 (2003).

[7.10] E. Ibe, H. Kameyama, Y. Yahagi, K. Nishimoto, Y. Takahashi, "Distinctive Asymmetry in Neutron-Induced Multiple Error Patterns of 0.13um process SRAM," *RASEDA2004, Tsukuba, October 6-8, 2004,* pp.19-23 (2004).

[7.11] E. Ibe, S.S. Chung, S.J. Wen, H. Yamaguchi, Y. Yahagi, H. Kameyama, and S. Yamamoto, "Multi-Error Propagation Mechanisms Clarified in CMOSFET SRAM Devices under Quasi-Mono Energetic Neutron Irradiation," *NSREC2006, Ponte Vedra Beach, Florida*, No. PC-6 (2006).

[7.12] E. Ibe, Y. Yahagi, H. Yamaguchi, "Monte-Carlo Simulation for the Effects of Composite Materials on Single Event Effects of Sub-100 nm Semiconductor Devices," *Proc. 2004 Symposium on Nuclear Data, November 11-12, 2004, JAERI, Tokai, Japan, Vol. JAERI-Conf 2005-003,* No. 2.14, pp.100-105 (2005).

[7.13] E. Ibe, Y. Yahagi, H. Yamaguchi, H. Kameyama, "SEALER: Novel Monte-Carlo Simulator for Single Event Effects of Composite-Materials Semiconductor Devices," *2005 RADECS, Sep.19-23, Palais des Congres, Cap d'Agde, France*, No. E-4 (2005).

[7.14] Y. Yahagi, E. Ibe, Y. Takahashi, Y. Saito, A. Eto, M. Sato, H. Kameyama, M. Hidaka,, K. Terunuma, T. Nunomiya, and T.Nakamura,, "Threshold Energy of Neutron-Induced Single Event Upset as a Critical Factor," *2004 IRPS, April 25-29, Phoenix, Arizona*, , pp.669-670(2004).

[7.15] G. Gasiot, V. Ferlet-Cavrois, J. Baggio, P. Roche, P. Flatresse, Guyot, A. Morel, P. Bersillon, O. du Port de Pontcharra, J., "SEU Sensitivity of Bulk and SOI Technologies to 14-MeV Neutrons," *IEEE Trans. Nucl. Sci.,* Vol. 49, No. 6, pp.3032-3037 (2002).

Chapter 8

[8.1] JEDEC, "Measurement and Reporting of Alpha Particles and Terrestirial Cosmic Ray-Induced Soft Errors in Semiconductor Devices: JESD89," *JEDEC STANDARD, No. 89*, pp.1-63 (2001).

[8.2] JEITA (Japan Electronics and Information Technology Industries Association), "JEITA SER Testing Guideline," *JEITA SER Testing Guideline, Vol. EIAJ, No. EIAJ EDR-4705*, pp.1-62 (2005).

[8.3] JEDEC, "Measurement and Reporting of Alpha Particles and Terrestirial Cosmic Ray-Induced Soft Errors in Semiconductor Devices: JESD89A," *JEDEC STANDARD, No. 89A*, pp.1-99 (2006).

[8.4] E. Ibe, H. Kameyama, Y. Yahagi, and H. Yamaguchi, "Single Event Effects as a Reliability Issue of IT Infrastructure," *3rd International Conference on Information Technology and Applications, July 3-7, 2005, Sydney,* Vol. I, pp.555-560 (2005).

[8.5] Y. Yahagi, E. Ibe, Y. Takahashi, Y. Saito, A. Eto, M. Sato, M. Kameyama, M. Hidaka, K. Terunuma, T. Nunomiya, and T. Nakamura, "Threshold Energy of Neutron-Induced Single Event Upset as a Critical Factor," *2004 IEEE International Reliability Physics Symposium, April 25-29, Phoenix, Arizona*, pp.669-670 (2004).

[8.6] P.W. Lisowski, "The Los Alamos National Laboratory Spallation Neutron Sources," *Nucl. Sci.e . Eng.*, Vol. 106, pp.208-218 (1990).

[8.7] E. Ibe, Y. Yahagi, F. Kataoka, Y. Saito, A. Eto, M. Sato, H. Kameyama, and M. Hidaka, "A Self-Consistent Integrated System for Terrestrial-Neutron Induced Single Event Upset of Semiconductor Devices at the Ground," *2002 ICITA, Buthurst, Australia,* No. 273-21 (2002).

Chapter 9

[9.1] J. JEITA, "JEITA SER Testing Guideline," JEITA SER Testing Guideline, Vol. EIAJ, No. EIAJ EDR-4705, pp.1-62 (2005).

[9.2] JEDEC, "Measurement and Reporting of Alpha Particles and Terrestirial Cosmic Ray-Induced Soft Errors in Semiconductor Devices: JESD89," JEDEC STANDARD, No. 89, pp.1-63 (2001).

[9.3] Ibe, E., Chung, S.S., Wen, S., Yamaguchi, H., Yahagi, Y., and H. Kameyama, S. Yamamoto, and T. Akioka, "Spreading Diversity in Multi-cell Neutron-Induced Upsets with Device Scaling," *The 2006 IEEE Custom Integrated Circuits Conference, September 10 - 13, 2006, San Jose, CA.* , No. 10.4 (2006).

[9.4] Schwank, J.R., Dodd, P.E., Felix, J.A., Sexton, F.W., Hash, G.L., and M.R. Shaneyfelt, J. Baggio, V. Ferlet-Cavrois, P. Paillet, and E. Blackmore, "Effects of Particle Energy on Proton and Neutron-Induced Single-Event Latchup," *2005 IEEE Nuclear and Space Radiation Effects Conference, Seattle, Washington, July 11-15, 2005,* No. I-4 (2005).

[9.5] R. Koga, S. Penzin, K. Crawford, and W. Crain, "Single Event Functional Interrupt (SEFI) Sensitivity in Microcircuits," RADECS 1997, pp.311-318 (1997).

[9.6] Ibe, E., Chung, S.S., Shijie, W., Yahagi, Y., Kameyama, H. and S. Yamamoto, T. Akioka, and H. Yamaguchi, "Valid and Prompt Track-down Algorithms for Multiple Error Mechanisms in Neutron-Induced Single Rvrnt Effects of Memory Devices," *Radiation Effects on Components and Systems, Athens, Greece, September 27-29, 2006* , E-2 (2006).

[9.7] Ibe, E., Chung, S.S., Wen, S.J., Yamaguchi, H., Yahagi, Y., H. Kameyama, and S. Yamamoto, "Multi-Error Propagation Mechanisms Clarified in CMOSFET SRAM Devices under Quasi-Mono Energetic Neutron Irradiation," *Int'l Nuclear and Space Radiation Effects Conference, Ponte Vedra Beach, Florida, July 17-21, 2006* , No. PE-4 (2006).

[9.8] D. Ball, K. Warren, R. Weller, R. Reed, L. Massengill, and a. R.Schrimpf, "Simulating Nuclear Events in a TCAD Model of a High-Density SEU Hardened SRAM Technology," 2005 RADECS, Sep. 19-23, Palais des Congres, Cap d'Agde, France, No. C1-2 (2005).

[9.9] N. Seifert, X. Zhu, and L. Massengill, "Impact of Scaling on Soft-Error Rates in Commercial Microprocessors", IEEE Transactions on Nuclear Science, Vol. 49, No. 6, pp.3100-3106 (2002).

[9.10] P. Shivakumar, M. Kistler, W. Keckler.S, D. DougBurger, and L., "Modeling the Effect of Technology Trends on the Soft Error Rate of Combinational Logic," *Int'l Conf. on Dependable Systems and Networks*, pp.389-398 (2002).

[9.11] G. Gasiot, V. Ferlet-Cavrois, J. Baggio, P. Roche, P. Flatresse, A. Guyot, P. Morel, O. Bersillon and J. du Port de Pontcharra, "SEU Sensitivity of Bulk and SOI Technologies to 14-MeV Neutrons," *IEEE Transactions on Nuclear Science*, Vol. 49, No. 6, pp.3032-3037 (2002).

[9.12] P.D. Bradley, and E. Normand, "Single Event Upsets in Implantable Cardioverter Defibrillators," *IEEE Transactions on Nuclear Science*, Vol. 45, No. 6, pp.2929-2940 (1998).

[9.13] M. Bruggenmann, H. Schmidt, F. Gliem and F.J Ombeck, "Single Event Errors of FIFO Memory Devices," *The 6th International Workshop on Radiation Effects on Semiconductor Devices for Space Application, Tsukuba, October 6-8, 2004*, pp.115-118 (2004).

[9.14] J. Benedetto, P. Eaton, K. Avery, D. Mavis, M. Gadlage, and T. Turflinger, "Heavy Ion Induced Digital Single-Event Transients in Deep Submicron Processes," 2004 Nuclear and Space Radiation Effects Conference, Atlanta, Georgia, July 20-24, No.E-5 (2004).

[9.15] Makihara, A., Midorikawa, M., Shindou, H., Kuboyama, S., Hirao, T., S. Onoda, T. Ohki, and Y. Takahashi, "Optimization for SEU/SET Immunity on 0.15 μm Fully Depleted CMOS/SOI Digital Logic Devices," *Int'l Nuclear and Space Radiation Effects Conference, Ponte Vedra Beach, Florida, July 17-21, 2006*, No. PD-3 (2006).

[9.16] A. Brown and C. Lin, "Limitations of Software Solutiuons for Soft-Error Detection," *The Second Workshop on System Effects of Logic Soft Errors, Urbana-Champain, IL, April 11-12, 2006* (2006).

[9.17] A. Makihara, H. Shindou, H., N. Nemoto, S. Kuboyama and S. Matsuda, "Analysis of Single-Ion Multiple-Bit Upset in High-Density DRAMs," *Proc. the 4th Int'l Workshop on Radiation Effects on Semiconductor Devices for Space Application , Tsukuba, May 15-18, 2005*, pp.103-107 (2000).

[9.18] J. Maiz, S. Hareland, K. Zhang and P. Armstrong, "Characterization of Multi-bit Soft Error events in advanced SRAMs," *2003 IEEE International Electron Devices Meeting, Washington, DC, December 7-10, 2003*, No. 21.4 (2003).

[9.19] M. Baze, J. Wert, J. Clement, M. Hubert and A. Witulski, "Propagating SET Characterization Technique for Digital CMOS Libraries," *Int'l Nuclear and Space Radiation Effects Conference, Ponte Vedra Beach, Florida, July 17-21, 2006*, No. E4 (2006).

[9.20] S. Mitra, M. Zhang, S. Waqas, N. Seifert, B. Gill and K.S. Kim, "Combinational Logic Soft Error Correction," *The Second Workshop on System Effects of Logic Soft Errors, Urbana-Champain, IL, April 11-12*, 2006 (2006).

[9.21] A. Wood, R. Jardine, and W. Bartlett, "Data Integrity in HP Nonstop Servers," SELSE II, Urbana-Champain, IL, April 11-12, 2006 (2006).

[9.22] N. Wang, and S. Patel, "Sympton Based Redundant Multithreading," SELSE II, Urbana-Champain, IL, April 11-12, 2006 (2006).

Appendices

[A.1] E. Ibe, H. Kameyama, Y. Yahagi, K. Nishimoto, and Y. Takahashi, "Distinctive Asymmetry in Neutron-Induced Multiple Error Patterns of 0.13μm process SRAM," *The 6th International Workshop on Radiation Effects on Semiconductor Devices for Space Application, Tsukuba, October 6-8, 2004*, pp.19-23 (2004).

[A.2] E. Ibe, Y. Yahagi, H. Yamaguchi, and H. Kameyama, "SEALER: Novel Monte-Carlo Simulator for Single Event Effects of Composite-Materials Semiconductor Devices," *2005 Radiation and Its Effects on Components and Systems, Sep.19-23, Palais des Congres, Cap d'Agde, France*, No. E-4 (2005).

[A.3] http://www.nea.fr/janis/welcome.html.

[A.4] M.B. Chadwick, and E. Normand, "Use of New ENDF/B-VI Proton and Neutron Cross Sections for Single Event Upset Calculations," *IEEE Trans. Nucl. Sci.*, Vol. 46, No. 6, pp.1386-1394 (1999).

[A.5] C.M. Perey, "Compilation of Phenomenological Optical-Model Parameters 1954-1975," *Atomic Data and Nuclear Data tabeles*, Vol. 17, No. 7, pp.1-101 (1976).

[A.6] S. Furihata, "Parameters Used in GEM," Development of a Generalized Evaporation Model and Study of Residual Nuclei Productions, Thesis for PhD, Tohoku Univ. pp.18-20 (2004).

[A.7] J.F. Ziegler, "The Stopping of Energetic Light Ions in Elemental Matter," *J. Appl. Phys.*, Vol. 85, pp.1249-1272 (1999).

TERMS AND DEFINITIONS

Albedo

The original meaning is the intensity ratio of reflected light to incident light from the sun on the earth. Here, it means the neutron components backscattered from the earth by incident cosmic-ray neutrons.

Ambient dose equivalent

One of the units expressing the strength of a radiation field. It expresses the field weighted by the magnitude of biological effects.

ASIC (Application-Specific Integrated Circuit)

A chip designed for a particular application (as opposed to the integrated circuits that control functions such as RAM in a PC). ASICs are built by connecting existing circuit building blocks in new ways. Since the building blocks already exist in a library, it is much easier to produce a new ASIC than to design a new chip from scratch. ASICs are commonly used in automotive computers to control the functions of the vehicle and in PDAs.

ATE

Automatic test equipment.

Atmospheric depth

Atmospheric shielding at a given altitude which is determined by the mass thickness per unit area of the air above.

Attenuation length

The cosmic-ray neutron variation with altitude shows an exponential attenuation, $\exp(-D/\lambda)$, as a function of the atmospheric depth, D, with an attenuation length, λ, in g/cm^2.

AVF cyclotron

Azimuthally-Varying-Field cyclotron. A cyclotron with a magnetic field which is alternatively varying in the azimuth direction. It is a powerful technique to accelerate ions beyond the limit energy in a normal cyclotron due to relativistic effects.

Beam dump

A beam stop for a beam which passes through a target with no interactions.

Cascade maximum

A region of the air above about 20 km where the number of secondary cosmic-ray nucleons produced by cascade reactions of primary cosmic-ray particles with atoms in the air are at a maximum.

CERN (Conseil Européen pour la Recherche Nucléaire)

A research laboratory headquartered in Geneva, Switzerland, and funded by many countries. Most work there deals with nuclear and elementary particle physics.

C.L. (Confidence level)

The confidence level is the probability value 1-α associated with a confidence interval, where α is the level of significance. It can also be expressed as a percentage 100(1-α)% and is then sometimes called the confidence coefficient.

CMOS (Complementary Metal Oxide Semiconductor)

The most widely used integrated circuit design. It is found in almost every electronic product from handheld devices to mainframes. CMOS uses PMOS and NMOS transistors wired together in a balanced fashion that causes less power to be used than NMOS or PMOS transistors by themselves. The first transistors were bipolar, which are still used when higher power is required.

Collected charge

The charge collected by a particular device node after the passage of a particle.

NOTE: The collected charge is dependent on the geometry and doping of the node, the particle mass, energy, and trajectory, and the density and type of material in the volume being irradiated.

Component

A packaged die (silicon chip) or integrated circuit.

NOTE: This could be either a test vehicle or an actual product.

Concrete attenuation length

High-energy neutrons (tens of MeV and above) are attenuated in building materials, but can still penetrate the building. The absorption of neutrons in ordinary concrete can be roughly estimated with an empirical attenuation factor of 3.3 for every meter of concrete, or attenuation = exp (-1.17*t) where t is the concrete thickness in meters.

CORIMS (COsmic-Ray IMpact Simulator)

A SER simulator originated by Hitachi Production Engineering Research Laboratory.

CPU (Central Processing Unit)

The part of the computer that contains the microprocessor, power supply, hard drive, and disk drives.

Critical charge

The minimum amount of collected charge that will cause a device node to change state. This concept applies most directly to dynamic devices such as DRAM memory cells.

Cross-section

The probability of the reaction per incident particle and target nucleus, expressed in a unit of cm^2.

Cut-off rigidity

The geomagnetic field deflects low-momentum primary cosmic particles back into space, lowering the neutron flux produced in the atmosphere. The minimum momentum per unit charge (magnetic rigidity) that an incident (often, vertically incident) particle can have and still reach a given location above the earth is called the geomagnetic cutoff rigidity for that point.

CYRIC

CYclotron and RadioIsotope Center at Tohoku University, Sendai, Japan.

DIMM (Dual Inline Memory Module)

A device with a number of memory components that are attached to a board.

DRAM (Dynamic Random Access Memory)

Device used for the main memory in subsystems. The elementary cells must be periodically refreshed.

DUT

Devices Under Test. Devices undergoing irradiation tests with beams from accelerators, reactors and so on.

Dynamitron

A trade mark of the accelerator developed by Radiation Dynamics Inc., USA, on the basis of the electric circuit of Schenkel type. A high beam current is available.

ECC (Error-Correcting Code)

ECC allows data that is being read or transmitted to be checked for errors and, when necessary, corrected on the fly. It differs from parity-checking in that errors are not only detected but also corrected. ECC is increasingly being designed into data storage and transmission hardware as data rates (and therefore error rates) increase.

EEPROM (Electrically-Erasable PROM)

A device which is similar to ROM, but with the capacity of selective erasure and programming through electrical stimulus.

Excited state

The energy levels of atoms and nuclei with energies higher than the ground states.

Fabrication

In semiconductor manufacturing, fabrication usually refers to the (front-end) process of making devices in semiconductor wafers, but usually does not include the packaging, assembly, and test (back-end) stages.

Failure Rate

A rate of failure per unit time. Semiconductors are usually measured in failures per 1,000 hours.

FBGA (Fine Ball-Grid Array)

A FBGA features an array of metallic balls, of which the grid pitch is 0.8 mm or lower and the stand-off height is more than 0.1 mm, on the underside of the package.

FF (Flip Flop)

An electronic circuit that can assume either of two stable states.

Fission chamber

An ionization chamber for neutron detection containing a fission foil as a material to convert neutrons into charged-particles.

FIT

Failure in time; the number of failures per 10^9 device-hours.

Fluence

Fluence is originally defined as the time-integrated quantity of flux. NOTE: fluence is usually expressed in particles per unit area.

Fluence-to-dose conversion factor

The conversion factor from fluence to dose defined by the International Commission on Radiological Protection (ICRP) and the International Commission on Radiation Measurements and Units (ICRU). Different values in units of nSv per n/cm^2 are given for effective dose, ambient dose equivalent and personal dose equivalent.

Flux

The time rate of fluence. Flux is usually expressed in particles per unit area per unit time.
NOTE: The term flux is used in this standard way, whereas other standards might use terms such as flux density or fluence rate for the same meaning.

FNL

Fast Neutron Laboratory at the Faculty of Engineering, Tohoku University, Sendai, Japan.

FN (Fowler-Nordheim) tunneling

Cold electron or Fowler-Nordheim tunneling is used for program/erase of EEPROM and flash memory devices. It is a quantum-mechanical effect that allows electrons to pass through a gate oxide and to be trapped in a floating gate.

Forbush decrease

Secondary cosmic-ray intensity in the atmosphere surrounding the earth decreases in some cases during a couple of days after a large solar flare creates a significant geomagnetic disturbance.

Gas target

A target employing a foil to enclose a gas with window foils separating the target from the accelerator vacuum.

Geomagnetic latitude

The geographic latitude is converted into a geomagnetic latitude corresponding to the iso-contour of the geomagnetic field strength.

GLE (Ground Level Enhancement)

An enhancement of the ground-level neutron flux caused by a large solar flare which emits a large amount of high-energy protons (above several hundreds of MeV) that can reach the earth.

GPS (Global Positioning System)

A system of satellites, computers, and receivers that is able to determine the latitude and longitude of a receiver on Earth by calculating the time difference for signals from different satellites to reach the receiver.

Ground state

The energy levels of atoms and nuclei with the lowest energy.

Hard error

An irreversible change in operation that is typically associated with permanent damage to one or more elements of a device or component (e.g., gate oxide rupture, destructive latch-up events).
NOTE: The error is "hard" because the data is lost and the component or device no longer functions properly, even after a power reset and re-initialization.

Heliocentric potential and Solar deceleration potential

Heliocentric potential and Solar deceleration potential are the indices of the solar modulation, which varies with the 11-year solar cycle. Heliocentric potential is deduced from neutron monitors on the ground, and Solar deceleration potential is deduced either from the sunspot index or from neutron monitors.

ICITA

International Conference on Information Technology and Applications.

Ionosphere

A layer of ionized air in the atmosphere extending from almost 80 km above the Earth's surface to altitudes of 600 km and more. Technically, the ionosphere is not another atmospheric layer. It occupies the same region of the upper atmosphere as the thermosphere.

JAXA

Japan Aerospace eXploration Agency.

JEDEC

The **J**oint **E**lectron **D**evice **E**ngineering **C**ouncil. The semiconductor engineering standardization body of the Electronic Industries Alliance (EIA).

JEITA

Japan Electronics and Information Technology Industries Association.

J-PARC

Japan Proton Accelerator Research Complex. An accelerator system for multi-purpose utilization of intense proton beams, including an intense spallation neutron source similar to SNS. Presently, under construction in Tokai-mura, Japan, under a joint project between JAEA and KEK, Japan.

K number

A parameter which gives maximum energy of cyclotrons T; $T = K \cdot Z^2/A$, where Z and A are, respectively, the charge number and the mass number of the ion to be accelerated.

LANSCE

Los Alamos Neutron Science Center, Los Alamos National Laboratory, U.S.A.

Lethargy flux

Flux per lethargy, $\phi(u)$, which is equivalent to energy flux, $E\phi(E)$, where E is energy and u is lethargy, $u = \mathit{ln}\ E/E_0$.

LSI

Large Scale Integrated circuit.

Maxwellian distribution

The formula which expresses the energy distribution of thermal neutrons in thermal equilibrium with surrounding materials.

MCBI

Multi-Coupled Bipolar Interaction.

MCU: (Multiple-Cell Upset)

A single event that induces several bits on the same IC to fail at one time.
NOTE: The error bits are usually, but not always, adjacent.

Mesosphere

The third highest layer in our atmosphere, occupying the region 50 km to 80 km above the surface of the Earth, above the troposphere and stratosphere, and below the thermosphere. It is separated from the stratosphere by the stratopause and from the thermosphere by the mesopause.

Momentum transfer

Momentum transferred from the initial state to the final state of the reaction. In this book, it corresponds to the difference of the momentum of neutrons before and after scattering.

Monoenergetic neutrons

Neutrons which have single energy.

MONOS (Metal-Oxide-Nitride-Oxide-Semiconductor) type EEPROM

The storage EEPROM device using MONOS structure can be composed of a coupler or a non-volatile memory cell.

Multiple-bit upset (MBU)

A multiple-cell upset in which two or more error bits occur in the same word.

NOTE: An MBU cannot be corrected by the most common form of ECC that only has the capability to correct for a single bit in a word.

Multiplexer

A device that performs multiplexing is called a multiplexer (MUX), and a device that performs the reverse process is called a demultiplexer (DEMUX).

Variable-bit-rate digital bit streams may be transferred efficiently over a fixed bandwidth channel by means of statistical multiplexing, for example, packet mode communication. Packet mode communication is an asynchronous mode time-domain multiplexing, which resembles, but should not be considered, as time-division multiplexing.

NASA (National Aeronautics and Space Administration)

An independent agency of the United States government responsible for aviation and space flight.

NMOS

N channel transistor. The source and drain of the transistor are N doped. The transistor is on when the gate voltage is increased.

NSREC

IEEE Nuclear and Space Radiation Effects Conference.

Nuclear evaporation

A process in which a nucleon is emitted from the equilibrium state of a exited nucleus.

Nuclear cascade and spallation

A high-energy nucleon or nucleus hits the target nucleus and produces a number of nucleons through nuclear collisions.

Package

The protective container for an electronic component, with terminals to provide electrical access to the components inside.

Parasitic-neutron

Background neutrons produced by reactions of incident particles with elements other than target.

PCSE

Power Cycle Soft-Error.

PMOS

P-channel transistor. The source and drain of the transistor are P doped. The transistor is on when the gate voltage is decreased.

Process

A combination of people, procedures, methods, machines, materials, measurement equipment, and/or environment for specific work activities to produce a given product or service.

NOTE: For the purposes of this book, the process specifically refers to the manufacturing steps and methodologies used to fabricate an IC.

Product

A component or service sold to satisfy a particular customer application.

NOTE: For the purposes of this book, a product is a complete integrated circuit sold to satisfy a particular customer application.

Proton-recoil counter

A counter for neutron measurement that uses a hydrogen-containing material/gas as a neutron-to-charged particle converter.

Pulse shape discrimination

The technique which neutron pulses are discriminated from photon and muon pulses from the scintillator by using the difference of their pulse shapes. The output pulses from neutron-induced charged particles are slower than those from photon-induced electrons and muons.

Proton-recoil telescope

A proton recoil counter which uses several detectors operating in coincidence to clearly define the solid angle for detection.

Quasi-monoenergetic neutron

Neutrons which have additional energies besides the major single energy.

RADECS

Radiation Effects on Components and Systems Workshop.

RASEDA

International Workshop on Radiation Effects on Semiconductor Devices for Space Application.

RCNP

Research Center for Nuclear Physics at Osaka University, Osaka, Japan.

Reaction threshold

The lowest energy of particles where the reaction starts to occur.

Real-time SER: Real time soft error rate (RTSER)

A soft error rate measurement technique under a naturally occurring alpha particle and/or neutron environment using a large number of devices to obtain a statistically significant error count.
NOTE: The former term was System soft error rate (SSER).

Response function

The output pulse height distribution or detection efficiency per unit flux of radiation incident to a detector as a function of its incident energy.

Radiation

Energy emitted in the form of electromagnetic waves or moving nuclear particles.

NOTE: For purposes of this book, the primary radiation of concern is ionizing radiation and includes protons, alpha particles and nuclear reaction products.

Ring cyclotron

A type of cyclotron designed to accelerate ions beyond the relativistic limit present in conventional designs. The magnet yoke is divided into several pieces as an extension of the AVF cyclotron.

Room-scattered neutrons

Background neutrons which are scattered by walls, floors and objects inside the room.

SAA (South Atlantic Anomaly)

A region where the earth's radiation belt (Van Allen Belt) makes its closest approach to the surface of the earth. SAA is produced by a "dip" in the earth's magnetic field, caused by the fact that the center of the Earth's magnetic field is offset from its geographic center.

SEALER

Single Event Adverse and Local Effects Reliever.

SECIS

SElf-Consistent Integrated System for terrestrial neutron-induced soft-error.

Sensitive volume

A region, or multiple regions, containing nodes whose states can be changed by incident radiation.

Single-Bit Upset (SBU)

A single event that induces a single bit on the same IC to fail once.

Single-event effect (SEE)

Any measurable or observable change in the state or performance of a microelectronic device, component, subsystem, or system (digital or analog) resulting from the passage of a particle.

NOTE: Single-event effects include single-event burnout (SEB), single-event functional interrupt (SEFI), single-event gate rupture (SEGR), single-event latch-up (SEL), single-event upset (SEU), multiple-bit SEU (MBU), multiple-cell upset (MCU), single-event hard error (SHE) and single-event transient (SET).

Single-event functional interrupt (SEFI)

A soft error that causes a component to reset, lock-up, or otherwise malfunction in a detectable way.

NOTE 1: In contrast to an SEU or MBU, where one or more memory bits may be affected but the device may still operate normally, a SEFI necessitates a restart or reboot operation.

NOTE 2: A SEFI does not usually result in a loss of functionality requiring a power cycling the device to restore operability, unlike the single-event latch-up (SEL). Nor does a SEFI result in permanent damage, as does a single-event burnout (SEB).

Single-event latch-up (SEL)

An abnormal high-current state produced in a CMOS device when the passage of a single particle induces the creation of a parasitic bipolar shorting of power to ground.

NOTE: SEL may cause permanent damage to the device. If the device is not permanently damaged, power cycling of the device (off and back on) is necessary to restore normal operation.

Single event transient (SET)

A momentary voltage excursion (voltage spike) at a node in an integrated circuit caused by the passage of a single particle.

Single-event upset (SEU)

A soft error caused by a transient signal induced by the passage of a single particle.

Signal-to-noise ratio

A ratio of the number of signals of interest to that of unwanted noise. It represents the quality of the signal.

SNS

An intense Spallation Neutron Source under construction in the Oak Ridge National Laboratory, USA, to be used for material and life sciences. The planed maximum beam power is 1.4 MW for the first stage of operation.

Soft error

A data error that can be removed or corrected by rewriting the device.
NOTE 1: As commonly used, the term refers to an error caused by an upset of the data state by the passage of a particle, not a hard or intermittent error associated with a physical defect introduced during the manufacturing process.
NOTE 2: Soft errors can be generated from SEU, SEFI, MBU, MCU, or SET. This was the original term used when the phenomenon was discovered (1979) and is still widely used in industry. The more specific terms SEU, SEFI, etc. are typically used by the avionics, space and military electronics communities.

Soft error rate (SER)

The rate at which soft errors occur.

Solar flare

A phenomenon of an explosion in the solar atmosphere which takes place in the solar corona and chromosphere, and accelerates electrons, protons and heavier ions near the speed of light.

Solar modulation – solar maximum and solar minimum

The varying magnetic field carried outward from the sun by the solar wind plasma that permeates the solar system also reduces the cosmic-ray intensity at earth. The cosmic-ray induced neutron flux is highest when solar activity is at a minimum (solar minimum), and lowest during solar maximum.

Spallation reaction

Nuclear reaction which is induced by high energy particles and results in production of various fragments. It produces a large number of neutrons per incident particle with a very high efficiency.

SRAM (Static RAM)

A type of semiconductor memory in which a flip-flop stores the information. The information is stored as long as the power voltage is applied.

SSRAM (Synchronous SRAM)

A high-speed type SRAM that uses a clock to synchronize a signal when outputs change states.

Tail correction

Correction of data for the effect of tail components in a quasi-monoenergetic neutron source, like the ^{7}Li(p,n) reaction. Usually it is done on the basis of neutron spectrum data and the cross sections for energies lower than the point of interest.

Terrestrial neutrons

Neutrons produced in the terrestrial environment by nuclear reactions between the atmosphere and cosmic rays incident on the earth.

Test vehicle

A circuit designed for the purpose of evaluating the soft error sensitivity of a particular process.
NOTE: This test vehicle is typically not a product but is a dedicated component designed to be used to extrapolate the product's SER.

TFBC

Thin Film Breakdown Counter, Fast timing gas counter for neutron detection using a ^{235}U fission foil for neutron converter, developed at TSL, Uppsala University.

TFT (Thin Film Transistor)

The TFT cell is a type of SRAM cell that consists of a polysilicon load acting as a PMOS transistor.

Thermal neutrons

Neutrons which are energetically equilibrated with atoms in matter. They have energy spectrum with a Maxwellian shape.

Three-body phase-space distribution

A distribution function which provides the energy spectrum of particle 1 that is produced from a three-body simultaneous breakup reaction: A + B → 1 + 2 + 3. It is derived under the assumption of no interaction among particles (1, 2 and 3) and the constancy of the matrix element which describes the interaction causing the breakup. The energy of particle 1 relative to the center-of-gravity of particles 2 and 3 is given simply as

$$\frac{d^2\sigma}{d\Omega_{1-23}dE_{1-23}} = C[E_{1-23}(E_{tot}^{c} - E_{1-23})]^{\frac{1}{2}}.$$

Time-of-Flight (TOF)

A method to measure particle energy. The energy is obtained by measuring the time required to travel a known distance.

TNS

Transactions on Nuclear Science.

Transistor

An active semiconductor device with three electrodes that is mainly used as a switch.

Tritium-loaded titanium target

A tritium target consisting of titanium metal absorbing tritium atoms.

TRIUMF

Tri-University Meson Facility in Vancouver, Canada.

TSL

Theodore Svedberg Laboratory at Uppsala, Sweden.

Unfolding method

The method used to get the integrand of an energy spectrum, $\phi(E)$, by solving the integral equation of detector counts and the detector response function through an inverse transformation (deconvolution), with a computer program.

Wall effect

The effect occurs when secondary charged particles produced in a detector by incident particles escape from the detector without complete energy loss.

VLSI (Very Large Scale Integration)

ICs that contain 100,000 or more transistors, but less than 10 million transistors.

Wafer

A thin disk of semiconductor material (usually silicon) on which many separate chips can be fabricated.

INDEX